RADAR ANTI-JAMMING TECHNIQUES

ЗАЩИТА ОТ РАДИО-ПОМЕХ

ПОД РЕДАКЦИЕЙ   М. В. МАКСИМОВА

МОСКВА «СОВЕТСКОЕ РАДИО» 1976

RADAR ANTI-JAMMING TECHNIQUES

A translation from the Russian of
Zaschita Ot Radiopomekh

M.V. Maksimov, M.P. Bobnev,
L.N. Shustov, B.K. Krivitskiy,
G.I. Gorgonov, V.A. Il'in,
and B.M. Stepanov

Artech House Books 1979

CONTENTS

1 INTRODUCTION

NATURAL RADIO INTERFERENCE 1

1.1   Radio Interference from Active Sources 1
    1.1.1   Internal Radio Receiver Noise 1
    1.1.2   Antenna System Noise 7
    1.1.3   Atmospheric Noise 10
    1.1.4   Industrial Noise 12
1.2   Passive Radio Interference 13
    1.2.1   Reflected Interference from the Earth's Surface 13
    1.2.2   Interference from Meteorological Formulations 25
    1.2.3   Reflected Interference from the Water Surface 27

2 MANMADE RADIO INTERFERENCE 29

2.1   Active Jamming 29
    2.1.1   Description of Continuous Noise Jamming 30
    2.1.2   Kinds of Continuous Jamming 34
    2.1.3   Description of Random Pulse 42
    2.1.4   Description of Bursts of Regular Pulses 45
2.2   Active Deception Jamming 46
    2.2.1   General Information 46
    2.2.2   Deception Jamming of Pulse Detection Radars 47
    2.2.3   Deception Jamming Against Automatic Tracking Radars 48
    2.2.4   Deception Jamming of Automatic Angle Tracking Systems 48
    2.2.5   Jamming Automatic Range Tracking Systems 62
    2.2.6   Jamming Automatic Velocity Tracking Systems 63
2.3   Passive Jamming 63
    2.3.1   Effective Scatter Area 64
    2.3.2   Bandwidth Properties of Chaff 65
    2.3.3   Dispersion of Chaff Cloud 66
    2.3.4   Spectra of Signals Reflected by Chaff Cloud 68
    2.3.5   Chaff Corridors 70
2.4   Passive Deception Jamming 74

3 MUTUAL INTERFERENCE AND ELECTROMAGNETIC COMPATIBILITY OF ELECTRONIC SYSTEMS 77

3.1   General Information on Mutual Interference and Electromagnetic Compatibility 77
3.2   Electromagnetic Situation and EMC Parameters 78
3.3   Secondary Radiations of Radio Systems 80
    3.3.1   Harmonic Emissions 81
    3.3.2   Subharmonic Emissions 82

3.3.3   Parasitic Emissions 82
3.3.4   Cross-Product Emissions 83
3.3.5   Intermodulation Emissions 84
3.3.6   Out-of-Band Emissions 84
3.4   Receiver Heterodyne Emissions 86
3.5   Secondary Receiving Channels 86
3.6   Industrial Interference 88
3.6.1   Ignition Noise 88
3.6.2   Power Transmission Noise 89
3.6.3   Interference from Arc Welding Machinery 89

4   GENERAL DESCRIPTION OF RADIO INTERFERENCE CONTROL TECHNIQUES 91

4.1   Jamming Immunity of Radio Systems and Methods of Analyzing It 91
4.1.1   Quantitative Description of Jamming Immunity 91
4.1.2   Methods of Analyzing Jamming Immunity 92
4.2   Security Enhancement Techniques 102
4.3   Noise Immunity Enhancement Techniques 104
4.3.1   General Information 104
4.3.2   Prevention of Receiver Overload 105
4.3.3   Jamming Cancellation 106
4.3.4   Primary Selection 108
4.3.5   Secondary Selection 112
4.3.6   Functional Selection 113
4.3.7   Adaptation, Complete Utilization of Information, Utilization of Radio Interference 114
4.4   Signal Processing Optimization 115
4.4.1   General Information 115
4.4.2   Signal Detection 117
4.4.3   Filtering 133

5   PROTECTION OF RECEIVERS FROM OVERLOADS AND CANCELLATION OF RADIO INTERFERENCE 147

5.1   Protection of Receivers from Overloads 147
5.1.1   Logarithmic Amplifiers 149
5.1.2   Basic Dynamic Feature of AGC Systems: Response to Low-Level Interference 156
5.1.3   Response of AGC Receiver to Pulsed Jamming 161
5.1.4   Some Ways of Protecting Receiver from Temporary Overloads and Temporary Sensitivity Loss 168
5.2   Jamming Cancellation with Extra Receiver 172
5.2.1   General Information 172
5.2.2   Amplitude Jamming Cancellation 172
5.2.3   Coherent Jamming Cancellation 176

5.3   Interperiod Clutter Cancellation 189
    5.3.1   Interperiod Clutter Cancellation Signals Reflected from Fixed Clutter 189
    5.3.2   Interperiod Cancellation of Chaff 201
5.4   Jamming Cancellation with Dual Balanced Mixers 205
    5.4.1   Design of Dual Balanced Mixer and Principle of Jamming Cancellation 205
    5.4.2   Influence of Noise on Single Balanced and Dual Balanced Mixers in Radars with Continuous Double Frequency-Modulated Transmission 208

6 SPATIAL, POLARIZATION, FREQUENCY AND PHASE SELECTION 221

6.1   Spatial Selection and Jamming Suppression with Shaped Radiation Patterns 221
    6.1.1   Choice of Amplitude and Phase Distributions Over Antenna Aperture 221
    6.1.2   Nonlinear Processing Antennas 223
6.2   Polarization Selection 224
    6.2.1   Basic Definitions 224
    6.2.2   Parameters of Elliptical Polarization 225
    6.2.3   Polarization Selectors 227
    6.2.4   Reception Polarization Coefficient 231
6.3   Change of Operating Frequency of Radio Systems 237
    6.3.1   Effectiveness of Change of Carrier Frequency 237
    6.3.2   Effectiveness of Changing Radar Pulse Repetition Frequency 243
    6.3.3   Changing the Radar Pulse Repetition Period as a Means of Combatting Jamming 246
6.4   Frequency Diversity Radar 247
6.5   Application of Frequency and Phase Selection Systems 254
    6.5.1   Application of Frequency Stabilization and Automatic Frequency Tracking Systems 254
    6.5.2   Application of Phase Locked Loops 264
    6.5.3   On Application of Tracking Receivers for Improving Noise Immunity of FM Receiver 269
    6.5.4   Frequency Selection in Automatic Frequency Tracking Systems 271

7 TIME AND AMPLITUDE SELECTION 275

7.1   Time Selection of Pulsed Signals 275
    7.1.1   Time Position Pulse Selection 275
    7.1.2   Pulse Repetition Frequency Selection 280
    7.1.3   Pulse Duration Selection 282
7.2   Amplitude Selection 285
    7.2.1   Selection of Bottom-Clipped Signals 285
    7.2.2   Pulse Level Selection 286

7.2.3   Storage 287
7.2.4   Angle Gating 292

8 FUNCTIONAL, STRUCTURAL AND COMBINED SELECTION 305

8.1   Functional Selection 305
8.2   Structural Selection 310
    8.2.1   Structural Selection without Feedback 310
    8.2.2   Structural Selection with Feedback 317
8.3   Amplitude-Frequency Selection 320
8.4   Space-Time Processing of Signals 337
    8.4.1   General Information 337
    8.4.2   Holographic Processing of Radio Signals 344
    8.4.3   Synthetic Aperture Radar [47, 136] 356
    8.4.4   Spatial Filtering in Video Channel 362

9 COMBINED COORDINATE SENSORS 367

9.1   Design Principles and Applications of Combined Sensors 167
9.2   Combined Systems with Independent Sensors 374
9.3   Systems with Corrected Electronic Tracking Sensors 389
9.4   Combined Navigation Systems with Mutually Correcting Sensors 400

BIBLIOGRAPHY 405

Preface

The basic characteristics of various types of natural interference, crosstalk, and jamming are presented and methods of protecting radio systems from radio interference are described in the book. In contrast to previously published books, much attention is devoted in this book to methods of protecting systems from jamming.

Methods of protecting receivers from overloads, cancellation of radio interference, spatial, polarization, frequency, phase, time, amplitude, structural, amplitude-frequency and space-time selection of signals in a noise background, and the complete utilization of information for enhancing noise immunity are analyzed.

The book is intended for the large community of electronic experts engaged in the development and operation of various kinds of electronic equipment and may also serve as a textbook for students enrolled in advanced radio engineering courses at higher institutes of learning.

The book includes 170 figures, 1 table and 214 bibliographic references.

INTRODUCTION

Protecting various kinds of electronic systems from radio interference is one of the most important problems encountered in the development and operation of such systems. This problem arose primarily in connection with the ever-expanding development of radio eletronics and its applications, accompanied by a sharp increase in the level of mutual interference. Interference is also produced by industrial enterprises, household appliances, etc. To ensure the normal operation of radio equipment under such conditions, it has become necessary in recent years to undertake the solution of the problem of electromagnetic compatibility on an ambitious scale. The problem of protecting electronic systems from radio interference also is associated with the rapid and effective development of electronic countermeasure systems.

Many journal articles and conference and symposium reports have been published on aspects of interference control. Individual aspects of interference control are examined in many textbooks, training manuals and monographs. N.D. Papaleksi's book *Radio Interference and Its Control,* published in 1942, was the first book published in the Soviet Union on this subject. The Soviet scientist Ye. G. Momot [115], who successfully analyzed synchronous reception techniques, is given much credit for increasing the noise immunity of radio communications systems, particularly in relation to site radio interference. V.A. Kotel'nikov's outstanding work, *Theory of Optimum Noise Immunity,* a classic study of the noise immunity problem, came to light in 1946.

A.A. Kharkevich offers a compact presentation of noise immunity in his monograph *Combatting Interference* (Fizmatgiz, 1963), which is accessible to research engineers and engineering technicians.

Works of other Soviet scientists [8, 52, 77, 159, 167, etc.] also represent a significant contribution to the solution of the problem of protecting electronic systems for interference. Many books and periodicals [9, 20, 26, 73, 79, 147] attest to the attention that is being devoted to this problem abroad.

Until the late 1950's the technical literature dealt primarily with the control of natural and mutual interference. Many articles have been published in journals during the last 10 to 15 years specifically on anti-jamming measures. There are also books that examine individual aspects of electronic countermeasures and counter-countermeasures [24].

There is a need today to generalize the results of research on radio interference control by way of examination of aspects of interference control from unified points of view, to the extent that that is possible. The solution of this problem is the subject of this book. In writing the manuscript the authors considered that the developing theory of optimum reception most often provides recommendations on the assumption that useful signals are accompanied by additive noise. This assumption is usually valid in relation to receiver noise and certain kinds of man made interference. However, the application of optimum reception is more a means of exploiting the capabilities of radio electronic systems than a means of increasing their noise immunity under conditions of jamming. The quest for optimum anti-jamming measures, which are extremely diversified in terms of structure and parameters, is made difficult by the fact that optimum reception systems depend to a great extent on the statistical properties of interference and useful signals. Therefore, anti-jamming techniques are analyzed in this book in accordance with the specified structural and parametric differences that describe useful signals and interference.

The contents of this book may be divided into two related parts. The first part deals with natural, intentional, and inadvertent mutual interference. A rather complete picture of the variety of known kinds of radio interference is presented from the standpoint of structure and parameters in an effort to determine the attributes that identify useful signals and interference in the interest of developing countermeasures.

The features of natural interference are described briefly in Chapter 1. Chapter 2 combines materials on active and passive jamming, which are divided into two groups: concealment and deception. The former makes a noise background in which it is difficult to identify a useful signal and also suppresses that signal in nonlinear receiver elements. Deception interference is a forgery of useful signals in relation to one or more parameters.

Concealment jamming includes continuous noise, random intermittent noise, bursts of deterministic radio pulses, and chaff. Active deception jamming is analyzed in application to radar systems that detect and automatically track targets by angle, range, and speed. Passive deception jamming is examined in the form of false targets, radar decoys, and clouds of chaff.

Chapter 3 contains general information about electromagnetic compatibility and mutual interference, which may be concealment or deceptive jamming, and also deals with problems of radiation on harmonics and subharmonics, cross-products, intermodulation and out of band radiation, as well as secondary reception channels.

The protection of radio electronic systems from radio interference is discussed in the second part of the book. Chapter 4 is an introduction to the entire second part, since it presents a general description of noise control techniques. The jamming immunity of any radio system is assumed to be determined by its security, i.e., by its capacity to prevent an enemy from jamming the receiver input, and by its noise immunity, which is its capacity to function properly under conditions when its input is acted upon by both natural and man made radio interference.

The necessary initial concepts and definitions are followed in Chapter 4 by the basic and secondary quantitative characteristics and brief descriptions of methods of analyzing the jamming immunity of various kinds of radio electronic systems. Also presented in Chapter 4 is basic information about the construction of mathematical models for radio electronic systems under conditions of radio interference. Security enhancement techniques are discussed in only one section, primarily because of a lack of sufficient information in this area. The remainder of the chapter is devoted to a general description of noise immunity (interference control) enhancement techniques in relation to both natural and intentional radio interference, and to the optimization of radio signal processing.

The classification of jamming immunity enhancement techniques is based on signal and noise attributes (polarization, pulse repetition frequency, etc.), in accordance with which the target signal can be selected (identified) and active interference can be counteracted or used in the interests of the system. Also examined are methods based on the principles of adaptation and the combined utilization of information from different sources.

Since signals and interference can be described as functions at least of time or frequency, noise immunity enhancement techniques are not always uniquely determined. In the familiar moving target indicator (MTI) radar systems, for example, moving target signal selection may be regarded as a means of counteracting random signals reflected by fixed objects, and at the same time as a technique based on the spectral difference of the signals that characterize fixed and moving targets. The authors are perfectly aware of the stated ambiguity, but they do not discuss it specifically in this book.

Signal processing optimization is examined in application to signal detection and filtering. Of primary importance here is a discussion of ways to exploit the capabilities of a system in consideration of its performance under conditions of white noise. The latter may be regarded as a satisfactory model for internal receiver noise and especially for intentional nonselective interference. The advantages of optimum techniques are noted wherever possible. By and large, optimum processing theory is reduced in this book not to a set of procedures for finding the best systems, but to the development of a general guideline for seeking the optimum structures of radio systems. Indeed, the development of modern noise-immune systems obviously will continue to be for a long time largely an art in which the developer must exhibit both intuition and skill to combine heuristic techniques with recommendations derived from optimum signal processing theory.

All of the material in Chapters 5 through 9 is composed and written in such a way that the reader, after analyzing Chapter 4, will immediately be able to understand in detail the interference control technique in which he is most interested, without necessarily having to read the entire book word for word.

Methods of protecting radio receivers from overloads under conditions of interference (the application of logarithmic amplifiers and AGC systems) and countermeasures are discussed in Chapter 5.

Materials on spatial, polarization, frequency, and phase signal selection on a noise background are presented in Chapter 6; Chapter 7 pertains to time and amplitude selection.

Functional, structural and combined selection are examined in Chapter 8. Functional selection is defined as the selection of a signal using several independent receiving channels and subsequent processing of the entire set of signals that act upon the system. Of the many possible combined selection methods, only the best known are analyzed in detail, namely amplitude-frequency selection, based on the use of a "wide-band-limiter-narrow-band" system, and space-time selection, which is used in holographic signal processing.

Finally, Chapter 9 presents a comparatively detailed analysis of an interference counteraction technique that is used when electronic and non-electronic coordinate and time derivative measurement instruments are combined.

The main results of the theoretical analysis presented in this book are illustrated by numerical examples, experimental research data, graphs and tables. All the numerical characteristics were borrowed from the open Soviet and foreign literature or were determined by the authors as the result of calculations based on existing theoretical formulas. The initial data for the calculations are hypothetical.

The mathematical tools employed in the book, as well as the body of preliminary information on radio electronics necessary for its comprehension, do not go beyond the scope of the radio engineering programs of higher institutes of learning. The reader is assumed to be sufficiently well versed in the fundamentals of the theory and technology of radar, telecontrol, radio navigation and telephone-telegraph radio communications.

The book is intended for the large community of radio engineers, graduate students, and scientists engaged in the development and operation of radio engineering installations and systems. It may also serve as a textbook for students enrolled in various programs of the radio engineering departments of higher institutes of learning.

The authors gratefully acknowledge the reviewers, professors, and doctors of engineering sciences, B.V. Vasil'yev and N.M. Tsar'kov, for their esteemed advice which promoted an improvement of the book.

Work on the writing of the manuscript was distributed among the authors as follows: M.P. Bobnev—Chapter 9 and §8.4 (jointly with L.N. Shustov);

G.I. Gorgonov — §4.4; V.A. Il'in — Chapter 3 and §2.3; B. Kh. Krivitskiy — §1.1, 5.1, 6.5 and 7.1; M.V. Maksimov — Introduction, §4.1-4.3, 5.2-5.4, 7.2.3 and 8.2; B.M. Stepanov — §2.1, 6.2-6.4, 7.2.1, 7.2.2 and 8.3; L.N. Shustov — §1.2, 2.2, 2.4, 6.1, 7.2.4, 8.1, 8.4 (jointly with M.P. Bobnev).

Chapter 1

Natural Radio Interference

1.1 RADIO INTERFERENCE FROM ACTIVE SOURCES

1. Internal Radio Receiver Noise

Internal radio receiver noise is caused by voltage and current fluctuations in amplifiers (tubes and transistors) and by electrical fluctuations in resistors and active components of resistors. Noise in the input stages of a radio receiver is most important, since it is subjected to the strongest amplification.

Any active resistor, R (just like the active component of a compound resistor), as is known, is a source of wide band Gaussian noise [38, 94, 114]. The voltage variance, σ_n^2 of that noise in the equivalent (energy) band, B_n is

$$\sigma_n^2 = 4kTRB_n \qquad (1.1.1)$$

Here $k = 1.38 \cdot 10^{-23}$ J/deg is the Boltzmann constant, and T is absolute temperature, $^\circ$K.

An active resistor may be represented as a source of noise, as an equivalent voltage generator, u_n with internal (quiet) resistor, R (Figure 1.1a), or current generator, i_n with internal conductance $Y = 1/R$ (Figure 1.1b); here the variance of this voltage is expressed by Equation (1.1.1), and the variance σ_{in}^2 of current, i_n is

$$\sigma_{in}^2 = 4kTYB_n \qquad (1.1.2)$$

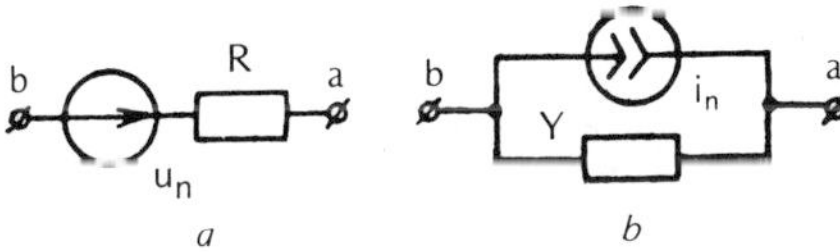

Figure 1.1

An active resistor may be characterized as a source of noise by the nominal (or available) power

$$P_N = kTB_n = \sigma_n^2/4R \tag{1.1.3}$$

It is often convenient to examine, instead of σ_n^2 and σ_{in}^2, the spectral densities, G_u and G_i of noise voltages and currents, equal to

$$G_u = \sigma_n^2/B_n = 4kTR \tag{1.1.4}$$

$$G_i = \sigma_{in}^2/B_n = 4kTY \tag{1.1.5}$$

The spectral densities are constant in an extremely wide frequency range, from the lowest to frequencies on the order of $f = 10^{12}$ Hz. At higher frequencies, G_u is a function of frequency, f and is determined by the following equation:

$$G_u = \frac{4kTRhf}{kT} \left[\exp\left(\frac{hf}{kT}\right) - 1 \right]^{-1} \tag{1.1.6}$$

Here $h = 6.62 \cdot 10^{-34}$ J $\cdot$ s is the Planck constant. The analogous expression for G_i is written by substituting R for Y.

A receiver is characterized as an amplifier in relation to internal noise by the noise coefficient (noise factor). The noise factor is the ratio

$$F_n = \frac{(S/N)_i}{(S/N)_o} \tag{1.1.7}$$

where $(S/N)_i$ and $(S/N)_o$ are the ratios of the available (nominal) powers of the signal and noise at the input and output of the receiver, matched with the source of the signal and noise. Here and henceforth in this section a radio receiver is defined as the linear part of it (from the input to the detector).

If a receiver is quiet, then the output and input signal-to-noise ratios are identical and $F_n = 1$. The excess of F_n over 1, i.e., the value $F_n - 1$, is called the excess noise factor, which gives an indication of the noise power contributed by the receiver itself. We denote the power gain of a receiver as

$$k_p = S_o/S_i$$

Then Equation (1.1.7) may be written as

$$F_n = N_o/k_p N_i \tag{1.1.8}$$

This relation may be used as another definition of the noise factor. Since a receiver is a linear system, the output power, N_o may be broken down into two components: $k_p N_i$, corresponding to the amplified input noise, and N_e, corresponding to internal receiver noise. Hence,

$$F_n = (N_e/N_i k_p) + 1 \tag{1.1.9}$$

It follows from this relation that the noise factor uniquely describes the internal receiver noise only if N_i is "standardized." For such a noise source we may use an active resistor, R with temperature, $T_0 = 293°K$. Then

$$F_n = 1 + (N_e/k_p N) \tag{1.1.10}$$

where $N = kT_0 B_n$ is the available noise power of resistor, R, the magnitude of which has no influence on N. The receiver energy band is

$$B_n = \frac{1}{2\pi} \int_0^\infty |K(j\omega)|^2 \, d\omega/K^2(\omega_i) \tag{1.1.11}$$

Here $|K(j\omega)|$ is the amplitude-frequency response of the receiver and $K(\omega_i)$ is the gain of the receiver on the nominal intermediate frequency, ω_i. Substituting P_{No} into Equation (1.1.10), we obtain

$$F_n = 1 + (N_e/kT_o B_n k_p) \tag{1.1.12}$$

Hence

$$N_e = (F_n - 1) kT_o B_n k_p \tag{1.1.13}$$

Set noise may be expressed not only in terms of the output, but also in terms of the input of the receiver. Then instead of Equation (1.1.13), we have

$$N_{ei} = (F_n - 1) kT_o B_n \tag{1.1.14}$$

The noise factor, F_n of a receiver may be expressed through the noise factors of individual stages in accordance with the following equation [85]:

$$F_n = F_{n1} + \frac{F_{n2} - 1}{k_{p1}} + \frac{F_{n3} - 1}{k_{p2}} + \ldots \tag{1.1.15}$$

Here $F_{n1}, F_{n2}, \ldots; k_{p1}, k_{p2}, \ldots$ are the noise factors and gains of the first, second, etc., stages, respectively. (When this formula was written it was assumed that the frequency response of each stage was nearly rectangular, although the formula may also be generalized for other kinds of characteristics.)

The noise components that are inherent to an electronic tube as as electronic device consist of shot noise, distribution noise, induced noise and jitter noise [85, 94, 181].

Shot noise is attributed to the discrete nature of electrical current: the number of electrons reaching the anode per unit time is not constant and fluctuates rapidly relative to the average number.

Current distribution noise occurs in multiple grid radio tubes and is the result of the random distribution of electrons between the electrodes (primarily the anode and screening grid) of a tube. Both of these kinds of noise are wide band and Gaussian. They may be characterized as random EMF, u_{sh} (Figure 1.2), which may be ascribed to some equivalent resistor, R_{sh}, operating at a prescribed (standard) temperature, T_0. The variance of this noise is

$$\sigma_{sh}^2 = 4kT_0 R_{sh} B_n \tag{1.1.16}$$

The size of noise resistor R_{sh} depends on the type and parameters of the tube. For a triode, $R_{sh} \approx 2.5/S$, where S is the anode current transconductance, and for pentodes [94]

$$R_{sh} = \left(\frac{2.5}{S} + \frac{20 I_{po}}{S^2} \right) \frac{I_{ao}}{I_{ao} + I_{po}}$$

where I_{ao}, I_{po} are the constant components of the anode and plate currents of a tube.

Induced noise current is also wide band and is induced by electrons in the grid-cathode part of a tube; its magnitude depends on resistance, R_{g-c} in the grid circuit. This noise is conveniently characterized as current generator, i_{ind} (Figure 1.2) with variance

$$\sigma_{iind}^2 = 4kT_{ind} Y_{g-c} B_n \tag{1.1.17}$$

where $Y_{g-c} = 1/R_{g-c}$ is the conductance of the grid-cathode part, determined at that temperature, T_{ind} which is responsible for the noise variance, equal to that which actually takes place.

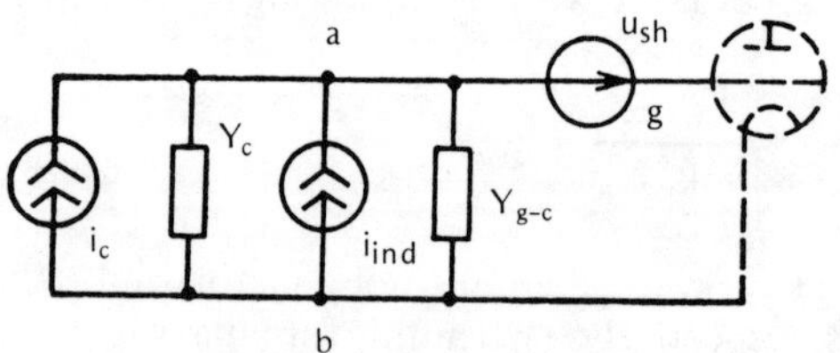

Figure 1.2

Noise of the input circuit is characterized as current generator, i_c (Figure 1.2) with conductance, $Y_c = 1/R_r$ and temperature, T_c, at which the circuit operates (here R_r is the resonance impedance of the circuit), and the variance of the noise current from both generators is

$$\sigma_{ia,\,b}^2 = \sigma_{ind}^2 = 4k\,[T_c Y_c + T_{ind}\,Y_{g-c}]\,B_n$$

Current generators, i_c and i_{ind} may be replaced with one noise current generator, i_{ne} with conductance, $Y_{ne} = Y_{g-c} + Y_c$ (Figure 1.3), and, in consideration of Equation (1.1.18), we may write one voltage variance, $\sigma_{a,\,b}^2$ as

$$\sigma_{a,\,b}^2 = \frac{\sigma_{iab}^2}{Y_{ne}^2} = 4k\,B_n\,\frac{Y_c T_c + Y_{g-c}\,T_{ind}}{(Y_c + Y_{g-c})^2}$$

$$= 4k\,B_n\,\frac{R_c^{\,2}\,R_{g-c}^{\,2}}{(R_c + R_{g-c})^2}\left(\frac{T_c}{R_c} + \frac{T_{ind}}{R_{g-c}}\right) \tag{1.1.18}$$

Variance, σ_{g-c}^2 of voltage, u_{g-c} between the grid and cathode is

$$\sigma_{g-c}^2 = \sigma_{a,\,b}^2 + \sigma_{sh}^2 \tag{1.1.19}$$

Voltage u_{g-c} is then amplified, which causes noise to appear at the receiver output.

The grid circuit of a tube is connected to the antenna, which is the source of the signal and noise. Therefore, a wide band noise circuit must include a source of current i_{na}, which characterizes the noise properties of the antenna, and which has "quiet" conductance, Y_{na}. The variance, σ_{ina}^2 of current i_{na} is

$$\sigma_{ina}^2 = 4k\,T_{ii}\,Y_{na}\,B_{ii} \tag{1.1.20}$$

As a result, the circuit shown in Figure 1.3 is changed to the one shown in Figure 1.4.

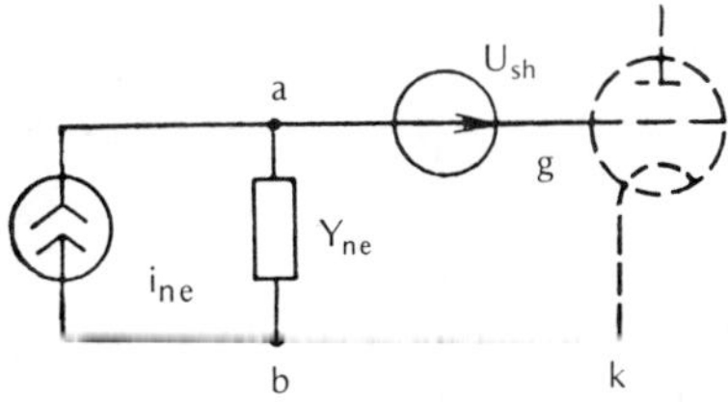

Figure 1.3

Tube jitter noise diminishes in inverse proportion to frequency and is significant only for low frequencies on the order of 10-100 Hz. This noise need be considered only in certain special cases of radio reception.

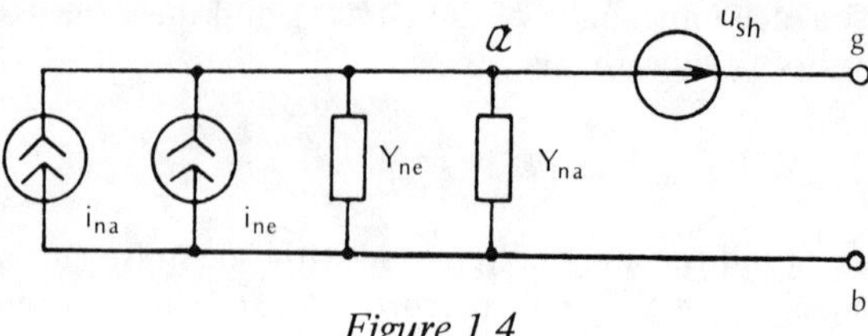

Figure 1.4

Noise produced by semiconductor devices is similar to tube noise [181].

Low-frequency noise (in the range of frequencies below 500-20,000 Hz) is caused by processes that occur on the surface of a semiconductor. The intensity of such noise diminishes in proportion to $1/f^2$ and depends on the design and production technology of a semiconductor device.

Wide band fluctuation noise is produced as a result of the shot effect due to the inclusion of active resistors in emitter, base and collector circuits; base resistance noise is the strongest of these types.

A general simplified noise T-circuit of a transistor is illustrated in Figure 1.5 [19, 181]. Here the generator of EMF u_{nb} with variance $\sigma_{nb}^2 = 4kT_0 R_b B_n$ characterizes the noise of base resistance, R_b. The other two generators of noise, u_{ne} and i_{nc} are attributed to the shot type of injection of carriers through the emitter and collector junctions.

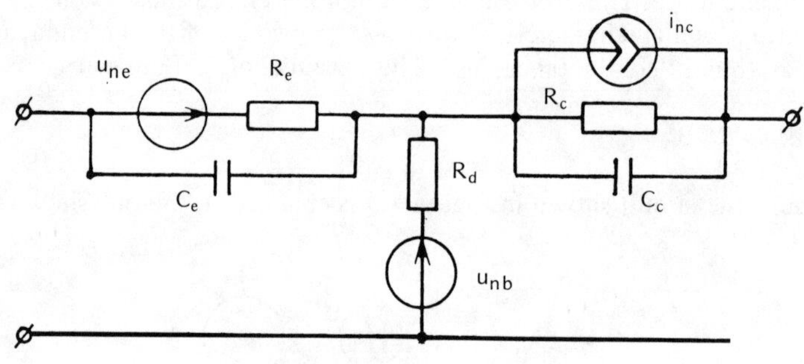

Figure 1.5

Emitter junction noise is caused by the shot type of injection of carriers through the emitter junction. The variance of this noise current, $\sigma_{in\,e}^2 = 2eI_{eo} B_n$, where I_{eo} is the constant component of the current and e is the electron charge. Since $I_{eo} = kT_0 Y_e/e$ (Y_e is the conductance of the emitter circuit), this noise may be considered also as a generator of control voltage, u_{ne} with variance

$$\sigma_{ne}^2 = 2kT_0 R_e B_n$$

Statistically independent sources of noise operate in the collector junction. This noise is produced by the emitter circuit and by current distribution. The former correlates with the emitter noise current and is considered as a noise generator, u_{ne} in the emitter circuit (Figure 1.5).

Current distribution noise is characterized as the generator of driving noise current, i_{nc} with variance, $\sigma_{inc}^2 = 2eI_{0e}\alpha_0 (1 - \alpha_0) B_n$, where α_0 is the emitter-collector current transconductance.

Equivalent noise circuits of tubes and transistors (if the stage is an input stage, then antenna noise also must be taken into account) can be used to calculate the noise factors of stages in various versions of amplifier circuitry [181, 150].

2. Antenna System Noise

An antenna system is acted upon by electromagnetic radiation of natural origin in addition to man made signals in the form of radiation from various radio stations or industrial noise from operating electrical systems.

Natural noise components include the following [19]:
(a) internal antenna loss resistance noise, which is of the nature of active resistance thermal noise;
(b) noise produced by radiation from extraterrestrial sources, or extra-terrestrial noise;
(c) noise attributed to the fluctuating character of the absorption of electro-magnetic radiation in the earth's atmosphere;
(d) noise caused by the thermal radiation of the earth;
(e) noise generated by lightning discharges in the atmosphere.

The first three kinds of noise have an extremely wide spectrum. The spectral densities of such noise are constant within the passband of radio receivers and may be considered as white noise. Atmospheric noise is more narrow banded, and for this reason it is convenient to examine it separately from antenna noise.

The spectral density of the noise produced by the antenna loss resistance, $G = 4kT_{10}R_{10}$, and the variance of the EMF $\sigma_{10}^2 = 4kT_{10}R_{10}B_n$. In these equations, T_{10} is the temperature at which the antenna operates and R_{10} is the loss resistance. In practice, it is more convenient to use, instead of R_{10}, the expression for the total resistance of an antenna, $R_a = R_{10} + R_\Sigma$, where R_Σ is the radiation resistance.

Since antenna efficiency is

$$\eta_a = R_\Sigma/R_a = R_\Sigma/(R_\Sigma + R_{10}),$$

we obtain for R_{10}

$$R_{10} = R_a(1 - \eta_a)$$

and consequently

$$\sigma_n^2 = 4kT_{10}R_a (1 - \eta_a) B_n \tag{1.1.21}$$

Noise components from external radiation (noise from space, from absorption in the atmosphere, and from the earth's thermal radiation) are statistically independent, in view of which their spectral densities are added

$$G_{ex\ rad} = G_{sp} + G_{ab} + G_{ea}.$$

It is convenient to assume that each component of this noise is produced by an equivalent resistance, which is equal to the radiation resistance that occurs at the temperatures T_{sp}, T_{ab}, T_{ea}, respectively. The temperatures are given by the equations

$$T_{sp} = G_{sp}/4kR_\Sigma; T_{ab} = G_{ab}/4kR_\Sigma; T_{ea} = G_{ea}/4kR_\Sigma,$$

and the overall noise is produced by the resistance that occurs at the temperature $T_{ex\ rad} = G_{ex\ rad}(4kR_\Sigma)^{-1}$. Therefore,

$$T_{ex\ rad} = T_{sp} + T_{ab} + T_{ea}. \tag{1.1.22}$$

The variance of the noise EMF of an antenna from external sources is

$$\sigma_{ex\ rad}^2 = 4kT_{ex\ rad}R_\Sigma B_n. \tag{1.1.23}$$

Considering that $R_\Sigma = \eta_a R_a$, we obtain from (1.1.21) and (1.1.22) the variance of antenna noise:

$$\sigma_a^2 = \sigma_{10}^2 + \sigma_{ex\ rad}^2 = 4kR_a B_n [T_{10}(1 - \eta_a) + T_{ex\ rad}\eta_a].$$

Thus, an antenna system in relation to noise may be represented as an equivalent system, consisting of noise EMF source u_{na} and series-connected resistor $R_a = 1/Y_{na}$, which operates at the temperature

$$T_a = T_{10}(1 - \eta_a) + T_{ex\ rad}\eta_a. \tag{1.1.25}$$

The stated equivalent circuit is used to determine the noise transmitted by the antenna into the input circuits of the receiver, and to calculate the receiver noise factor.

Extraterrestrial noise depends on the angular position of the main lobe. There is background noise and noise from discrete sources of radiation ("radio stars"). Background noise is characterized by the temperature $T_{sp\ b}$, which is a function of the angular coordinates ϕ and θ of the point of the celestial sphere. $T_{sp\ b}$ is virtually constant for directional antennas within the pattern.

The background temperature depends on frequency. As can be seen in Figure 1.6, where the typical function $T_{sp\ b}(f)$ is shown, the background temperature is low for frequencies above 3-5 GHz and may be ignored [94, 147, 19].

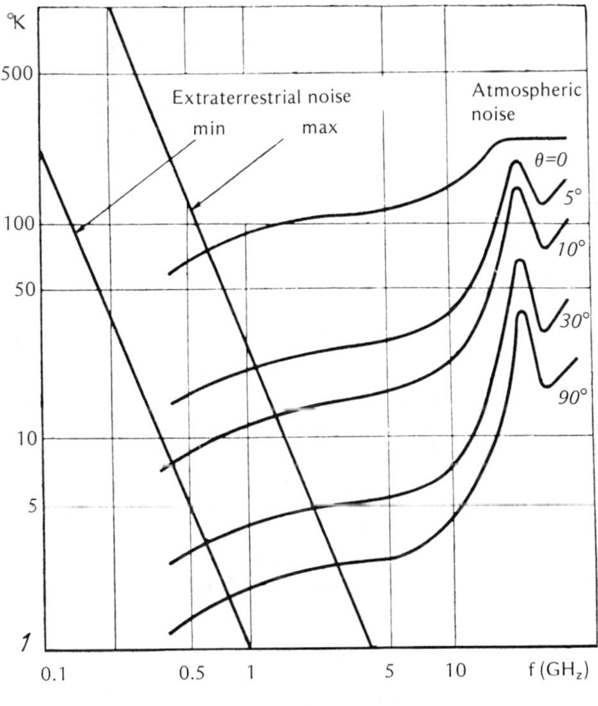

Figure 1.6

The background temperature of discrete sources, with angular dimensions not exceeding the angular dimension of the main lobe Ω_a, is

$$T_{sp\ di} = T_{so}\Omega_{so}/\Omega_a, \tag{1.1.26}$$

where T_{so} is the effective temperature of the source and Ω_{so} is its angular size.

Examples of discrete sources are the sun ($T_{sp\ di} = (10^4 - 10^6)°K$) and the moon ($T_{sp\ di} = (150-250)°K$). We note that the angular dimensions of these sources are about the same and $\Omega = 0.5°$.

Thermal noise of the atmosphere is caused by fluctuation scattering of electromagnetic waves by oxygen and moisture in the atmosphere. Atmospheric thermal noise T_{ab} is strong in the region of very high frequencies (from 0.5 GHz and above) and depends on the orientation of the antenna, increasing as the main lobe approaches the horizon. Graphs of the dependence of atmospheric temperatures on frequency are presented in Figure 1.6 for angles of elevation from $0 = 90°$ (zenith) to $0°$ (horizon). The lowest level of extraterrestrial and atmospheric thermal noise occurs in the range of from 2-3 to 10-15 GHz, and the mean total temperature of the noise is on the order of $10\text{-}20°\text{K}$ if the angle of elevation is not less than $10°$ [147].

An antenna is also acted upon by terrestrial radiation, since the warm surface of the earth is a source of electromagnetic noise radiation. Temperature, T_{ea} will be high for an antenna aimed at the earth (for example, at a search radar station, for a Doppler aircraft velocity meter and radio altimeter) and low for an antenna whose side lobes only are aimed at the earth. Thus, T_{ea} may vary in a rather wide range.

3. Atmospheric Noise

Sources of atmospheric noise are numerous lightning discharges that occur simultaneously in various regions of the earth. The number of such discharges may reach several thousands. If local storms are ignored the atmospheric noise level is quasistationary in nature. It depends on the geographic coordinates of the point of reception and changes comparatively slowly during the day and from season to season. This makes it possible to forecast the atmospheric noise level. Such forecasts are summarized in *Special Report No. 322* [59] of the 10th plenary assembly of the International Radio Consultative Committee (IRCC).

IRCC has proposed two values for atmospheric noise: the effective noise factor, which represents the energy characteristic, and the amplitude probability distribution (abbreviated APD in the report), which in most cases satisfy practical needs.

The amplitude probability distribution is important in cases when it is necessary to know the probability that noise will exceed a certain level (for example, in the case of radio communications using binary amplitude transmissions).

The effective atmospheric noise factor is determined by the relation

$$f_{atm} = P_n/kT_0 B_n.$$

Here P_n is the noise power (in watts), obtained from a short equivalent antenna without losses (i.e., efficiency $\eta_a = 1$) above a perfectly conducting earth's surface.

It is convenient to express the available power, P_n, yielded by an antenna to a matched load, through the relation

$$P_n = kT_a B_n \tag{1.1.27}$$

where T_a is some equivalent noise temperature, characterizing the intensity of noise. This relation is not rigorous, since atmospheric noise is considerably narrower in bandwidth than other antenna noise components. However, the noise spectrum usually may be assumed uniform within a comparatively narrow receiver band, and consequently, the coefficient f_a may be determined as the ratio of noise temperatures:

$$f_a = T_a/T_0 . \tag{1.1.28}$$

The abovementioned IRCC report contains tables for f_a, in decibels; $F_a = 10 \log f_a$ (dB). The relation

$$E_a = F_a - 65.5 + 20 \log f \tag{1.1.29}$$

is used for calculating the mean square field intensity, E_{atm} of atmospheric radio noise for a receiver band of 1 kHz [59]. Here E_a is given in decibels relative to 1 μW/m for a 1 kHz band, and f is frequency in MHz.

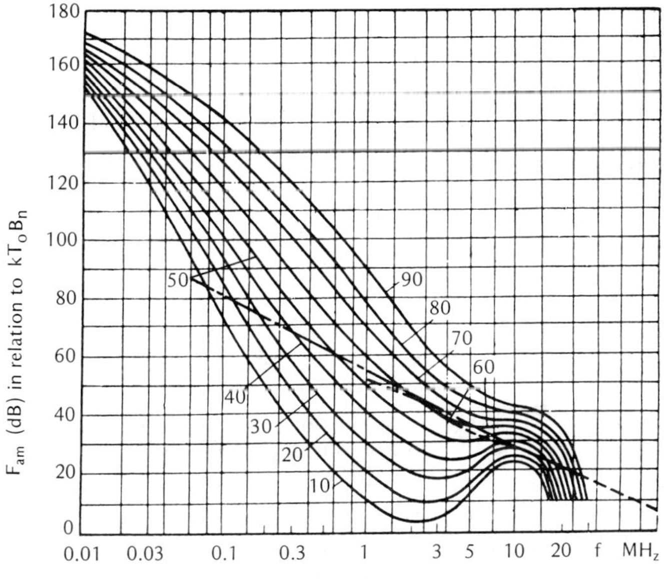

Figure 1.7

The noise field intensity for any band, B_n (Hz), other than 1 kHz, is found by adding $10 \log B_n - 30$ to E_a. Maps of the earth, on which are inscribed the average lines of identical levels of F_{am} for the wavelength $\lambda = 300$ m (f = 1 MHz), have been compiled for the median value $F_a - F_{am}$ of the noise field (in decibels relative to $kT_0 B_n$). Here the year is broken down into four seasons (winter, spring, summer, autumn), and days into six 4-hour segments. To convert F_{am} values to other frequencies, the graph of the dependence of the average F_{am} values on frequency, presented in Figure 1.7, is placed in correspondence with each map. (The numbers that identify the curves correspond to the F_{am} values for f = 1 MHz.) Noise power diminishes sharply with frequency, and only in the range of frequencies of 4-20 MHz do the curves rise a little. Atmospheric noise may be ignored on frequencies exceeding 20-30 MHz.

The broken line in Figure 1.7 shows the expected industrial noise level in "quiet" regions, far from major cities and large population centers, transmission lines and concentrated industrial noise sources. The curve of extraterrestrial background noise is indicated in Figure 1.7 by the dotted line.

The other characteristic, APD, shows the fraction of time, t(E) during which a given noise level is not exceeded. It has been established that t(E) obeys a distribution law that is approximated by the two following laws [67]:

$$t(E) = \Phi \left[0.9 \frac{E-E_m}{D_u} \right] + 1, \quad E \leqslant E_m \qquad (1.1.30)$$

$$t(E) = \Phi \left[0.9 \frac{E-E_m}{D_\ell} \right] + 1, \quad E > E_m \qquad (1.1.31)$$

Here $D_u = E_{0.9} - E_m$; $D_\ell = E_m - E_{0.1}$; $E_{0.9}$; $E_{0.1}$ are the values of E that are not exceeded for 90% and 10% of the time of each 6-hour interval; E_m is the median noise level;

$$\Phi(z) = \frac{2}{\sqrt{\pi}} \int_0^z e^{-x^2} dx.$$

E_m is connected to the median value, F_{am} by the relation shown in Equation (1.1.29).

The main parameter for D_ℓ and D_u is frequency, f. Families of D_u and D_ℓ for all for all regions of the world that correspond to each family of F_{am} are given in reference [59]. Also given in the cited work are other graphs that describe various parameters of atmospheric noise.

More detailed analyses of the laws of distribution of noise amplitudes for frequencies of 20-30 MHz and for very low frequencies are presented in references [72, 138, and 139].

4. Industrial Noise

Industrial noise is produced by operating electrical machinery, automotive ignition systems and power transmission lines. The industrial noise spectrum and intensity depend not only on the nature of the sources, but also on the nature and effectiveness of measures employed to localize them (screening of local sources). The highest industrial noise level is characteristic of large cities, industrial centers and moving objects (automobiles, airplanes, etc.). With the exception of noise from power transmission lines, the industrial noise level and spectrum are difficult to forecast and calculate. These data are monitored experimentally for each case with special test receivers. Some data on the industrial noise level at points distant from their sources are presented in IRCC *Report No. 322.* There are standards that restrict the levels of industrial radio interference, whose sources are located in cities and population centers.

1.2. PASSIVE RADIO INTERFERENCE

1. Reflected Interference from the Earth's Surface

General Description of Reflected Interference.
Radio signals radiated by both the main and side lobes of the radiation pattern of a transmitting antenna are reflected by the earth's surface and, when reaching the input of a receiver, interfere with its operation. Reflections from the earth's surface exert an increasingly strong influence on the performance of land-based radar systems as the height of a radar antenna increases. Airborne radar systems are more susceptible to reflected interference than surface radar systems. This is related to the motion of the airplane and the character of reflection from the earth at different grazing angles.

The intensity of reflected interference depends on many factors, the most important of which are wavelength, polarization of the signal, the structure, physical and chemical properties of the reflecting surface, grazing angle, etc. The reflection of waves from the earth's surface may be specular and diffuse. Pure specular reflection occurs only from perfectly smooth (mirror) surfaces, which may be, for instance, concrete-paved roads, airfields and asphalt highways. Real earth surfaces (grasslands, forests and grazing lands) produce both specular and diffuse reflections. The specular component is often called coherent and the diffuse component, incoherent.

The radiation pattern of the specular component has the same shape as when waves are reflected from a perfectly smooth surface [167]. The radiation pattern of the diffuse component approximates a sphere, tangent to the earth's surface. The size of the diffuse component depends basically on the degree of unevenness (roughness).

The kind of reflection (diffuse or specular) influences radar performance in various ways. For example, specular reflection breaks up the radiation pattern, produces false targets, or "ghosts," and often results in systematic coordinate measurement errors. Specular reflections sometimes have the same effect as deception jamming. The strength of image interference is estimated through the reflection coefficient, ρ, which is the ratio of the reflected power, P_{ref} to the incident power, P_{inc}.

The reflection coefficient, ρ, which determines the proportion of the power of the reflected wave, depends strongly not only on the kind of surface and the angle, θ_{rec} between the line perpendicular to the exposed surface and the line to the radar, but also on the polarization of the signal.

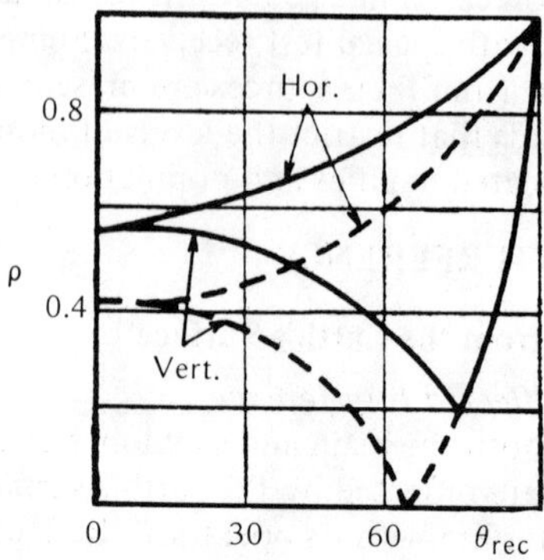

Figure 1.8

Figure 1.8 shows the curves of the reflection coefficient, ρ, as a function of θ_{rec} for radio waves with horizontal and vertical polarization [148, 74]. These curves were plotted for 10 cm (continuous curve) and 1 cm (broken curve) wavelengths for a smooth surface. At certain angles θ_{rec}, called Brewster angles, the phenomenon of perfect refraction occurs and the reflected wave either disappears or has extremely low energy. The physical reason for this is that when an electromagnetic wave from an optically less dense medium enters a medium that is more dense, the wave, when refracted at the interface between the two media, enters the denser medium (the earth) and is absorbed there. This interesting phenomenon can be used in radar to eliminate reflected interference by selecting the right radar site.

Diffuse reflection, which produces a random jamming noise signal, is most often encountered in radar. The jamming effect of diffuse reflection is estimated through the surface reflectivity, σ^0 of the reflecting part of the earth's surface. The reflectivity, σ^0 expresses the effective scatter area per unit area of the reflecting surface, and is determined experimentally. The most complete information on measurements of σ^0 is assembled in reference [65]. The characteristics of the reflectivity of certain types of surfaces at vertical incidence angles are presented in Table 1.1 [21].

The signal ξ, reflected from the earth's surface, may be represented as an additive mixture of the coherent (specular) component $\xi_c = A_m \cos \omega_s t$ and narrowband normal noise $\xi_d = A(t) \cos [\omega_s t + \phi(t)]$, produced by the diffuse component, i.e.,

$$\xi(t) = \xi_c + \xi_d = V(t) \cos [\omega_s t + \psi(t)]$$

where $V(t)$ is the envelope and $\psi(t)$ is a random phase.

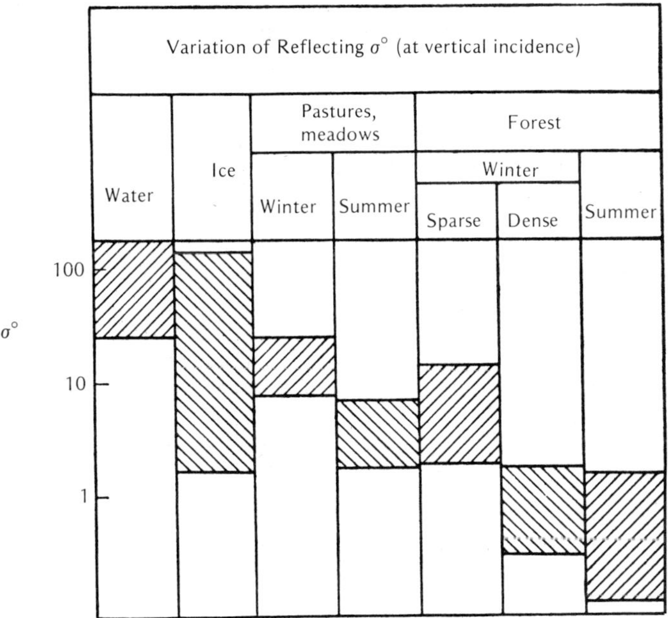

Table 1.1

In the general case, the one-dimensional amplitude distribution is Rician [65]:

$$w(V) = \frac{V}{\sigma_{\xi d}^2} \, \exp\left(\frac{V^2 + A_m^2}{-2\sigma_{\xi d}^2}\right) \, I_0\left(\frac{V A_m}{\sigma_{\xi d}}\right)$$

where $\sigma_{\xi d}^2$ is the variance of the noise component, and I_0 is the zero order Bessel function of an imaginary argument.

If the surface is extremely rough and the coherent component is small ($A_m \approx 0$), then the distribution of the envelope is close to Rayleigh

$$w(V) = \frac{V}{\sigma_{\xi d}^2} \, \exp\left(\frac{-V^2}{2\sigma_{\xi d}^2}\right) .$$

When roughness is small, i.e., when the coherent component predominates ($A_m \gg \sigma_{\xi d}$), the distribution becomes normal:

$$w(V) = \frac{1}{\sqrt{2\pi}\,\sigma_{\xi d}} \, \exp\left(-\frac{(V - A_m)^2}{2\sigma_{\xi d}^2}\right) .$$

The distribution of phase ψ of the reflected signal also depends on the ratio of the intensity, $A_m/\sigma_{\xi d}$ of the coherent and incoherent components. In the general case the probability density for phase ψ is expressed by the following equation [172]:

$$w(\psi) = \frac{1}{2\pi} \exp\left(-\frac{1}{2}\frac{A_m^2}{\sigma_{\xi d}}\right)\left\{1 + \sqrt{2\pi}\;\frac{A_m}{\sigma_{\xi d}}\cos\Phi\left(\frac{A_m}{\sigma_{\xi d}}\cos\psi\right)\times\right.$$

$$\left.\exp\left[\frac{1}{2}\left(\frac{A_m}{\sigma_{\xi d}}\cos\psi\right)^2\right]\right\}, \quad -\pi \leqslant \psi \leqslant \pi.$$

When $A_m/\sigma_{\xi d} \gg 1$ the distribution density, $w(\psi)$ is nonuniform and is maximum when $\psi = 0$. As roughness increases distribution, $w(\psi)$ approaches uniformity. In the centimeter band the coherent component often may be ignored. In this case the amplitude distribution may be assumed to be Rayleigh, and the phase distribution, uniform.

The kind of correlation function of the reflected signal or the shape of its spectrum is also important during analysis of reflected interference.

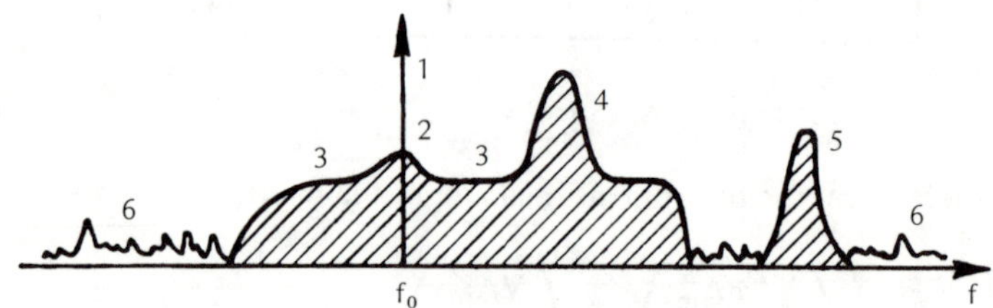

Figure 1.9

The relative motion of a radar and a target causes the spectrum of the interfering signal at the radar receiver input to be distorted considerably in comparison with the spectrum of the useful signal. Part of a spectrum corresponding to a pulse signal from an airborne radar and adjacent to the carrier frequency, f_0 of the transmitted signal is depicted in Figure 1.9 [147]. The shape of the spectrum of the entire signal may easily be determined by considering the symmetry of the components $f_0 \pm kF_p$ ($k = 1, 2, 3, \ldots$), where F_p is the pulse repetition frequency.

The transmitted signal, part of which passes directly into the receiving channel, has a discrete component (1), which is surrounded by peaked trace (2) of the signal from the earth's surface, located directly beneath the airplane. Spectral components (3) correspond to reflections from local objects, picked up by the front and rear sidelobes. The asymmetry of that part (3) of the spectrum relative to frequency, f_0 is attributed to the presence of different Doppler frequency shifts of the signals received by the antenna sidelobes from different directions.

The interfering reflection from the earth's surface, picked up by the main lobe, produces peaked trace (4). Useful signal (5) from the target, which approaches the radar, is observed in this case on an internal noise background (6). The reflecting surface may also influence the size of the systematic Doppler frequency shift. In the case of reflection from the sea surface, wave action causes a systematic frequency shift, amounting to fractions and even units of percent relative to the average Doppler frequency [65].

There are several factors by which the earth adversely affects radar performance (detection range, resolution, precision, etc.), including:

- the breaking up of the radiation patterns of radar antennas in the region of small angles of elevation, which reduces the range of detection of low-flying targets;
- the presence of local features that reduce range due to the masking of targets;

- the multiple beam nature of propagation, which increases the target elevation tracking error (the target is divided into the true target and its image);
- interference of the direct and reflected beams in conical scanning radar, which causes beat interference of the received signal and results in considerable angle errors; and
- the roughness of the earth's surface, which, when radio waves are reflected from it, substantially distorts the spectrum of the signal, interfering with the performance of moving target indication and automatic velocity tracking systems.

By analyzing these interfering factors in detail it is possible to determine rational (in a certain sense) ways and means of counteracting interference produced by natural earth factors.

Low-Flying Target Detection Range
The influence of the earth at low angles of elevation imparts multiple lobes to the pattern of a radar antenna. The bottom lobe is raised from the earth by an angle β, the size of which is related to antenna elevation, H_a and wavelength, λ by the following equation [166]:

$$\beta = \lambda/4\, H_a$$

The strength of the signal received by the radar system in consideration of the earth's influence, is written in the form [166]:

$$S = \frac{P_t\, G_t^2\, \sigma\, \lambda^2}{(4\pi)^3\, r_t^4}\, F^4(\beta) \tag{1.2.1}$$

where P_t is the radar transmitter power, G_t is the radar antenna gain, σ is the radar cross section of the target, r_t is target range, and $F(\beta)$ is the pattern-propagation factor, which for small angles of elevation, β is

$$F(\beta) = 4\pi H_t H_a / \lambda r_t \tag{1.2.2}$$

where H_t is the altitude of the target relative to the earth's surface.

If the sensitivity of the radar receiver is $S = S_{min}$ then, from Equations (1.2.1) and (1.2.2), we find the maximum range of a radar system for detecting low-flying targets

$$r_{t\,max} = \sqrt[8]{\frac{P_t G_t^2\, 4\pi\, H_t^4 H_a^4 \sigma}{S_{min}\, \lambda^2}} \tag{1.2.3}$$

In view of the fact that the maximum range of the radar in free space is

$$r_o = \sqrt[4]{\frac{P_t G_t^2\, \lambda^2 \sigma}{(4\pi)^3 S_{min}}} \tag{1.2.4}$$

we obtain from Equation (1.2.3) and (1.2.4):

$$r_{t\,max} = C \sqrt{r_o} \tag{1.2.5}$$

where

$$C = \frac{4\pi H_t H_a}{\lambda}$$

Equations (1.2.3) and (1.2.5) were derived by linearizing the function $F(\beta)$, and therefore, they are valid for the condition

$$r_{t\,max} > 10\pi\, H_t H_a / \lambda$$

Equation (1.2.5) shows that the detection range for low-flying targets is reduced substantially. Increasing the energy potential $P_t G_t$ has even less influence on the maximum detection range (eighth root) in comparison with the detection of high-flying targets (fourth root).

Obstacles also affect radar detection range because of masking (shadowing) of targets. The field in the shadow (behind an obstacle) is substantially weakened. The attenuation factor depends on the geometric size and shape of an obstacle. The attenuation factor and the corresponding reduction of radar range can be calculated in accordance with a method described in [57]. It is important to note that even comparatively small obstacles (hills, trees, etc.) can reduce detection range several fold.

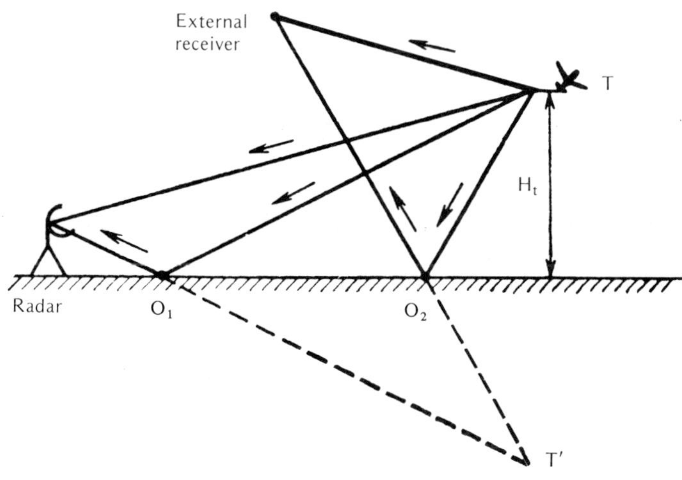

Figure 1.10

Fluctuations of Effective Reflection Center of Low-Flying Targets.
The multibeam propagation of radio waves near the earth makes a single low-flying target appear as a group. A single target may be represented with acceptable approximation as a double target (true target, T and its image, T'), as illustrated in Figure 1.10.

There are three basic cases of the formation of an image. The first is characterized by the fact that a reflection of radio waves takes place in a region adjacent to the radar antenna site (point O_1 in Figure 1.10). The resulting double target TT' exerts an interfering action on the ground radar, whose antenna tracks target T with an error that is especially appreciable in elevation. The effective center of the double target TT' wanders around in elevation and may even extend below the horizon.

The second case characterizes the reflection of radio waves in the vicinity of the target (point O_2 in Figure 1.10). This case is unfavorable for bistatic radar systems, in which the transmitting and receiving systems are separated by some distance. An example of such a system is a semiactive missile homing system of the surface-to-air type.

Finally, in the third case reflection occurs in the areas of the target and radar (points O_1 and O_2). The double target exerts an interfering influence on the performance of combined and bistatic radar systems. In application to a semiactive missile homing system, reflections exert an interfering influence both on the surface radar and on the radar homing head of the missile.

An expression was derived in reference [120] for the probability density, w(x) of the tracking error of the geometrical center of fluctuating sources TT′

$$w(x) = \frac{2\,q_s^{\,2}}{\sqrt{1 + q_s^{\,2}}\,[(1+x^2) + q_s^{\,2}\,(1-x^2)]^{3/2}} \tag{1.2.6}$$

where $x = \xi/H_t$ is the relative centroid tracking error of the paired sources (Figure 1.11); H_t is the flight altitude, ξ is linear error; and q_s^2 is the power ratio of the signal reradiated by the target to that from the earth.

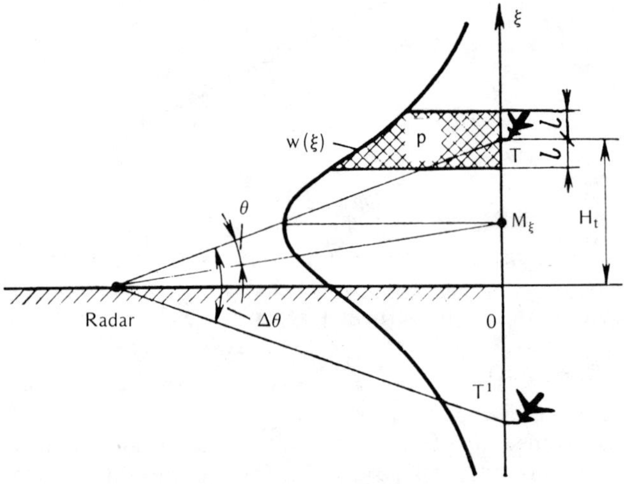

Figure 1.11

The mean relative error is

$$M_x = \frac{M_\xi}{H_t} = \int_{-\infty}^{\infty} x\,w(x)\,dx = \frac{q_s^2 - 1}{q_s^2 + 1}$$

Hence,

$$M_\xi = H_t\,\frac{q_s^2 - 1}{q_s^2 + 1} \tag{1.2.7}$$

Equation (1.2.7) shows that the mathematical expectation of the linear error of deviation of the equal signal line, $\ell_{esl} = H_t - M_\xi$ in low-flying target tracking cannot be greater than the flight altitude, i.e., $\ell_{esl} \leqslant H_t$.

If target T has linear dimensions 2ℓ (Figure 1.11), then the probability that the sighting line during tracking will lie within the target area is determined by the formula

$$p = \int_{\xi_1}^{\xi_2} w(x)\,dx = \Phi(\xi_1) - \Phi(\xi_2)$$

where

$$\Phi(\xi) = \frac{(1+q_s^2)\,\xi + (1-q_s^2)}{2\sqrt{1+q_s^2}\;[(1+\xi)^2 + q_s^2\,(1-\xi)^2]}$$

and

$$\xi_1 = (H_t + \ell)/H_t; \quad \xi_2 = (H_t - \ell)/H_t$$

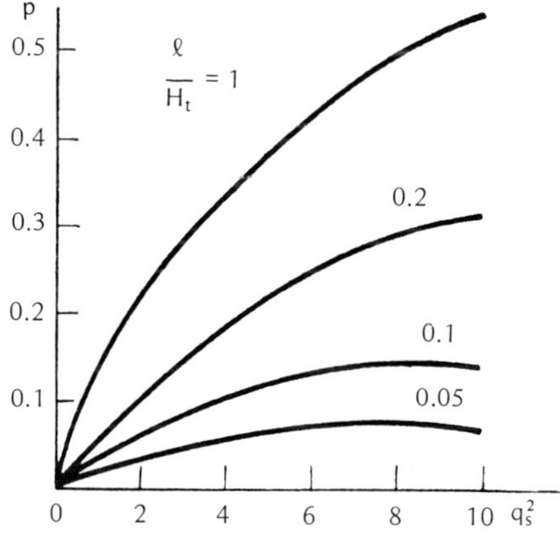

Figure 1.12

The dependence of probability, p on q_s^2 is depicted in Figure 1.12 for several ratios of ℓ/H_t. The curves illustrate the strong influence of a reflected signal on low-flying target tracking precision. The substantial deterioration of homing accuracy, even when reflections from the earth are weak, substantially reduces the effectiveness of electronic systems.

Influence of Interference Between Direct and Reflected Signals on Accuracy of Automatic Angle Tracking.

As an airplane flies over a radar station, low frequency beats appear at the input due to interference of the signals that are reradiated by the air target (direct signal) and the earth [64]. A second interfering signal also appears due to the mutual interference of the signals reflected by local objects (clutter). Interference caused by the mutual interference of the direct and reflected signals is stronger. The interfering reflection will exert a strong effect if the Doppler spectral shift, F_d is close to the radar antenna scanning frequency.

To esimate low-flying target tracking errors, we will assume that the radar operates in the passive direction finding mode with a continuous transmitted signal. Under these conditions the radar antenna input will receive the signals

$$U_1 = V_1 \cos \omega_1 t \qquad\qquad U_2 = V_2 \cos \omega_2 t$$

where U_1 is the amplitude of the useful signal, U_2 is the amplitude of the interfering signal, produced by reflections from the earth,

$$\omega_1 = \omega_0 \left[1 + (v_1/c) \right], \qquad \omega_2 = \omega_0 \left[1 + (v_2/c) \right],$$

ω_0 is the carrier frequency, and v_1 and v_2 are the radial approach speeds of the target, T and its image, T'.

For combined signal for $U_1 = U_2$ is

$$u = u_1 + u_2 = 2U \cos \frac{\omega_1 - \omega_2}{2} t \cos \frac{\omega_1 + \omega_2}{2} t.$$

If the direction finder has a quadratic detector, then its output voltage will be

$$u_t = c \left[1 + \frac{1}{2} \cos 2\omega_1 t + \frac{1}{2} \cos 2\omega_2 t + \cos (\omega_1 + \omega_2)t + \cos (\omega_1 - \omega_2)t \right]$$

$$(1.2.8)$$

where c is a constant coefficient.

Analyzing Equation (1.2.8), we notice that the detected voltage has the beat frequency component $\Omega_b = \omega_1 - \omega_2$. If the radial velocities v_1 and v_2 are such that beat frequency Ω_b comes close to scanning frequency Ω_{sc}, i.e.,

$$\Omega_b = \omega_1 - \omega_2 = \frac{2\omega_0}{c} (v_1 - v_2) \approx \Omega_{sc} \qquad\qquad (1.2.9)$$

then interference, which causes disturbance in the tracking circuit, is formed at the scanning frequency.

To determine tracking error, θ for target, T we find the generalized direction finding characteristic for the case of the direction finding of paired moving sources (Figure 1.10).

As a result of the action of signals u_1 and u_2 we obtain at the output of a radar receiving scanning antenna with pattern $f(\beta)$

$$u = U_1 L_1(\theta) [1 + m_1(\theta) \cos \Omega_{sc} t] \cos \omega_1 t +$$
$$U_2 L_2(\theta, \Delta\theta) [1 + m_2(\theta, \Delta\theta) \cos \Omega_{sc} t] \cos \omega_2 t$$

$$(1.2.10)$$

where

$$L_1(\theta) = \frac{1}{2} [f(\theta_0 - \theta) + f(\theta_0 + \theta)] \tag{1.2.11}$$

$$L_2(\theta, \Delta\theta) = \frac{1}{2} [f(\theta_0 - \theta + \Delta\theta) + f(\theta_0 + \theta - \Delta\theta)] \tag{1.2.12}$$

$$m_1(\theta) = \frac{f(\theta_0 - \theta) - f(\theta_0 + \theta)}{f(\theta_0 - \theta) + f(\theta_0 + \theta)} \tag{1.2.13}$$

$$m_2(\theta, \Delta\theta) = \frac{f(\theta_0 - \theta + \Delta\theta) - f(\theta_0 + \theta - \Delta\theta)}{f(\theta_0 - \theta + \Delta\theta) + f(\theta_0 + \theta - \Delta\theta)} \tag{1.2.14}$$

θ_0 is the displacement of the peak of the antenna beam relative to the equal signal line, and $\Delta\theta$ is the angular distance between sources T_1 and T_2.

Passing signal u from there through a quadratic detector and selective amplifier, tuned to the frequency Ω_{sc}, and assuming $\omega_1 - \omega_2 = \Omega_{sc}$, we obtain at the phase detector output, the voltage u_{pd}, equal to

$$u_{pd} = U_1^2 L_1^2(\theta) m_1(\theta) + U_2^2 L_2^2(\theta, \Delta\theta) m_2(\theta, \Delta\theta) +$$

$$U_1 L_1(\theta) U_2 L_2(\theta, \Delta\theta) + \frac{3}{4} U_1 U_2 L_1(\theta) L_2(\theta, \Delta\theta) m_1(\theta) \times$$

$$m_2(\theta, \Delta\theta)$$

$$(1.2.15)$$

Equation (1.2.15) represents the generalized direction finding characteristic. In the balanced state

$$u_{pd} = 0 \tag{1.2.16}$$

Equations (1.2.10) through (1.2.16) enable us to write an equation, with which we can find the angle tracking error for target, T

$$f^2(\theta_0 - \theta) - f^2(\theta_0 + \theta) + q_n [f^2(\theta_0 - \theta + \Delta\theta) -$$

$$f^2(\theta_0 + \theta - \Delta\theta)] + \frac{q_n}{4} \{ f(\theta_0 - \theta + \Delta\theta) \times \qquad\qquad (1.2.17)$$

$$[7f(\theta_0 - \theta) + f(\theta_0 + \theta)] + f(\theta_0 + \theta - \Delta\theta) \times$$

$$[f(\theta_0 - \theta) + 7f(\theta_0 + \theta)] \} = 0$$

where $q_n = 1/q_s$.

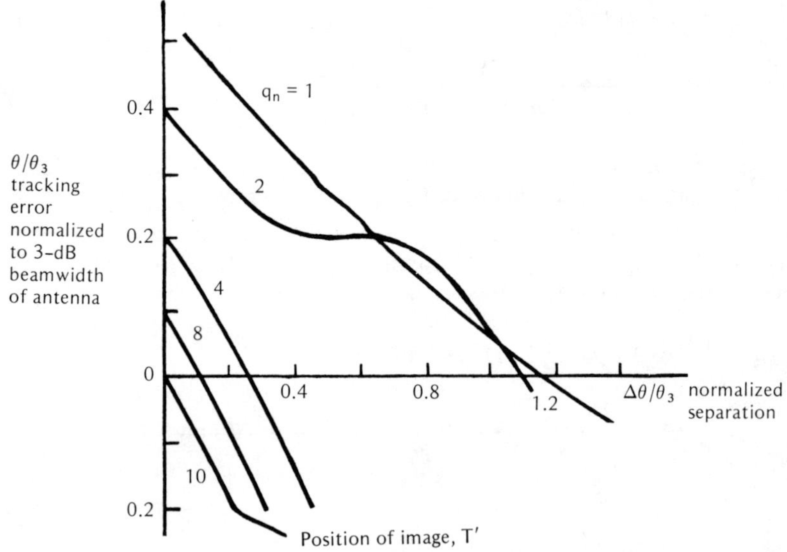

Figure 1.13

The numerical solution of Equation (1.2.17) yields the dependences of tracking error θ/θ_3 on angle $\Delta\theta$. These functions are given in Figure 1.13 for various $q_n = 1, 2, 4, 8, 10$. Analysis of the graphs shows that for comparable powers of the useful and interfering signals ($q_n \approx 1$) the effective center may be located above target T. As the power of the interfering signal increases (as q_n increases), the tracking error of target T decreases, and for small $\Delta\theta$ ($\Delta\theta/\theta_3 = 0.1$-$0.3$), the effective center shifts downward from the target.

Interfering disturbances occur whenever the beat frequency Ω_b becomes close to the scanning frequency or to its harmonics. The effectiveness of these disturbances can be reduced by changing the scanning frequency or by narrowing the passband of the tracking channel. The latter measure influences the time of action of the interference, i.e., the time during which the interfering disturbances of frequency $|\Omega_b - \Omega_{sc}|$ are located in the passband of the tracking circuit.

2. Interference from Meteorological Formulations

In certain cases interference may occur as a result of the inhomogeneous structure of the medium in which radio waves propagate. Inhomogeneities of a medium are usually related to the presence of various hydrometeorological formations in the form of rain, hail, thunder storms and cloud cover. This kind of interference may also be caused by abnormally large temperature field gradients in different regions of the atmosphere.

Reflected Interference from Clouds and Precipitation.
Many experimental investigations and observations, related to analysis of the kinds of interference that occur as a result of reflections of radar signals from clouds and precipitation, have been conducted and generalized. In application to long-range scanning radar systems with a high energy potential, these kinds of interference may be divided roughly into four major groups.

The first group includes interference which produces on PPI radar scopes bright and sharply outlined bands, the length of which in many cases reaches 200-300 km and longer, and the width several tens of kilometers. Sometimes the bright bands that light up the scope exhibit a focal type of structure. The horizontal dimensions of individual foci are estimated at 5-40 km, and their vertical dimensions reach 12-15 km.

Comparison of the results of radar observations with analysis of weather conditions in the corresponding regions suggests that the examined type of interference is caused by the passage of cold fronts, occlusions of the cold front type or strong secondary cold fronts. The bands of interference caused by cumulonimbus clouds are usually stable and appear on radar scopes for several hours.

Cold fronts may produce interference which appears on PPI radar scopes as a deformed, wavy bright band or as individual, randomly oriented bright regions on the scope, corresponding to 5-10 km long sections of the terrain. Interference of this kind is caused by cumulonimbus and large cumulus clouds and by showers in main cold fronts, located at great distances from the radar.

The second group includes interference that produces a large number of individual, unconnected and randomly scattered bright spots all over the scope. These spots are usually quite bright, and their boundaries are sharp. Interference of this kind is caused by showers and thunder storms.

The third group includes interference caused, as a rule, either by warm fronts or by occlusions of the warm front type, associated with nimbostratus clouds and continuous precipitation, covering large areas. In this connection interference produced by such atmospheric formations causes large bright regions to appear on the scope. These large bright spots are approximately uniform in the middle, gradually dim toward the periphery and do not have distinct boundaries. The size of the bright part of the scope corresponds to regions with linear dimensions reaching tens and hundreds of kilometers.

Interference of the fourth group shows up on radar scopes operating in tropical and subtropical regions. This interference is caused by clouds and precipitation produced by tropical storms (typhoons). The bright portions of PPI radar scopes caused by image interference usually appear as circular or spiral bands, converging near the center of a storm.

Attenuation in the Troposphere.
When the space between a radar and a target is filled with a mixture of various gases, water vapor and sometimes clouds and precipitation zones, it is necessary when determining the strength of the received signal to take into consideration additional attenuation. This attenuation is caused by absorption and molecular scattering of the energy of a signal by the gases that comprise the atmosphere, and also by the absorption and scattering of that energy by particles of clouds and precipitation.

The attenuation of a reflected signal is taken into consideration in calculations by introducing the attenuation factor into the corresponding formulas. The value of this factor is influenced by the distance between the radar station and target and by the properties of the matter that fills a medium in which radio waves propagate. Of the gases that make up the earth's atmosphere, oxygen and water vapor cause the greatest attenuation of electromagnetic waves.

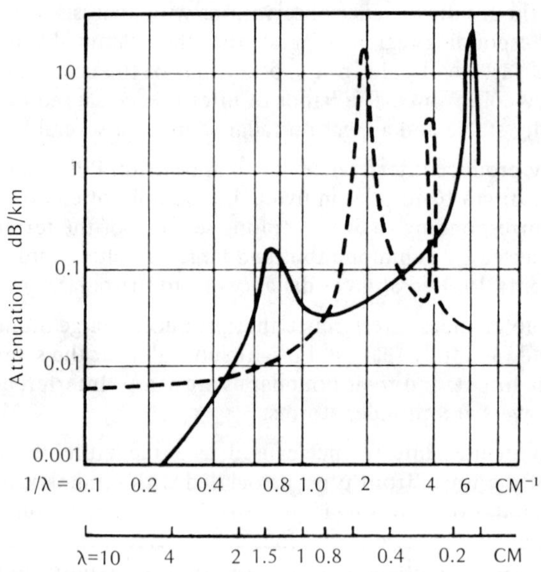

Figure 1.14

Without going into the details of the absorption mechanism, we shall simply
point out that it is related to resonance phenomena in dipolar gas molecules
and is consequently sharply frequency-dependent. The dependence of the
absorption coefficient on wavelength, λ is depicted in Figure 1.14 [4]. The
curves were plotted for the atmosphere at a pressure of 760 mm and temperature
of 20°C. As pressure decreases, for example, as altitude increases, the density of
air decreases and, consequently, so does the absorption of energy by oxygen
molecules. For example, absorption by oxygen molecules in the 0.7 to 10 cm
band decreases in proportion to the square of pressure [135, 4].

If clouds appear between a radar station and a target, attenuation of electromagnetic energy increases due to absorption and scattering by cloud particles.
This additional attenuation is characterized by the effective areas of absorption,
S_{ab} of cloud particles, and by the overall effective areas of scattering, S_{scat} of
these particles.

These areas are determined as the ratio of the power absorbed (scattered) by a
particle to the flux density, ρ_0 of the incident power:

$$S_{ab} = P_{ab}/\rho_0 \; ; \; S_{scat} = P_{scat}/\rho_0$$

Here P_{ab} and P_{scat} are the powers lost by an electromagnetic wave to absorption
and scattering.

3. Reflected Interference from the Water Surface

As of this time there is no simple analytical means of analyzing the conditions of
the surface of a body of water. This not only impedes the theoretical analysis of
the reflection of electromagnetic energy from water, but also hampers the
generalization and systematization of the results of experimental studies.

The status of any portion of the surface of a body of water, which in each specific experiment is a reflecting surface, yields only to rough, primarily descriptive
approximation. To generalize measurement results, therefore, it is necessary to
work with values that are averaged on the basis of a vague set of parameters, and
only the qualitative dependences of the properties of a reflected signal on the
parameters of the incident wave and on the type of reflecting surface can be
considered reliable. Quantitative forecasts of reflections in each specific case
may be only grossly approximate.

In most cases of practical importance emphasis is placed on the charácteistics of
backscatter, i.e., of the signal that is reflected toward the source of the radiation.

It is assumed at the present time, during theoretical determination of the
characteristics of such a signal, that a theoretical model, representing a super-
position of two statistical rough surfaces, most often corresponds to the conditions of the scattering of radio waves by a water (for example, sea) surface. One
of the mentioned surfaces is large-scale and "smooth." Its statistical parameters
are the dispersion of the random height of surface irregularities and the spatial

radius of their correlation. These parameters must be selected in accordance with the parameters of heavy sea wave action. The parameters of the other surface are assigned such that they may be viewed as small deviations from the average values, corresponding to points of the former surface. These parameters correspond to "gentle" wave action, caused, for example, by local breezes, and representing small ripples on the slopes and crests of waves.

By using the model described above, it is possible to obtain results that satisfactorily agree with experimental results in most of the frequency bands used in radar.

Analysis of existing data indicates that for grazing angles smaller than $60°$, most of the energy of a reflected signal is attributed to reflections from small-scale resonance ripples. Here the strength of an echo depends on the polarization of the transmitting and receiving antennas and is greater for vertical polarization than for horizontal.

The large-scale statistical roughness of water is determined by the mean square slope tangent of large waves. To describe small-scale surface roughness (ripples) the mean square deviation of such a surface from the plane that forms angle, α with the horizontal plane, and the relative spatial correlation radius, $k\ell$, where $k = 2\pi/\lambda$ is the wave number, and ℓ is the absolute value of the spatial correlation radius, are employed.

Theoretical analysis indicates that when grazing angles, θ_s are small (smaller than $60°$), the determining contribution to backscatter comes from resonance ripples, i.e., small irregularities of the sea surface; the wave numbers, κ of these are related to the wave number, k of the irradiating electromagnetic field by the equation $\kappa = 2k \cos\theta_s$. The height and character of ripple waves depend on the velocity of the wind above the water surface. Therefore, it is hypothesized that a direct relationship exists between local wind velocity over a given section of water surface and the strength of the signal reflected by that section of the surface in the "reverse" direction. This hypothesis is confirmed by the findings of a number of direct and indirect experiments.

Chapter 2

Manmade Radio Interference

2.1 ACTIVE JAMMING

Radio signals produced by special radio transmitters and intended for interfering with or precluding the normal operation of an enemy radio system are called active jamming. Active jamming produces at the input of a victim system a background which impedes the detection and recognition of useful signals and determination of their parameters. Jamming interference is usually linearly added to the signal at the receiver input and is therefore called additive interference.

Active jamming may be divided into three classes: continuous noise, random pulse noise and regular pulse noise. Any type of jamming reduces the probability of correct signal detection, increases the probability of false alarm and reduces measurement accuracy. For example, in a pulsed search radar system with a cathode ray display as the output system, a single blip appears on the scope when signals are received from one target in the absence of interference. Interference produces flickering (fluctuating) blips along the entire range scan, which interfere with the detection of the signal (Figure 2.1b).

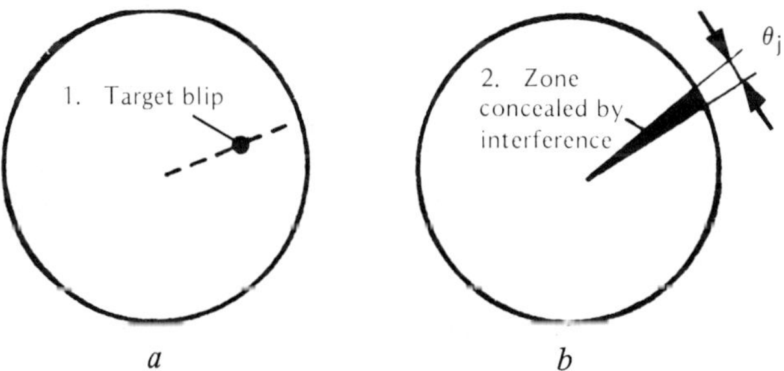

Figure 2.1

When an automatic range tracking radar system is used, the position of the gates of the range tracker is determined uniquely by the position of the pulse signals on the time axis. Interference disrupts the normal operation of the system, since the random change of the amplitude of the pulse signals across the entire range scale makes it impossible to firmly establish the tracking gates in a certain position.

The same thing happens when a radio telegraph or radio telephone transmission channel is jammed. The interference, combined with the useful signals, distorts the latter at the receiver output, which reduces the probability of correct recognition of the transmitted signals.

The effectiveness of jamming depends on many factors, especially on the time and frequency structure both of the noise and of the signal, and also on the energy ratio of the interference and signal at the input of the receiver of the victim system.

1. Description of Continuous Noise Jamming

This kind of interference is most common. It may be used successfully to overwhelm all kinds of radio systems in all operating modes [5, 14, 24, 79, 156, 203]. Continuous noise jamming can conceal useful signals on the time and frequency axes, and also their directions of arrival.

The interference examined here consists of continuous high-frequency waves, one or several parameters of which (amplitude, frequency, phase) randomly change. The systems that produce this kind of interference are characterized by the following features [79, 121, 153, 154, 203, 206]:

- the basic parameters (for example radiated power, bandwidth, etc.) of the jamming transmitter in a selected mode remain constant;
- the transmission spectrum is relatively narrow;
- the spectral density of the interference is comparatively uniform.

It may be assumed in the first approximation that the voltage of the interference $u_j(t)$ at the receiver input is a stationary ergodic narrowband random process, in which the instantaneous values have a normal distribution and the frequency spectrum is uniform within the band B_j. Such a process is called band-limited white Gaussian noise [129, 172].

Thus, interference voltage $u_j(t)$ is a random process which depends only on one real parameter, namely time, t. The specific form of a random process in some observation time is called a sample function. The form of a sample function changes from test to test. The set of all possible sample function is called a random process.

A random process is called stationary (in the broad sense) if its mathematical expectation remains constant in time and its correlation function is a function only of the time shift between two examined moments of time $\tau = |t_1 - t_2|$. In view of the property of ergodicity, averaging of the process $u_j(t)$ in time yields the same result as statistical averaging for a given moment of time for all possible sample functions of a process.

Random processes whose spectral width is many times smaller than their center frequency, f_0, i.e.,

$$B_j \ll f_0 \qquad (2.1.1)$$

are called narrowband processes. For definition we shall assume in the following discussion that the energy spectrum, $G_j(f)$ of examined random process, $u_j(t)$ has constant spectral density, N_j within the band, B_j.

The unidimensional probability density of a process is Gaussian

$$w(u_j) = \frac{1}{\sigma_j \sqrt{2\pi}} \exp\left[-\frac{(u_j - M_j)^2}{2\sigma_j^2} \right] \qquad (2.1.2)$$

where M_j is the mathematical expectation, and σ_j^2 is the variance of process $u_j(t)$, equal to the average power released per ohm of impedance by the variable component of voltage $u_j(t)$. Mathematical expectation M_j usually may be assumed to be zero.

If the interference power at the receiver input is σ_j^2, then its spectral density is

$$G_j = \sigma_j^2 / B_j \qquad (2.1.3)$$

If the spectral density of process $G_j(f)$ is symmetrical with respect to the central frequency, f_0, the correlation function of the narrow band process takes the form [129]:

$$R(\tau) = \sigma_j^2 \, \rho(\tau) \cos \omega_0 \tau = \cos \omega_0 \tau \int_{-\infty}^{\infty} G_j(f_0 - v) \cos 2\pi v \tau \, dv \qquad (2.1.4)$$

where $\omega_0 = 2\pi f_0$, and $\rho(\tau)$ is the correlation coefficient which is a slowly changing function in comparison with $\cos \omega_0 \tau$.

If the spectral density of a narrow band process is constant within the band, B_j, then

$$R(\tau) = G_j B_j \frac{\sin \pi B_j \tau}{\pi B_j \tau} \cos \omega_0 \tau \qquad (2.1.5)$$

The function $R(\tau)$, determined by Equation (2.1.5), is illustrated in Figure 2.2. The correlation time of a narrow band process is determined by the relation

$$\tau_c = \int_0^{\infty} \rho(\tau) \, d\tau \qquad (2.1.6)$$

It is often assumed in practice that correlation vanishes outside of interval τ_c.

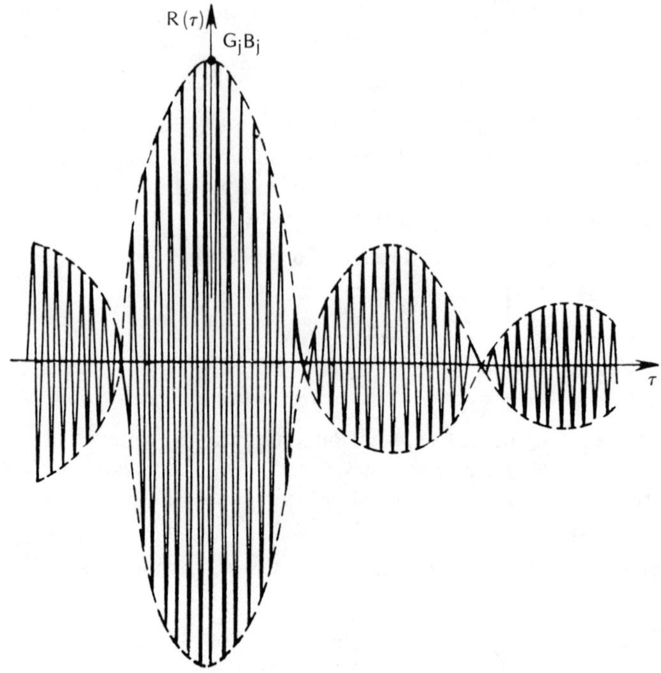

Figure 2.2

For a narrow band process with uniform spectral width, B_j, the correlation time
is

$$\tau_c = 1/2B_j \tag{2.1.7}$$

Since the sample functions of a narrowband process resemble a modulated
sinusoidal wave, it is customary to use an approximate representation of such
a process in the form

$$u_j(t) = U_j(t) \cos [\omega_0 t - \phi(t)] \tag{2.1.8}$$

where $U_j(t)$ and $\phi(t)$ are slowly changing (in comparison with $\cos \omega_0 t$) random
functions. The rate of change of the envelope and phase of wave $u_j(t)$ is inversely
proprtional to the spectral width of the process.

Real continuous jamming exerts less jamming action than Gaussian noise.
Specific examples will be presented below to explain why the effectiveness of
real jamming is diminished.

The effectiveness of jamming is characterized quantitatively as the jamming coefficient k_{jr}, which determines the minimum required input power ratio of jamming, J to signal, S in the passband, B_r of the linear part of a receiver for a given jamming power [5, 24], i.e.,

$$k_{jr} = (J/S)_{min} \tag{2.1.9}$$

A given jamming power for different systems produces different effects. For example, when two regular signals are used in a radio telegraph communications channel the jamming effect will consist in an increase of the general recognition error probability for these signals to a specified level. The required jamming effect in scanning radar consists either of an increase in the false alarm probability to the level where the system cannot function properly (the correct detection probability remains constant), or in a reduction of the correct detection probability below tolerance at a constant false alarm probability. The required jamming power for automatic target tracking radar systems may be expressed as the tracking interruption probability.

A system is considered to be jammed if the receiver input noise-to-signal ratio, throughout the time of jamming satisfies the inequality

$$(J/S) \geqslant k_{jr} \tag{2.1.10}$$

The smaller the jamming coefficient (under otherwise identical conditions), the more effective jamming is.

If, as a result of the action of Gaussian noise on a receiver, the jamming coefficient is k_j, and when real jamming is used it becomes k_{jr}, then the inequality

$$k_j < k_{jr} \tag{2.1.11}$$

is valid. The coefficient k_{jr} usually may be expressed through k_j as follows [24]:

$$k_{jr} = k_j \, (\beta_{cm}/\eta) \tag{2.1.12}$$

Here $\eta \leqslant 1$ is called the jamming quality coefficient, which is the power ratio of Gaussian noise to real jamming at the jammed receiver input for the same jamming effect; $\beta_{cm} \geqslant 1$ is the coefficient that characterizes the attenuation of jamming by special counter-countermeasures that can be incorporated in the receiver. Assuming that a receiver always has a filter that is matched with the signal, we may write for Gaussian noise $\beta_{cm} = 1$.

Some examples of continuous jamming and the dependences of η and β_{cm} on the characteristics of different types of jamming are explained below.

2. Kinds of Continuous Jamming

Direct-Noise Jamming.

Direct-noise jamming most closely approximates Gaussian noise. There are two ways to make direct-noise jamming. The first is to use an SHF noise generator. The signals produced at the output of such a generator are amplified and radiated into space. Gas discharge tubes may be used as primary SHF noise generators. A noise generator consists of a gas discharge tube, a segment of a high-frequency transmission line, and a matching system. Coaxial and wave guide generators are used, depending on the type of high-frequency line employed. Wave guide noise generators are built for 0.2 to 10 cm wavelengths, and coaxial for 10 - 12 to 120 - 140 cm wavelengths [168]. Gas discharge tubes are extreme wide-band sources of high-frequency noise and have a highly uniform spectrum.

The other way to generate direct-noise jamming is to use the heterodyning method to transfer the noise from a low-frequency oscillator to the high-frequency range. Primary noise sources on low frequencies are directly heated diodes, magnetic field thyratrons and photomultiplier tubes [5, 168].

The direct-noise quality coefficient would be equal to unity (for the usually valid condition $B_i > B_r$), but for amplitude limiting, which occurs in any feasible amplifier. Amplitude limiting results in a change of the jamming spectrum and of the distribution of its instantaneous amplitudes, with the result that the quality of the jamming is diminished.

If the spectrum of a high-frequency noise generator is assumed to have a rectangular envelope and the center frequency in the spectrum is f_0, then, after a broadband symmetrical limiter the noise spectrum will consist of an infinite series of components, grouped around the frequencies f_0, $3f_0$, $5f_0$, ... [174]. Broadening of the main part of the spectrum (near f_0), caused by the formation of beat frequencies from components of the input noise spectrum as the result of limiting, does not exceed 10%. Not less than 70% of the output noise power falls within the main part of the spectrum at the limiter output.

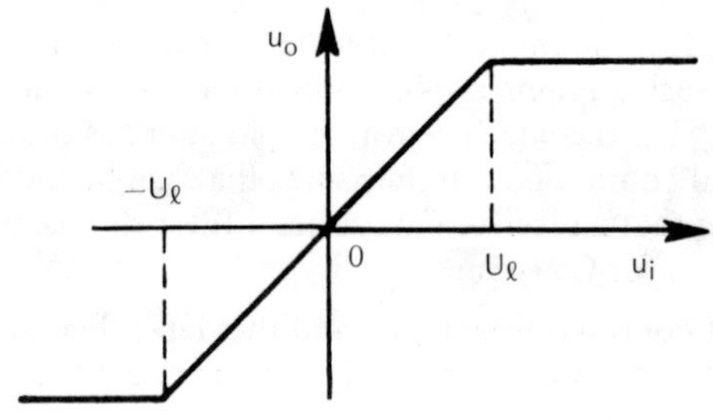

Figure 2.3

The probability density of the Gaussian process in Equation (2.1.2) is deformed at the output of a broadband clipper. Let the limiter have the response depicted in Figure 2.3, i.e.,

$$u_o = \begin{cases} -U_1 & \text{for } u_i < -U_1 \\ u_i & \text{for } -U_1 \leq u_i \leq U_1 \\ U_1 & \text{for } u_i > U_1 \end{cases} \qquad (2.1.13)$$

where U_1 is the limiting threshold.

In the interval $[-U_1, U_1]$ the conversion $u_o = f(u_i)$ is linear. Therefore, the probability density remains constant within that interval. The probability that $u_o < -U_1$ or $u_o > U_1$ is zero. All values of u_i, for which $u_i < -U_1$ is valid, are converted by the limiter to one output voltage $u_o = -U_1$; similarly $u_o = U_1$ for all $u_i > U_1$. Therefore, the probabilities

$$\rho_1 = \int_{-\infty}^{-U_1} w(u_i)\, du_i + \rho_2 = \int_{\infty}^{U_1} w(u_i)\, du_i$$

for u_o are transformed into delta functions, located at the points $u_o = -U_1$ and $u_o = U_1$; the factors of the delta functions are p_1 and p_2.

Thus the probability density of the limiter output noise is

$$w(u_o) = \begin{cases} w(u_i) + \rho_1 \delta (u_o + U_1) + \rho_2 \delta (u_o - U_1) \\ \quad \text{for } -U_1 \leq u_o \leq U_1, \\ 0 & \text{for } u_o < -U_1 \text{ and for } u_o > U_1 \end{cases} \qquad (2.1.14)$$

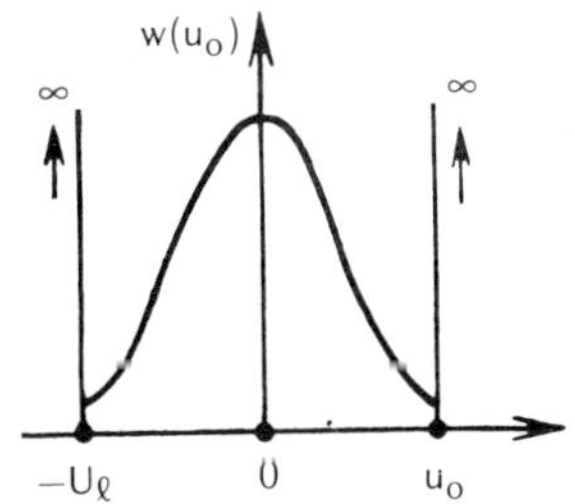

Figure 2.4

The graph of probability density $w(u_o)$ is shown in Figure 2.4.

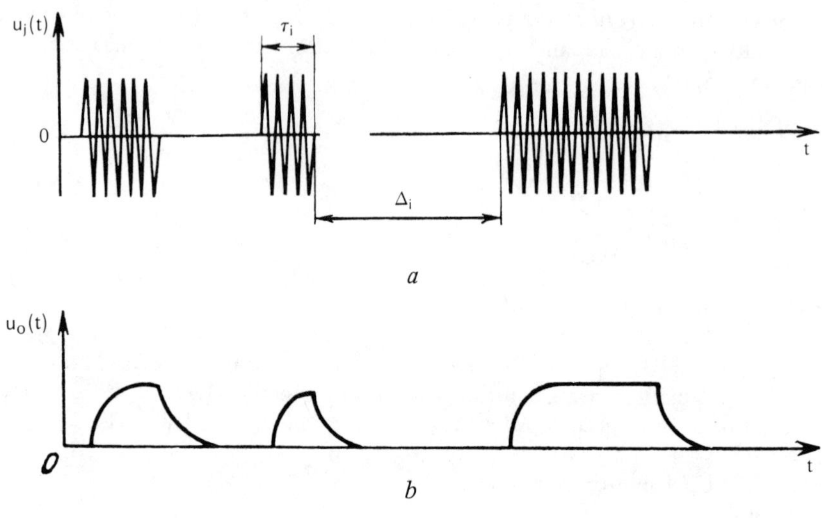

Figure 2.5

Amplitude limiting reduces the effectiveness of jamming, primarily because of the fact that its energy is redistributed through the spectrum. That, however, is not the most important reason. The reduced effectiveness of jamming is due largely to a change of its structure as a result of hard limiting.

If the mean square value of the noise is much higher than the limiting threshold $(\sigma_j \geqslant U_l)$, then jamming will degenerate to pulses, $u_j(t)$ with approximately constant amplitude and with randomly changing durations, τ_i and intervals, Δ_i (Figure 2.5a). Assuming that the spectral width of the jamming is matched to the passband of the receiver of the victim system $(B_j \simeq B_r)$, we arrive at the conclusion that the amplitude of the victim system pulses is also the amplitude of the jamming maintained constant at the receiver output (Figure 2.5b). Jamming of this kind has poor properties [24].

It is natural to ask why the ratio σ_j/U_l cannot be reduced in order to increase the effectiveness of jamming. Actually, when $(\sigma_j/U_l) \leqslant 1$ limiting in the amplifier has no effect on jamming and its properties remain optimum, but the output power amplifiers of the jamming transmitter in this case would operate in an extremely poor mode and transmitter efficiency would be exceedingly low. A reasonable compromise can be found by selecting $(\sigma_j/U_l) \simeq 1$. In this case the quality of the jamming would be somewhat poorer than that of Gaussian noise, but would be a minor sacrifice; the quality coefficient of the real jamming would remain close to unity.

According to existing data [5, 79, 156], direct-noise jamming is a promising technique.

Amplitude-Modulated Noise Jamming.
Amplitude-modulated noise consists of continuous sine waves, amplitude-modulated by noise. The noise signal at the receiver input may be written as follows:

$$u_j(t) = U_j \left[1 + k_a \Delta U_m(t)\right] \cos \omega_0 t \tag{2.1.15}$$

Here k_a is the steepness of the modulation characteristic of the transmitter; $\Delta U_m(t)$ is the modulating voltage, which is supplied by a noise generator.

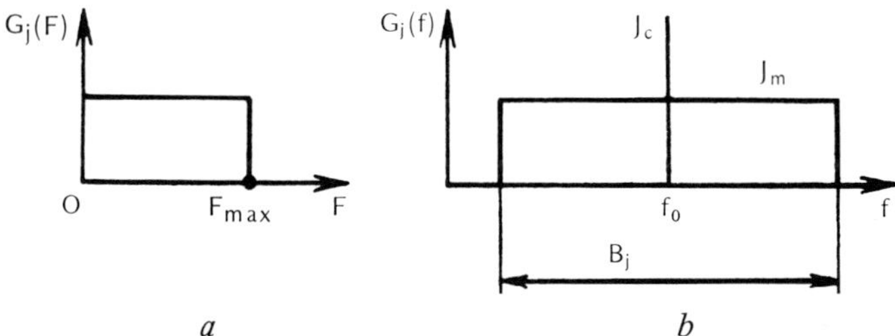

Figure 2.6

If the modulating noise has a constant spectral density within the frequency range of zero to F_{max} (Figure 2.6a), then the spectral density of the modulated wave will also be constant (Figure 2.6b), and spectral width is

$$B_j = 2F_{max} \tag{2.1.16}$$

The noise spectrum includes a wave at the carrier frequency and modulation components. It is easy to show that at 100% modulation of the sine wave by rectangular bipolar pulses with an average duty factor of $1/2$, the power ratio of the modulation components, J_m to the carrier power, J_c is equal to unity. Therefore, the inequality

$$J_m \leqslant J_c \geqslant 0.5 J \tag{2.1.17}$$

is valid, where J is the total jamming power.

Since only the modulation components of the spectrum produce a jamming effect, not more than 50% of jamming power is used for the direct jamming effect in amplitude modulation.

If the spectral width, B_j of the jamming exceeds the receiver passband, B_r of the victim system and the center frequency, f_0 of the jamming and the resonance frequency of the receiver are identical, then part of the power of the modulation components of the jamming spectrum, equal to

$$J_{mr} = J_m (B_r/B_j) \tag{2.1.18}$$

will pass through the receiver. Suppression of a signal in the absence of J_c requires at least that the condition

$$J_{mr}/S = k_j \tag{2.1.19}$$

be satisfied. In consideration of J_c we obtain for the jamming power, J that falls within the receiver passband

$$J = J_c + J_{mr}$$

and the real jamming coefficient is

$$k_{jr} = \frac{J}{S} = \frac{J_c + J_{mr}}{S} \tag{2.1.20}$$

Assuming $J_c = J_m$ and substituting Equations (2.1.18) and (2.1.19) into Equation (2.1.20), we obtain

$$k_{jr} = k_j [1 + (B_j/B_r)] \tag{2.1.21}$$

i.e., as the spectral width of the amplitude-modulated jamming increases in comparison with the receiver passband, the jamming coefficient increases nearly proportionately. The jamming quality coefficient in this case decreases correspondingly:

$$\eta = \frac{1}{1 + (B_j/B_r)} \tag{2.1.22}$$

Because of the fact that the spectrum contains a component at the carrier frequency, which does not produce a jamming effect, the jamming quality coefficient cannot be higher than 0.5 and decreases as B_j increases in comparison with B_r.

The quality of jamming is influenced by limiting, which occurs in any transmitter. For definition we shall examine limiting in a modulator. The characteristic dependence of the amplitude, U_w of high-frequency waves on the modulating voltage, U_m is illustrated in Figure 2.7. The operating point A is placed at the midpoint of that characteristic. The initial displacement, which determines the position of the operating point, is U_0. As the modulating voltage changes, the amplitude of the waves may change from 0 to U_{max}. The variable component of control voltage, $\Delta U_m(t)$ comes from the noise generator. If $|\Delta U_m(t)| > U_0$, then amplitude clipping of the jamming signal begins to take place. In Figure 2.7, limiting occurs in the intervals $t_1 - t_2$ and $t_3 - t_4$.

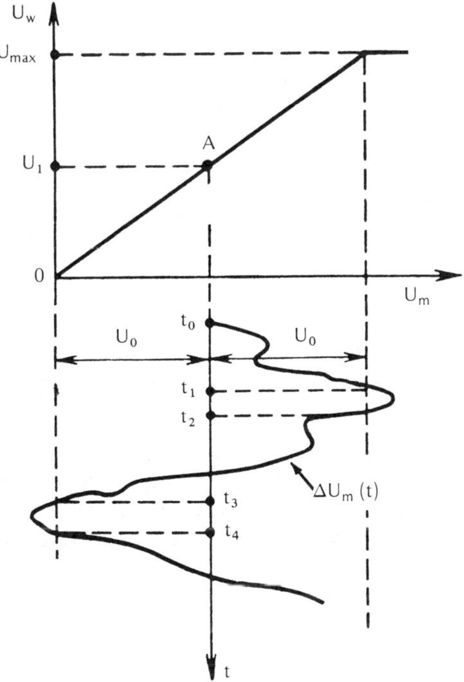

Figure 2.7

Voltage, U_0 may be called the limiting threshold. The relation between the effective modulating voltage, ΔU_e and the limiting threshold is

$$m_e = \Delta U_e / U_0 \qquad (2.1.23)$$

The effectiveness of jamming depends to a great extent on the choice of the value of m_e. The function $\eta = f(m_e)$ is illustrated in Figure 2.8. The character of this function is explained as follows. If $m_e \leqslant 1$, then the effect of amplitude limiting may be ignored, the jamming has good properties, but in this case the depth of modulation of high-frequency waves is very shallow. Consequently the power of the modulation components of the jamming spectrum is much lower than the power on the carrier and as a result the jamming quality coefficient is small.

The ratio J_m/J_c increases with m_e and becomes equal to unity when $m_e \to \infty$. Therefore, as m_e increases the jamming quality coefficient increases at first, but when $m_e \gtrsim 1$ the jamming effectiveness again becomes low because of a

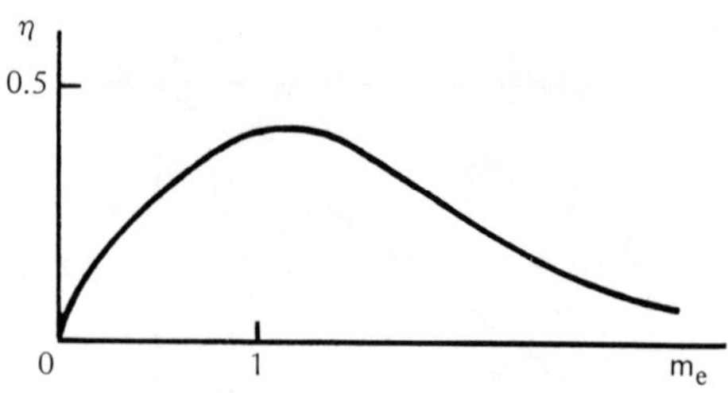

Figure 2.8

change of the structure of the jamming: the jamming transformed into pulses of approximately constant amplitude, which, as was explained above, exhibit poor properties.

To maximize the effectiveness of amplitude-modulated noise jamming, it is advisable to select $m_e \simeq 1$, and $B_j \approx B_r$. When these conditions are satisfied the jamming will differ little from Gaussian noise (with the exception of the presence of a carrier frequency), and it can be counteracted by optimum filtering methods.

Frequency-Modulated Noise Jamming.
The voltage of frequency-modulated noise at the input of a receiver may be written as follows:

$$u_j(t) = U_j\cos\left[2\pi f_0 t + 2\pi \int_0^t \Delta f(\xi)d\xi + \varphi_0 \right] \tag{2.1.24}$$

where U_j is the amplitude of the waves, f_0 is the center frequency, $\Delta f(t) = k_f \Delta U_m(t)$ is the random change of the frequency of the waves, and k_f is the steepness of the modulation characteristic.

One of the basic parameters of frequency-modulated waves is the effective frequency deviation index β_f, which is equal to the ratio of the effective frequency deviation, f_{de} to the effective spectral width of the modulating voltage, B_m, i.e.,

$$B_f = f_{de}/B_m$$

When narrow band noise is used for modulation, i.e., when $\beta_f \geqslant 1$, the spectral width of the jamming may be assumed equal to $2f_{de}$.

In the case of frequency modulation the frequency deviation, as is known, is proportional to the amplitude of the modulating voltage and does not depend on the modulation frequency. If the distribution $w(\Delta U_m)$ of the instantaneous values of the narrow band modulating voltage $\Delta U_m(t)$ is Gaussian, then the power spectrum $G_j(f)$ of the modulated waves [39] will also be Gaussian, i.e., the spectrum will be *a priori* nonuniform (Figure 2.9).

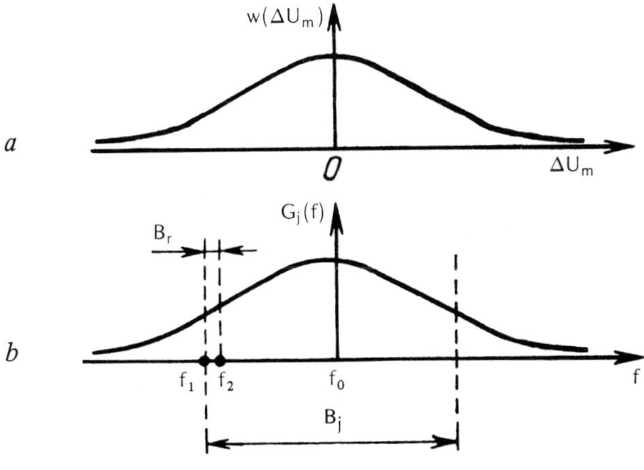

Figure 2.9

The fluctuation of the spectral density of jamming reduces the quality coefficient. If frequency-modulated jamming has a broad spectrum (selective barrage
jamming [24]), and the condition $B_j \gg B_r$ is satisfied, then the noise to signal
ratio will be substantially smaller at the edge of the spectrum (Figure 2.9b)
than in the middle. To satisfy these jamming conditions it is necessary that the
condition B_j be satisfied in the entire band

$$\frac{J}{S} = \int_{f_1}^{f_2} \frac{G_j(f)\, df}{S} = k_{jr}$$

If the condition is satisfied at the edge of the jamming spectrum, then the excess
of the jamming over the signal in the middle of the spectrum will be superfluous.

The quality coefficient of frequency-modulated noise jamming depends to a
great extent on the ratio between the spectral width of the jamming and the receiver passband of the victim system. For narrow band modulating noise, when
the conditions $f_{de} \gg B_m$ and $B_r \gg B_m$ are satisfied, the function $\eta = f(B_j/B_r)$ is
characterized qualitatively the same as the relation between η and m_e. To illustrate what we have said above, we shall examine the jamming of a pulse radar.
In the receiver of such a radar, frequency-modulated jamming produces the
greatest effect in the output, being converted only to an amplitude-modulated
wave at the output of a filter with bandwidth B_r. Figure 2.10a shows the fre-
quency response $|k_{ifa}(f)|$ of a receiver, acted upon by a jamming signal with
randomly changing frequency $f_0' + \Delta f(t)$. When $B_j \ll B_r$ the fluctuation component ΔU of the jamming is much smaller than its constant component
U_{max} (Figure 2.10b), i.e., the power of the jamming is utilized very inefficiently,
and consequently the jamming quality coefficient is small.

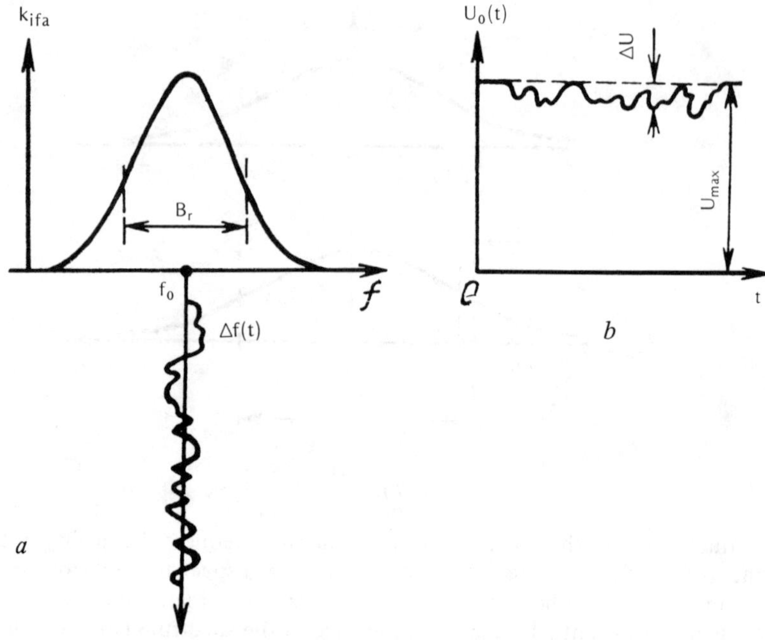

Figure 2.10

If $B_j \gg B_r$ then the effectiveness of the jamming decreases due to a change of its structure: at the output of the high-frequency filter the jamming will consist of radio pulses with approximately constant amplitude, and with randomly changing durations and intervals. The average duty factor of these pulses will increase in proportion to the ratio B_j/B_r.

The time structure of wide band frequency-modulated jamming at the output of a selective filter is a means of effectively utilizing special counter-counter-measure systems, for example, level selectors. It should be assumed that $\beta_{cm} \gg 1$ in application to wide band frequency-modulated jamming.

3. Description of Random Pulse

In general, jamming of this kind may be represented as a burst of radio pulses with a given duty factor, the amplitudes and durations of which, and also the spaces between adjacent pulses, change randomly. In practice such jamming is difficult to produce. It is considerably easier to generate a burst of radio pulses with a constant amplitude and with randomly changing duration and time spaces. The principle whereby such jamming is generated is explained in Figure 2.11. High-frequency waves of a given frequency are generated only at moments of time when voltage $u_n(t)$, supplied by the noise generator, exceeds threshold U_0. For example, consider a flip-flop, which may be used to control high frequency waves. When the noise voltage crosses the threshold from below,

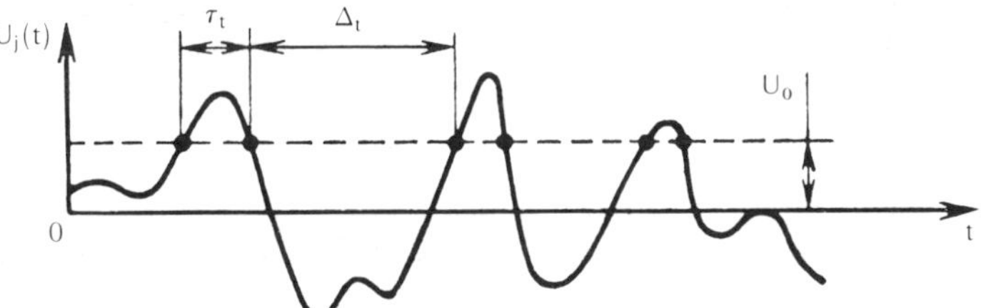

Figure 2.11

it is switched to one stable state, and when this threshold is crossed from above,
it is returned to the initial state. If the probability density of the instantaneous
values of the noise is Gaussian, with an average of zero, then the average pulse
durations M_τ, the spaces between them, M_Δ, and the number of crossings,
N_{av} of the threshold per unit time may be determined with the following
equations [129]:

$$M_\tau = \frac{\pi}{\sqrt{-\rho_0''}} \left[1 - \Phi\left(\frac{\gamma}{\sqrt{2}}\right) \right] \exp\left(\frac{\gamma^2}{2}\right)$$

$$M_\Delta = \frac{\pi}{\sqrt{-\rho_0''}} \left[1 + \Phi\left(\frac{\gamma}{\sqrt{2}}\right) \right] \exp\left(\frac{\gamma^2}{2}\right)$$

$$N_{av} = \frac{1}{\pi} \sqrt{-\rho_0''} \, \exp\left(-\frac{\gamma^2}{2}\right)$$

where

$$\rho_0'' = \left. \frac{d^2 \rho(\tau)}{d\tau^2} \right|_{\tau = 0}$$

$\rho(\tau)$ is the noise correlation coefficient of the generator

$$\Phi(\gamma) = \frac{2}{\sqrt{\pi}} \int_0^\gamma e^{-x^2} \, dx$$

is the probability integral, different from the one used in [129], $\gamma = U_0/\sigma_n$, and
σ_n^2 is the variance of the noise produced by the generator.

Since, for example, $B = F_{max} - F_{min}$ for wide band noise voltage characterized by uniform spectrum within the frequency band, the correlation coefficient is:

$$\sin (\pi B \tau)/\pi B \tau$$

In this case the average pulse durations, spaces between the pulses and the number of crossings of a given level are functions of threshold, U_0 and are written as follows:

$$M_\tau = \frac{\sqrt{3}}{B} \left[1 - \Phi \left(\frac{U_0}{\sqrt{2}\, \sigma_n} \right) \right] \exp \left(\frac{U_0^2}{2\, \sigma_n^2} \right)$$

$$M_\Delta = \frac{\sqrt{3}}{B} \left[1 + \Phi \left(\frac{U_0}{\sqrt{2}\, \sigma_n} \right) \right] \exp \left(\frac{U_0^2}{2\, \sigma_n^2} \right)$$

$$N_{av} = \frac{B}{\sqrt{3}} \exp \left(- \frac{U_0^2}{2\, \sigma_n^2} \right)$$

By changing U_0 it is possible to select the desired relation between M_τ and M_Δ. The average noise pulse repetition frequency is determined by the spectral width of the modulating noise.

By selecting the appropriate threshold, U_0 and values of M_τ and M_Δ, it is possible to make the average pulse duty factor equal to one half. Then the probability density for τ and Δ [129] is assumed to be determined by an exponential law:

$$w(\tau) = N_{av} \exp (-N_{av} \tau), \qquad \tau > 0$$

$$w(\Delta) = N_{av} \exp (-N_{av} \Delta), \qquad \Delta > 0$$

Spectral density, $G(\Omega)$ and the correlation function, $R(\tau)$ of a stationary sequence of independent rectangular pulses with constant amplitude, U_1 and $M_\tau = M_\Delta$ are

$$G(\Omega) = U_\varrho\, N_{av}/(\Omega^2 + 4\, N_{av}^2)$$

$$R(\tau) = (1/4)\, U_\varrho^2 \exp (-2\, N_{av}\, |\tau|)$$

Correlation time, τ_c, uniquely related to the spectral width of the process, is $\tau_c = 1/2N_{av}$.

It is possible to generate radio pulses with a given duty factor, constant amplitude and duration, but with a randomly changing space between the pulses.

To solve this problem it is sufficient in principle to establish each crossing of threshold level, U_1 from below by noise voltage with the aid of an electronic relay produces a voltage pulse with a certain amplitude and duration, which is used to control high-frequency waves.

It is important to note in relation to high-frequency random pulses that real opportunities exist for generating coherent bursts of noise pulses using systems for the long-term storage of the frequency of victim signal [24].

Random pulse jamming is effective against command guidance radio links, radio communications links and certain types of radars. Random pulse jamming produces a code barrage jamming effect against command links. It completely or partially jams transmitted commands, changes the subcarrier modulation parameters and introduces false commands. One of the most important indices for evaluating the effect of code barrage jamming on the performance of command links is the average number of noise pulses that reach the receiver input per unit time (0.5 N_{av}). The optimum value of N_{av} depends on the kind of useful signal. Another important characteristic is the pulse power ratio of the jamming and signal.

To ensure the effectiveness of random pulses when jamming radio telegraph and radio telephone communications lines it is also necessary to select the optimum durations of the jamming pulses and pauses between them. For example, to jam a radio telephone line it is desirable, depending on the energy spectrum of Russian speech, to use an average jamming pulse repetition frequency of 300-400 Hz with a duty factor of one half [24].

Random pulses may differ from useful signals in terms of a number of indices. There may be differences in the time structure. For example, random pulses are used to jam radio communications channels, which in many cases are characterized by a continuous signal, whereas the jamming is decidely pulse-like in character. There may be differences in the spectral widths of the signal and jamming. One of the major factors in relation to counter-countermeasures is the fact that the average frequencies of the jamming and signal are always different. The minimum jamming transmitter tuning error for active jamming is comparable with the receiver passband of the victim system. And if the receiver incorporates, for example, coherent signal processing, then the frequency difference between the signal and jamming may promote a substantial reduction of the jamming effectiveness.

The random position of jamming pulses on the time axis may also be of great importance for counter random pulses. For example, the use of pulse-to-pulse integration systems may substantially improve the signal to noise ratio.

4. Description of Bursts of Regular Pulses

An example of this kind of jamming is multiple synchronous pulses, which consists of a series of radio pulses, radiated in response to a signal received from a radar by an electronic countermeasure system. The jamming pulses match the useful signal in terms of shape and duration. The synchronism of the envelopes of jamming pulses relative to the radar timing is an important factor. At the same time the jamming differs in several ways from the useful signals. The jamming is usually considerably stronger than the signal in terms of amplitude (power). Consequently amplitude selection becomes very important in terms of ECCM.

Jamming pulses often have a constant amplitude and repetition frequency, which suggests the feasibility of using interperiod cancellation systems, for example, for countering jamming.

The frequency difference between the signal and jamming may be very important for countering this type of jamming. For example, when coherent processing is used, the frequency of the jamming after the phase detector may be far outside of the passband of its filter.

A change of the pulse repetition frequency from period to period or of the carrier frequency of radar signals is of considerable importance in terms of reducing the effectiveness of jamming. When these steps are taken, the jamming can conceal only regions that are separated from the radar by a distance greater than the distance to the jammer.

2.2 ACTIVE DECEPTION JAMMING

1. General Information

Active deception jamming is usually intended for inserting false information into a victim system. The corresponding information channels are occasionally overloaded by deception jamming. In the latter case the jamming causes the electronic system to operate at the limit of its transmission capacity, or more seriously, the design transmission capacity of the channel becomes inadequate for transmitting the required information.

To preclude the possibility of filtering, the deception jamming signal should not differ too much from the forged signal in terms of minor (secondary) parameters. For example, when imitating a false target on the same heading as the actual target, but at a different range, the jamming signal should have at least the identical polarization and carrier frequency as the useful signal. However, it differs from the useful signal in terms of the information parameter, i.e., the jamming signals are radiated with some lag in relation to the useful signals.

The information and secondary parameters of the jamming and useful signals are statistically related, and this relationship may become a functional relation in many cases.

Deception jamming is used to jam radars, radio communications links, command radio links, radio navigation systems, etc., depending on the purpose of the victim system.

The large variety of radar deception jamming can be divided into two groups: one for jamming detection systems, and the other basically for jamming automatic tracking systems.

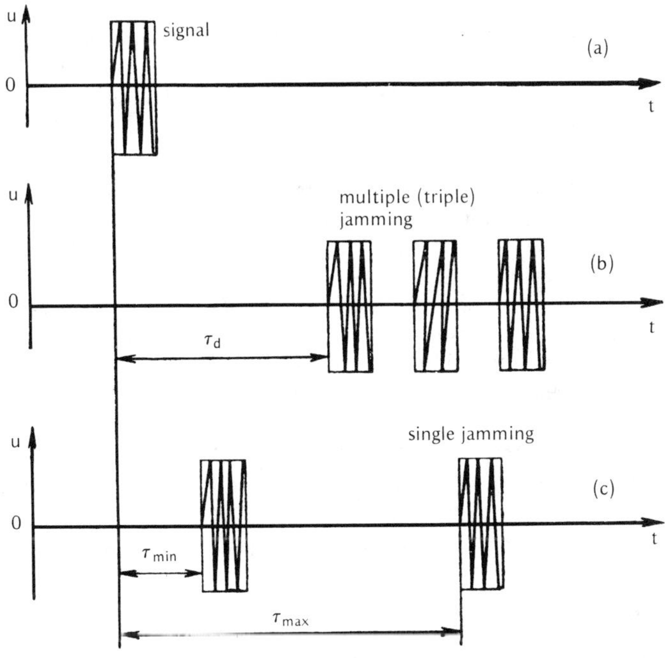

Figure 2.12

2. Deception Jamming of Pulse Detection Radars

Jamming of this group causes false blips, identical to the blips of real targets, to appear on the radar scope. By using such jamming, which, in turn, is broken down into multiple and single response jamming pulses, it is possible to confuse the radar operator and to overwhelm the data processing system.

Multiple response jamming pulses are a series of radio pulses, emitted in response to a signal received from victim radar (Figure 2.12b). The jamming and echo pulses are identical in terms of shape, duration and power. Multiple response jamming is classified as synchronous or asynchronous in relation to the repetition frequency of the transmitted radar pulses.

Asynchronous jamming pulses are generated by radiating a burst of radio pulses at arbitrary moments of time, which, in general, are not related to the time position of the radar pulses. Asynchronous interference pulses can be generated to simulate mutual interference in order to complicate the general situation and to confuse the people responsible for electromagnetic compatibility. A difference between the repetition frequency of asynchronous jamming pulses and useful pulse signal is used in the development of ECCM systems.

Synchronous pulse jamming is generated by multiple pulse jamming stations, designed on the principle of multiple retransmission of the pulse signals of the victim radar. The multiple response jamming causes series of blips, simulating nonexistent targets, to appear on radar scopes. The blips may be in front of or behind the actual target. The lag time of jamming pulses for making lead blips is about the same as the repetition period of the radar pulses.

Single response jamming is a radio pulse, radiated in response to the signal received from the radar with some delay, τ_d which varies in the range of τ_{min}-τ_{max} (Figure 2.12c). The delay time is usually changed in such a way as to imitate an actually moving target on the radar scope. The rate of change of the delay time $d\tau_d/dt$ corresponds to the speed of the imitated target (tank, ship, airplane). A sufficiently large jamming transmitter, by acting through the side lobes of the antenna, produces several false blips on the radar scope, which move at a certain speed, thereby confusing the operator or computer.

3. Deception Jamming Against Automatic Tracking Radars

Deception jamming used against an automatic tracking radar has the following effects:

- the information parameter is measured with unacceptably large errors;
- signals are incorrectly resolved;
- in many cases signal conversion characteristics are altered and the dynamic properties of a system deteriorate; and
- the operation of systems in the automatic tracking mode (in terms of range, speed and angular coordinates) is disrupted and tracking is interrupted.

4. Deception Jamming of Automatic Angle Tracking Systems

An automatic angle tracking channel, which performs angle selection of targets and measures angular coordinates and their derivatives, is one of the basic measuring channels of fire control radar. In most cases, a loss of information about the angular coordinates of the target (or the interruption of angle selection) causes the radar station to fail in the performance of its mission.

The methods used to jam an angle channel depend on the number of independent receiving channels (antennas). For example, the operation of a single-channel system, in which information about the angular coordinates of a target is obtained by scanning the antenna, can be interrupted by radiating jamming from one point (scan-frequency jamming). However, this jamming is completely eliminated in monopulse systems with two independent receiving channels (two separate receivers). Jamming generated at two and more points in space is used to jam two-channel systems.

It is important to note that jamming obeys a "priority" principle, i.e., the jamming radiated at two and more points in space is effective both for a two-channel (monopulse) system, and for a single-channel (conical scanning) system.

In some cases jamming of a "lower" class can jam systems with several independent receiving channels. For example, monopulse systems can be jammed with jamming radiated from one point, with orthogonal polarization relative to the useful signal. In this case the imperfection of antenna systems with different antenna patterns in the main and orthogonal polarizations is used.

Angle tracking systems can be jammed from one, two and more points in space.

Single-point jamming includes the following:
- scan-frequency selective jamming,
- scan-frequency barrage jamming,
- intermittent jamming, and
- cross-polarization jamming.

Jamming from two and more points includes:
- incoherent jamming,
- coherent jamming, and
- blinking jamming.

Scan-Frequency Selective Jamming.
This kind of jamming consists of an amplitude-modulated wave. The interference modulation frequency Ω_j is approximately the same as the antenna scanning frequency Ω_{sc}. Scan-frequency selective jamming can be used against a radar whose scanning frequency is either exactly known or can be determined during the process of electronic surveillance.

As a result of jamming the signal,

$$u = u_s + u_j = U_s \left[1 + m_s(0) \cos(\Omega_{sc}t + \varphi_s)\right] \cos\omega_o t +$$
$$U_j \left[1 + m_j \cos(\Omega_j t + \varphi_j)\right] \cos\omega_o t \tag{2.2.1}$$

appears at the radar input, where U_s, U_j are the amplitudes of the useful and interference signals, ϕ_s, ϕ_j are the phases of the envelopes of the useful and interference signals; $m_s(\theta)$, m_j are the useful and interference modulation coefficients. Modulation coefficient $m_s(\theta)$ depends on the angular target tracking error θ.

Jamming results in the deformation of the direction finding characteristic which for $\Omega_j = \Omega_{sc}$, $\phi_s - \phi_j = 180°$ and $\phi_s = 0$ is written for one direction finding channel in the form [24]:

$$u_{df} = k\left[m_s(\theta) - m_j(q_j/(2 + q_j))\right]$$

where $q_j = U_j/U_s$. In the absence of jamming, $q_j = 0$ and $u_{df} = km_s(\theta)$. The dependence of u_{df} on θ for $q_j = 0$ is illustrated in Figure 2.13 (curve 1).

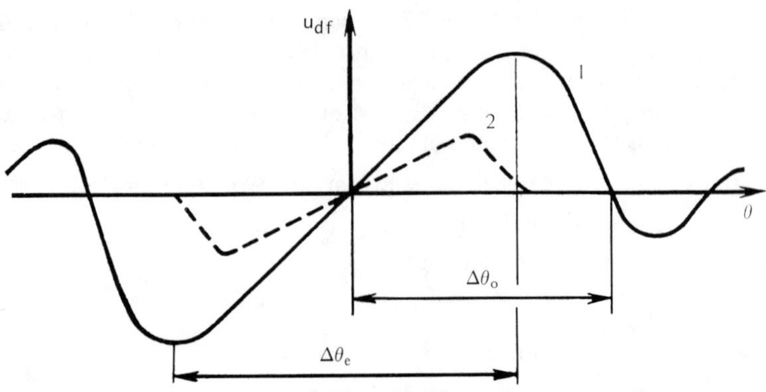

Figure 2.13

As m_j and q_j increase, the direction finding characteristic becomes asymmetrical. The stable equilibrium point is displaced to the right by the value θ, which is equal to the angular target tracking error. Error, θ can be found from Equation (2.2.2). If $u_{df} = 0$, then

$$m_s(\theta) = \frac{q_j}{2 + q_j}\, m_j \tag{2.2.3}$$

For small errors, θ we have

$$\frac{\theta}{\theta_{0.5}} = 0.36\, \frac{\theta_{0.5}}{\theta_0}\, \frac{q_j}{2 + q_j} \tag{2.2.4}$$

where θ_0 is the angle between the equal signal line and the maximum of the beam.

As follows from Equation (2.2.4), an error larger than the value

$$\frac{\theta}{\theta_{0.5}} = 0.36\, \frac{\theta_{0.5}}{\theta_0}\, m_j \tag{2.2.5}$$

cannot be attained, even with infinite jamming power.

The reason for this is that a jamming signal, radiated from a concealed object, itself carries information about its angular coordinates. Under the influence of jamming the radar antenna is deflected until the jamming modulation is counteracted by the useful modulation, which is caused by scanning the antenna, deflected from the target by angle θ.

Jamming also alters the dynamic characteristics of a system. As m_j and q_j increase, the transmission coefficient, k_{df} of the direction finding system decreases. To prove this we determine $k_{df} = du_{df}/d\theta$ at the stable equilibrium point. For simplicity, we approximate the direction finding characteristic with the aid of the function

$$u_{df} = km_s(\theta) = k\sin\theta \qquad (2.2.6)$$

Then, in consideration of Equations (2.2.6) and (2.2.3), we find the angles $\theta = \theta_{er}$, at which stable equilibrium occurs

$$\theta_{er} = \arcsin\frac{q_j}{2+q_j}\,m_j$$

The transmission coefficient of the direction finder, determined on the basis of Equations (2.2.2) and (2.2.6), for $\theta = \theta_{er}$ is

$$k_{df} = \frac{du_{df}}{d\theta}\bigg/_{\theta=\theta_{er}} = \frac{k\cos\left[\arcsin\left(\frac{q_j}{2+q_j}\,m_j\right)\right]}{\sqrt{1-\left(\frac{q_j}{2+q_j}\,m_j\right)^2}} \qquad (2.2.7)$$

A reduction of k_{df} as q_j and m_j increase may result in destabilization of the angle tracking system.

Scan Frequency Barrage Jamming.
When the scan frequency of a radar antenna is not known, scan frequency barrage jamming may be employed. There are two kinds of scan frequency barrage jamming: noise and sweep. The first kind of jamming is accomplished with a sinusoidal signal at a carrier frequency, amplitude-modulated by a low-frequency noise voltage with a uniform spectrum, overlapping the range of possible scan frequencies. If the carrier is amplitude-modulated with a sinusoidal voltage, the frequency of which changes periodically in a certain range, the result is sweep barrage jamming.

The jamming signal at the input of an angle tracking system may be represented for noise barrage jamming as

$$u_j(t) = U_j\,[1+m_j(t)]\,\cos\omega_0 t$$

where $m_j(t)$ is the amplitude modulation coefficient of the jamming signal.

When a tracking system is acted upon by low-frequency noise jamming in the absence of a signal, the variance of the angular error is determined by the formula

$$\sigma_\theta^2 = \frac{2}{k_m^2}\, G_m(f_s)\, \beta_n \tag{2.2.8}$$

where $k_m = 2.8\theta_0/\theta_{0.5}^2$, $G_m(f_s)$ is the spectral density of the random amplitude modulation coefficient of the jamming signal of the scan frequency, and β_n is the passband of the tracking system. Since $B_j \gg \beta_n$ we may assume

$$G_m(f_s) = m_{je}^2/B_j$$

where m_{je} is the effective amplitude modulation coefficient of the jamming signal. From Equations (2.2.8) and (2.2.9) we obtain

$$\sigma_\theta = \frac{\sqrt{2}}{k_m}\, m_{je}\, \sqrt{\frac{\beta_n}{B_j}} \tag{2.2.10}$$

As follows from Equation (2.2.10), the effective angle tracking error decreases as the spectral width B_j of the noise increases and the passband β_n of the tracker narrows.

The easiest way to determine the effective modulation coefficient is to represent modulating noise $m_j(t)$ as a random telegraph signal ("binary" noise). Then $m_{je} = 0.5$ [172].

Substituting $m_{je} = 0.5$ and $k_m = 2.8\,(\theta_0/\theta_{0.5}^2)$ into Equation (2.2.10), we obtain

$$\frac{\sigma_\theta}{\theta_{0.5}} = 0.25\, \frac{\theta_{0.5}}{\theta_0}\, \sqrt{\frac{\beta_n}{B_j}} \tag{2.2.11}$$

Comparison of Equations (2.2.11) and (2.2.5) shows that barrage jamming is $\sqrt{2\beta_n/B_j}$ times less effective than selective jamming.

Scan frequency barrage jamming, accomplished by changing the jamming modulation frequency $\Omega_j(t)$, causes periodic jamming at the moments of time when the jamming modulation frequency coincides with the scan frequency (to the accuracy of the passband of the tracking system).

The target angle tracking error depends on the time Δt_{ef} of the effective action of jamming on a tracking system, which is determined by the rate of change of modulating function $dF_j(t)/dt$. The effective jamming time for given tuning rate, $dF_j(t)/dt$ is

$$\Delta t_{ef} = \frac{\beta_n}{dF_j/dt} \tag{2.2.12}$$

In the case of slow tuning, the rate of change of the interference modulation frequency is selected from the condition $\Delta t_{ef} \geqslant 3\tau_n$, where

$$\tau_n = 1/\beta_n \qquad (2.2.13)$$

From Equations (2.2.12) and (2.2.13) we have

$$\frac{dF}{dt} < \frac{1}{3}\beta_n^2 \qquad (2.2.14)$$

Equation (2.2.14) shows that when slow tuning is employed, the jamming repetition period is rather long. It is determined by the formula

$$T = \frac{B_j}{dF_j/dt} \approx 3\,\frac{B_j}{\beta_n^2}$$

where B_j is the frequency control range of the jamming modulation oscillator.

In the case of high-rate interference modulation frequency control, the rate of change of the frequency is high:

$$\frac{dF_j}{dt} \gg \beta_n^2$$

The jamming time, Δt_{ef} is short ($\Delta t_{ef} \ll \tau_n$), but the jamming repetition period is close to the time constant of the system. In this case the jamming has about the same effect as barrage noise jamming [24]. The angle error also depends on the ratio of the equivalent passband to the frequency control band, β_n/B_j.

Scan frequency selective and barrage jamming also interferes with the operation of systems that measure the angular velocity of the line of sight of missile homing systems when the proportional guidance method is employed. The variance, σ_θ^2 of the measurement error of the angular velocity, θ of the sighting line is

$$\sigma_\theta^2 = \frac{1}{2\pi} \int_0^\infty |\Phi_\theta\,(j2\pi F)|^2 G_m(f_s)\,d2\pi F \qquad (2.2.15)$$

where $\Phi_\theta\,(j2\pi F)$ is the transfer function of the closed loop of the system that measures the angular velocity of the line of sight.

The distinguishing feature of an angular velocity estimator is the fact that the square of the modulus of its amplitude-frequency response, $\Phi_\theta\,(j2\pi F)$ is of the Rayleigh form. When $F = 0$, the equality $|\Phi_\theta\,(0)|^2 = 0$ is satisfied and the function $|\Phi_\theta(j2\pi F)|^2$ on some resonance frequency, F_m reaches the maximum

$$|\Phi_\theta(j2\pi F_m)|^2 = k_v^2$$

where k_v is the transmission coefficient of the angle-measuring channel in terms of speed.

The squared modulus of the transfer function $|\Phi_\theta(j2\pi F)|$ is satisfactorily approximated by the Rayleigh curve

$$|\Phi_\theta(2j\pi F)|^2 = k_1 F \exp(-k_2 F^2) \tag{2.2.16}$$

where k_1 and k_2 are approximation coefficients.

From Equations (2.2.15) and (2.2.16), and in consideration of Equation (2.2.9), we obtain

$$\sigma_\theta^2 = \frac{k_1}{2k_2} \frac{m_{je}^2}{B_j} [1 - \exp(k_2 B_j^2)] \tag{2.2.17}$$

Analysis of Equation (2.2.17) shows that for some optimum value of $B_{j\,opt}$ the variance σ_θ^2 of measurement of the angle error acquires the maximum value. The effectiveness of jamming decreases as the spectrum B_j narrows, just as in the case when it expands.

Incoherent Multipoint Jamming
Incoherent jamming is done with two or more spatially separated transmitters. There is no deterministic relation between the phases of the transmitted high-frequency waves. The radiated waves can be either modulated or unmodulated. The effectiveness of jamming depends to a great extent upon the form of modulating function.

In the simplest case, incoherent jamming is produced by two reflecting objects (for example, a pair of airplanes, ships, etc.). The action of such unmodulated interference generates target angle tracking errors.

If the angular distance between targets T_1 and T_2 is less than the beamwidth of the radar, then the equal-signal line is aimed at the effective center, 0 of the paired targets (Figure 2.14), the position of which depends on the power ratio $q_{12} = J_1/J_2$ of the jamming signals. The angle tracking error, θ of center 0 for small angular distances, $\Delta\theta$ between the sources is determined by the following formula [24]:

$$\theta = \frac{\Delta\theta}{2} \frac{q_{12} - 1}{q_{12} + 1} \tag{2.2.18}$$

If the targets are identical and $q_{12} \approx 1$, then $\theta = 0$ and the tracking errors of these targets are identical in terms of modulus and amount to $\Delta\theta/2$. An increase

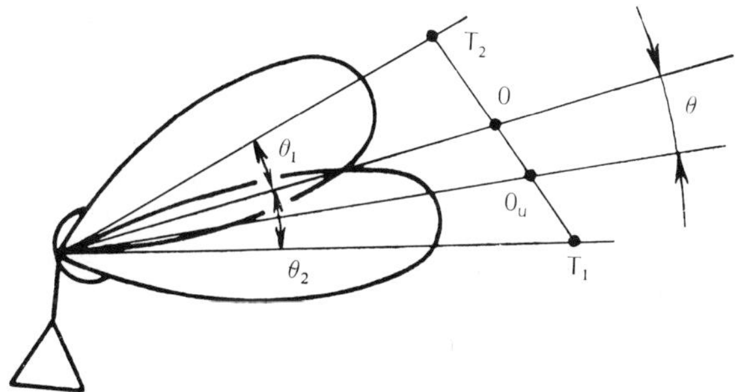

Figure 2.14

in the power of one of the sources results in a displacement of the equal-signal line toward the source with the larger power. When q_{12} = 4-5 the tracking error of the stronger source becomes negligible. This effect is incorporated in special angle tracking systems. In the Kuck system [101, 214] for example, the ratio $q_{12} \gg 1$ is achieved by "stressing" one target while tracking a group target with a transmitting antenna with an asymmetrical radiation pattern.

It is important to note that Equation (2.2.18) is valid for small angles, i.e., $\Delta\theta = (0.1\text{-}0.2)\theta_{0.5}$. As $\Delta\theta$ increases, the nonlinearity of the radiation pattern begins to exert an effect, and the behavior of the tracking system, while tracking a paired or compound target, must be analyzed with the aid of a family of generalized angle discriminator characteristics [24, 101].

Analysis of generalized angle discriminator characteristics shows that the resolution angle of unmodulated interference sources $\Delta\theta_r = (0.8\text{-}0.9)\theta_{0.5}$.

Blinking and Intermittent Jamming
Blinking jamming is done with two (or more) jamming transmitters by turning them on, one at a time. Foreign experts view blinking jamming as an effective electronic countermeasure against guided missiles [214, 101].

The important factor to remember during analysis of the effect of blinking jamming on an angle tracking system is the fact that one transmitter, T_1 or T_2 (Figure 2.14) is turned on at any given time. Thus, it is possible to evaluate the effectiveness of blinking jamming by analyzing the action of the interference signals of each source T_1 or T_2 on the system, using the angle discriminator characteristic of the system for a single source.

When subjected to blinking jamming, the control signal, u_{df} of an angle tracking system is a varying voltage with period, T_j (Figure 2.15b). The amplitude of the input signal is $\Delta\theta/2$. The slowly changing (constant) component, θ_0° determines the tracking error of the geometric center of a paired source.

The input jamming signal may be written as

$$\theta(t) = \theta_0^o + \sum_{k=1}^{\infty} a_k \cos(k\Omega_j t - \phi_k) \qquad (2.2.19)$$

where $\Omega_j = 2\pi/T_j$, and a_k and ϕ_k are the amplitude and phase of the k-th harmonic.

The output signal of the discriminator, which is assumed to be unsmoothed, is

$$u_{df}(t) = u^0 + \sum_{k=1}^{\infty} u_k \cos(k\Omega_j t - \phi_k) \qquad (2.2.20)$$

where u^0 is a slowly changing component, the magnitude and sign of which correspond to angle θ_0^o.

By analyzing the discriminator output signal, u_{df}, it is possible to identify two qualitatively different kinds of blinking jamming: slow and fast blinking.

Slow blinking is characterized by the fact that the blinking frequency, Ω_j falls within the passband of the tracking system, i.e., the condition

$$\Omega_j < \beta_n \qquad (2.2.21)$$

is satisfied.

In this case the components of frequency, Ω_j and of its individual harmonics pass through the system, with the result that the antenna alternately seeks the directions to targets T_1 and T_2. When the angle between the sources, $\Delta\theta$ exceeds the width of the main lobe of the discriminator characteristic, $\Delta\theta_0$, the blinking jamming sources are resolved. Consequently, the resolution angle for blinking jamming in the case of a low switching frequency is $\Delta\theta_r = \Delta\theta_0$.

For fast blinking and $\Omega_j > \beta_n$, the switching frequency falls outside of the passband of the tracking system. In other words, none of the harmonic components of control voltage, u_{df} passes through the system, and the control signal is written as

$$u_{con} = u^0 \qquad (2.2.22)$$

Thus, in the case when an angle tracking system is the target of fast blinking jamming, the control voltage is a slowly changing component which is proportional to angle θ_0^o. In this case, the action of a paired blinking target does not result in the oscillation of the antenna with frequency Ω_j. The system perceives the paired blinking target as a single target and tracks the center 0 of sources T_1 and T_2.

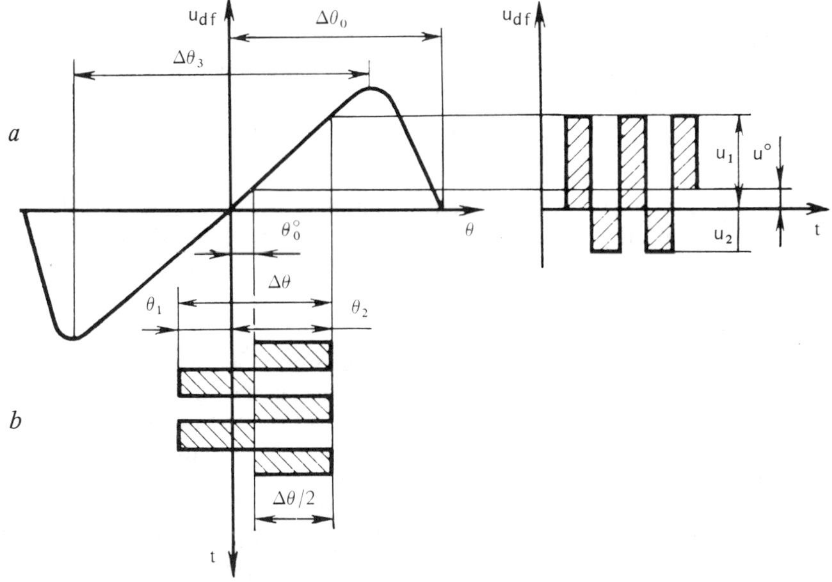

Figure 2.15

As can be seen in Figure 2.15, when θ_0° is small

$$u^0 = k_{df}\theta_0^\circ$$

where

$$k_{df} = \frac{du_{df}}{d\theta}\Bigg|_{\theta = \Delta\theta/2}$$

In tracking systems, unstable equilibrium occurs when $k_{df} \simeq 0$. By using the discriminator characteristic (Figure 2.15a), it is possible to determine the resolution angle of rapidly blinking sources for which $k_{df} = 0$. This angle is equal to the angular distance, $\Delta\theta_e$ between the extrema of the discriminator characteristic.

Intermittent jamming is a special case of blinking jamming, consisting of a periodic burst of strong radio pulses, radiated by the same jamming transmitter with duty factor, Q. The action of intermittent jamming on an angle tracking system with a scanning antenna is based on the use of transient processes which take place in an AGC system when its input is acted upon by strong pulse signals. Because of the time constant of the AGC system, the gain of the receiving channel cannot be changed abruptly. Moreover, the receiver dynamic range is constrained above and below. For these reasons intermittent jamming overloads the receiver and interrupts the transmission of information to the tracking channel as a result of the cutting off of the variable component with the scan frequency in time intervals next to the beginning and end of an interference pulse.

Intermittent jamming reduces the average gain of an angle tracking system. Its effectiveness depends primarily on the strength of the jamming signal, the duration of jamming pulses, their repetition period and the parameters of the AGC system [8].

Two-Point Coherent Jamming
The feasibility of jamming angle tracking channels with two spatially separated coherent jamming transmitters is based on the distortion of the phase and amplitude structure of the electromagnetic field in the aperture of the antenna of a tracking system. If the antenna is small, so that the amplitude distribution of the electrical field in the aperture may be considered uniform, then the angular coordinates of the jamming transmitter are determined by determining the spatial position of the line perpendicular to the plane of the different phases. For a single-point source this plane is a sphere, and a line perpendicular to any point on it coincides with the line to the source, i.e., the target.

For two coherent sources, T_1 and T_2, separated by distance d, the equiphase plane of the resulting wave is substantially deformed. Parts of the phase front are curved in some directions. When the amplitudes U_1 and U_2 of the signals radiated by coherent sources, T_1 and T_2 are equal $[q_{12} = (U_1^2/U_2^2) = 1]$ (Figure 2.16), the amplitude pattern is multilobed and the phase pattern experiences jumps by π in the directions corresponding to the absence of electromagnetic energy. The radiation pattern of two radiators with different amplitudes ($q_{12} \neq 1$) contains no zeros, and its phase characteristic at the boundary between adjacent lobes changes smoothly from some value of ϕ to $\phi \pm \pi$ in a range of angles $\Delta\theta'$ of finite width (Figure 2.16). The amplitude characteristic has a minimum in the range of angles $\Delta\theta'$ corresponding to a jump or a smooth change of phase from ϕ to $\phi \pm \pi$ (the phase inversion region).

At sufficiently great distances from sources T_1 and T_2, the linear dimensions of the phase inversion region may substantially exceed the dimensions of the radar antenna, so that the amplitude distribution in the aperture may be assumed to be constant and the phase distribution linear. In view of the operating principle of an angle system the equal signal line is automatically aimed perpendicular to the phase front of the wave. Therefore, if the phase inversion region is placed over the radar antenna, the system will track the center of paired targets with error θ.

The tracking error is determined in linear approximation by the following equation [54, 105, 24, 101]:

$$\theta = \frac{\Delta\theta}{2} \; \frac{1 - q_{12}}{q_{12} + 2\sqrt{q_{12}}\cos\psi + 1} \qquad (2.2.23)$$

where $\Delta\theta$ is the angle subtended by the jamming sources at the center of the radar antenna, and ψ is the phase difference of the waves, measured at the center of the aperture of the radar antenna.

Error, θ, becomes largest when $\psi = \pm\pi$ and $q_{12} \to 1$. If $\psi = \pi$, then

$$\theta = \frac{\Delta\theta}{2} \; \frac{1 + \sqrt{q_{12}}}{1 - \sqrt{q_{12}}} \qquad (2.2.24)$$

It is important to note that Equations (2.2.23) and (2.2.24) may be used only when $\Delta\theta \leqslant 0.05\,\theta_{0.5}$ and $q_{12} \geqslant 1.1$ or $q_{12} \leqslant 0.9$. The exact value of error, θ is determined by analyzing the family of generalized discriminator characteristics of the angle tracking system when it is acted upon by coherent jamming. It is necessary here to consider the amplitude and phase characteristics of the radar antenna. The analysis shows that the maximum error in the case of coherent jamming does not exceed the value $\theta_{max} \approx 0.7\,\theta_{0.5}$.

Coherent jamming is created naturally during the tracking of a compound target. In this case, the phase and amplitude patterns (Figure 2.16) rotate randomly around the effective target center. The tracking system senses "shock" type disturbances at the moments of time when phase inversion regions are placed over the radar antenna. This kind of jamming is sometimes called "angle" (or phase) noise.

Cross-Polarization Jamming
Most modern antennas radiate spurious lobes in orthogonal polarization in addition to emission in natural polarization. This emission is called cross-polarization. Cross-polarization emission is especially strong in reflecting antennas with doubly curved reflectors.

The radiation pattern of antennas in orthogonal polarization differs substantially from the main pattern. The distortion of the resulting radiation pattern of a radar antenna when irradiated with a field with orthogonal polarization is used to make cross-polarization jamming.

When a parabolic reflector is excited by an electric dipole, currents begin to flow on the surface of the paraboloid, and the distribution of those currents differs from linear distribution [2, 101]. Consequently, the field, E in the antenna aperture will have, in addition to the basic polarization $E_s = E_y$, a component with parasitic polarization $E_{int} = E_x$,

$$(E_{int}E_s) = (E_x E_y) = 0$$

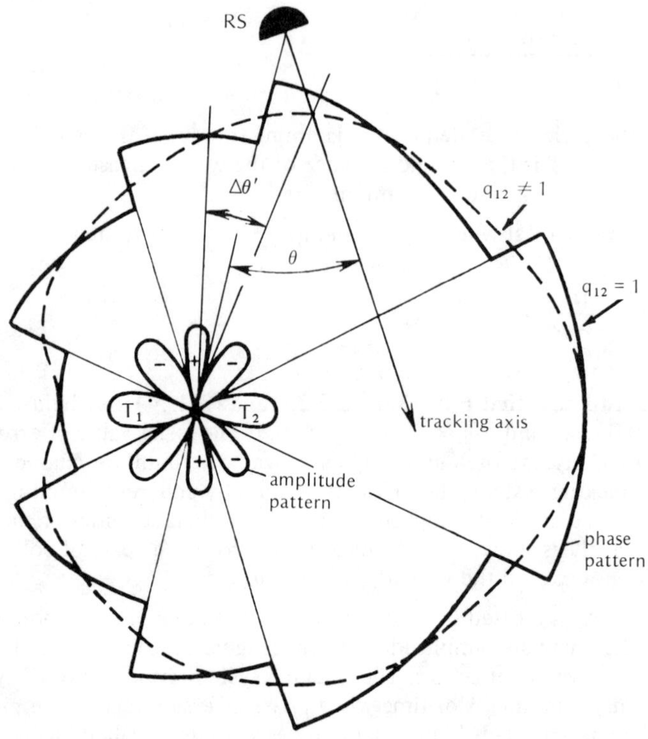

Figure 2.16

Analysis of the field distribution in the aperture of a parabolic antenna
(Figure 2.17) shows that cross-polarization emission does not exist in the
horizontal and vertical planes (the x and y axes). The maximum cross-polarization emission occurs in the planes passing through the axis of the paraboloid and
the axes ξ and η, which form an angle of 45° with the x and y axes. It coincides
with the line of the first minimum of the radiation pattern in the basic polarization. The structure of cross-polarization patterns depends on many factors,
primarily on the curvature of the reflector, type of radiator, distance between
the radiator and the focal point, and diffraction on the edges of the reflector,
on nearby objects, the radome, etc. The level of cross-polarization emission is
low. For reflector antennas it is –10 to –20 dB relative to the main maximum.

Polarization jamming can cause considerable angular coordinate measurement
errors. Detailed analysis of the action of polarization jamming on angle tracking
systems is done with the aid of a family of discriminator characteristics and is a
laborious task [101].

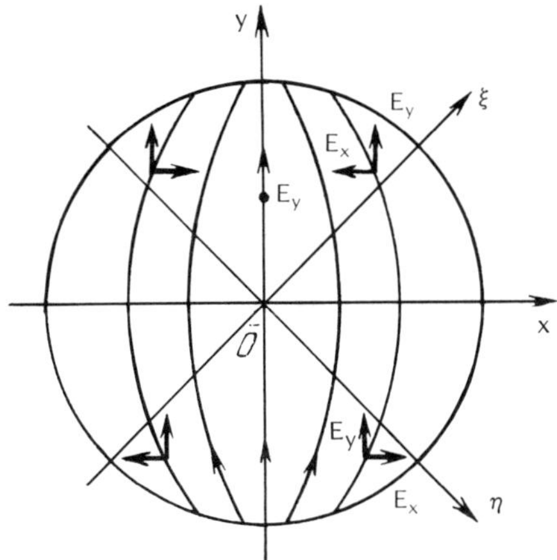

Figure 2.17

When cross-polarization jamming acts upon the simplest amplitude comparison tracker the following control signals appear on its output:

for the useful signal,
$$u_s = k\, U_s^2\, G_{mp}^2\, [F_{mp}^2\,(\theta_0 - \theta) - F_{mp}^2(\theta_0 + \theta)] \qquad (2.2.25)$$

for the interference,
$$u_j = k\, U_j^2\, G_{cp}^2\, [F_{cp}^2\,(\theta_0 - \theta) - F_{cp}^2\,(\theta_0 + \theta)] \qquad (2.2.26)$$

where k is a constant coefficient; G_{mp}, G_{cp} are the antenna gains in main and orthogonal polarization, respectively; and F_{mp}, F_{cp} are the standard radiation patterns in main and orthogonal polarization.

Equation (2.2.26) describes the direction finding characteristic in cross-polarization. The sum signal at the discriminator output is

$$u_{df} = u_s + u_j = k_1\, [L_{mp}\,(\theta) + k_p^2\, q_j^2\, L_{cp}(\theta)]$$

where $k_1 = kU_s^2 G_{mp}^2$; $k_p = G_{cp}/G_{mp}$ is the attenuation coefficient of polarization interference; $L_{mp}\,(\theta) = F_{mp}^2\,(\theta_0 - \theta) - F_{mp}^2\,(\theta_0 + \theta)$ is the discriminator characteristic for the signal; and $L_{cp}(\theta) = F_{cp}^2\,(\theta_0 - \theta) - F_{cp}^2\,(\theta_0 + \theta)$ is the discriminator characteristic for the jamming.

There are two qualitatively different cases of jamming. In the case of low-power jamming, when $k_p\, q_j \ll 1$, only the steepness of the discriminator characteristic decreases. As jamming power increases, the steepness of the discriminator characteristic steadily decreases and when $k_p q_j \gg 1$, it is substantially deformed: two stable equilibrium points appear, indicating a target tracking error in the system. The error caused by the jamming can be found from the equation $u_{df}(\theta) = 0$. It is $\theta \approx \theta_{0.5}$.

For angular misalignment, γ_j of the polarization planes of the jamming and useful signals, the electrical vector of the jamming signal, E_j may be broken down into two orthogonal components:

jamming, $E_j = E_{cp} \sin \gamma_j$

and useful, $E_s = E_{cp} \cos \gamma_j$

The value E_s acts on the intended polarization and will be the useful component. The jamming component attenuates by the factor k_p, and the signal-to-noise ratio at the antenna output

$$q_s = \frac{E_s^2}{k_p^2 E_j^2} = \frac{\cot^2 \gamma_j}{k_p^2} \qquad (2.2.27)$$

increases. Equation (2.2.27) indicates that the effectiveness of jamming depends heavily on the angle γ_j.

5. Jamming Automatic Range Tracking Systems

Range-gate pull-off is used for the deception jamming of automatic range tracking systems. This kind of jamming consists of a series of response pulses, which lag relative to the signal by a time interval that changes monotonically from zero to some fixed value, τ_{lag}.

At the moment the jamming transmitter is turned on, the lag time of the jamming signal $\tau_{lag} = 0$, so that the input of a ranging system simultaneously receives two pulses—signal and jamming. If the amplitude of the jamming pulse is greater than that of the signal, the range gate begins to be displaced by the stronger jamming signal as the lag increases. When τ_{lag} exceeds the aperture of the time discriminator characteristic, the system "loses" the target and begins to track the jamming.

However, the radar continues to receive information about the angular position of the target, in spite of the fact that the automatic range tracker tracks a false target, simulated by the jamming signal. The angle tracker functions normally, since the jamming signal carries information about the angular coordinates of the target, which carries the source of the jamming. In order to cause jamming of an angle channel, the jamming transmitter is turned off after the end of a series of jamming pulses and the ranging system switches to the search mode. During the search time, T_s the angle channel input receives no signal and information about the angular coordinates of the target is lost. After the ranging system locks onto the target a new pull-off cycle begins. The duty factor of the incoming information is determined by the expression

$$Q = T_p/(T_p + T_s)$$

where T_p is the range-gate pull-off time.

Thus, range pull-off jamming causes range measurement errors (and consequently speed measurement errors) and interrupts the flow of information in the angle channel.

6. Jamming Automatic Velocity Tracking Systems

Velocity gate pull-off jamming, which causes the velocity gate to drift and interrupts automatic Doppler frequency tracking of the useful signal[1], is also employed against automatic velocity tracking systems.

The feasibility of velocity gate pull-off is based on features of the action of two signals (useful and jamming), with different amplitudes and frequencies, on a frequency discriminator.

When the frequency discrimator of a velocity tracking system receives two signals, the stable zero of the generalized discrimination characteristic is displaced by the frequency of the stronger jamming. The carrier frequency of the jamming gradually changes and the velocity gate follows the frequency of the jamming. The system switches to automatic tracking of the jamming. During the pull-off cycle the velocity tracking channel provides false information about the velocity and acceleration of the target. Angular target tracking is not interrupted, since the signals from the velocity selector go to the input of the angle channel. When the pull-off cycle ends, the velocity gate signal disappears and the system switches to the search mode. During the pull-off cycle the radar receives false information about the velocity and acceleration of the target. In addition, the flow of information about the angular coordinates of the target is continued only during the time of action of the jamming signal. The radar loses the target during the time of search. The duty factor of the incoming information can be calculated for a given radar and for specific jamming. The duty factor is strongly influenced by the duration of the pull-off cycle and the search speed of the velocity tracker.

2.3 PASSIVE JAMMING

One of the most commonly used passive jamming techniques is to drop large quantities of chaff in the atmosphere. Chaff is most commonly used to jam radars.

Chaff consists of thin metallized passive resonators, the resonant frequency of which coincides with (or is close to) the carrier frequency of the victim radar. If the number of chaff particles that enter the pulse space of the victim radar is large and the chaff cloud is large enough the jamming signal will be substantially stronger than the echo signal from the target (an airplane, for example), traveling inside that cloud.

---

[1] In an automatic velocity tracking system, the frequency of the narrow band selective filter is not changed, and only the heterodyne is retuned. The concept "velocity gate" is introduced by analogy with the term "range gate" in ranging systems.

The techniques and characteristics of this type of jamming are steadily being improved. For example, the effective scatter area of one kilogram of chaff is $(2\text{-}3) \cdot 10^3 \, \text{m}^2$ [5].

Chaff must satisfy a wide range of sometimes contradictory requirements. Many of these requirements are related to the fact that jamming is usually done from airplanes or other types of aircraft. Modern chaff is made either from foil or from glass fiber coated with an extremely thin layer of zinc or aluminum. The cross sectional dimensions of the material are measured in tenths and sometimes hundredths of a millimeter [5]. The chaff particles must not bend after they are dropped into the atmosphere and must not break during storage and transportation. Furthermore, the surfaces must be treated in such a way as to prevent the particles from sticking together, even after long-term storage.

The basic properties of chaff are: effective scatter area, bandwidth, the character and time of development of a chaff cloud, the spectra of the signals reflected by the cloud and the width of the band that conceals the target.

1. Effective Scatter Area

The effective scatter area, σ of a single half-wave dipole reflector as a source of reradiated energy is determined analytically by the following relation:

$$\sigma = 0.86\lambda^2 \cos^4 \theta$$

Here λ is wavelength, and θ is the angle between the axis of the dipole and the electrical vector, E of the electromagnetic wave that irradiates the dipole reflector. When the axis of the dipole reflector and vector E are parallel, σ is maximum and is equal to $0.86\lambda^2$.

To create passive jamming, a large quantity of dipole reflectors is discharged into the atmosphere. These chaff particles are transported by the slip streams of the conveying aircraft and by the atmosphere's turbulent currents. As a result, the axes of the dipoles are randomly oriented relative to vector E.

For approximate calculations of the cross section of a chaff cloud, one may assume all spatial orientations of any given dipole to be equally probable and consider only the mathematical expectation, $\overline{\sigma}$ of its cross section. Calculations show that

$$\overline{\sigma} = 0.17\lambda^2 \tag{2.3.1}$$

The value of $\overline{\sigma}$ thus obtained is theoretical and is used to determine the necessary quantity of dipole reflectors. Dipole reflectors are usually placed in packages, which are dropped from an airplane (or other vehicle), the jamming conveyance. The number of dipoles per package varies from a few tens of thousands to several millions [5]. Entering the very turbulent slip streams of the aircraft, the dipoles are scattered, making "clouds" or corridors of dipoles (when packages of the reflectors are dropped periodically).

In view of the incoherence of the fields produced by the reflections from different dipoles, the mathematical expectation, σ_c for the ESA of a cloud consisting of N dipole reflectors is given by the formula

$$\sigma_c = 0.17\lambda^2 N\eta \tag{2.3.2}$$

where $\eta < 1$ is a coefficient that takes into consideration the effect of individual dipoles sticking together, breaking, etc., and the so-called scatter coefficient.

This expression gives only a comparatively rough estimate of the average cross section of a cloud because the assumption that all spatial orientations of the dipoles in a cloud are equally probable is by no means always valid, as will be demonstrated below. It is also important to remember that the preceding discussion covering the case when the polarization and location of the receiving and transmitting antennas coincide, which is characteristic of the overwhelming majority of radar systems.

2. Bandwidth Properties of Chaff

A serious deficiency of chaff as an electronic countermeasure is its relatively narrow bandwidth. The expressions presented above for determining the cross section of chaff pertain to the case when a dipole reflector is tuned to resonance with the frequency of the radar that irradiates it. In this case, the length of a dipole should be somewhat less than one half the wavelength of the radar. How much a particle can be shortened depends on the thickness of the dipole. Because of the fact that chaff is made as thin as possible in order to reduce the weight of the dipoles and the size of the packages, the permissible "shortening" of the dipoles is usually negligible. For example, if a segment of round wire with cross sectional radius, $r = 0.005\lambda$ is to be tuned to resonance, the length, ℓ of that wire should be made equal to 0.45λ.

A graph that characterizes the bandwidth of a dipole reflector in the mentioned case ($r = 0.005\lambda$) is shown in Figure 2.18. It is easy to see that the effectiveness of a reflector, characterized by its cross section, decreases sharply in the case when λ becomes larger than resonance. But if λ is smaller than the resonance wavelength of a dipole, the ESA of the dipole will decrease much more slowly.

The character of the dependence of σ_c on ℓ/λ, illustrated in Figure 2.18, is typical and changes little with the ratio r/λ. This should be taken into consideration during determination of the length of dipole reflectors in cases when the wavelength of the victim radar is only approximately known by the side that intends to jam it.

When the length of a dipole is increased significantly, its average cross section begins to increase again, and when $\ell \approx \lambda$ it even exceeds somewhat that cross section of a half-wave dipole. However, such an increase in the length of dipole reflectors results in a substantial increase in their weight and dimensions, which involves a reduction of the cross section of the reflectors per unit weight (1 kg, for example), and therefore finds no practical application.

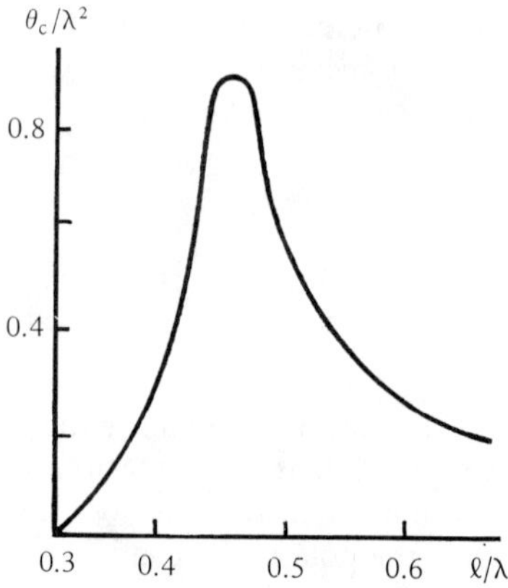

Figure 2.18

Dipoles of different lengths are assembled into a package in order to give a chaff cloud a wider band. However, this means of expanding the bandwidth of passive jamming involves a substantial increase in the weight of the dipoles necessary for making a cloud of reflectors with the required average effective reflecting area.

The problem of increasing the bandwidth of passive jamming can also be solved with equipment capable of cutting dipole reflectors of the required length right on board the jamming aircraft. To accomplish this, long bundles of thin strips of foil or metallized glass fiber are loaded into the containers of automatic chaff-dropping equipment. The bundles are the raw material for the production of dipole reflectors. The equipment of the jamming delivery system also includes an electronic surveillance receiver which determines the operating wavelengths of the victim radar stations.

3. Dispersion of Chaff Cloud

When an airplane delivers passive jamming, the packages of dipole reflectors that are dropped from the airplane enter slip streams, which have high turbulence. The dipoles, which are scattered under the influence of many strongly twisted local eddies, acquire substantial velocities and form relatively small clouds. Here both the distribution of the concentration of the dipoles and the distribution of their spatial orientation are random.

The assumption of an equally probable spatial orientation of dipoles in the cloud is considered to be most valid for a short period of time immediately following the opening of a package. Later, when the influence of the slip streams of the delivering aircraft becomes negligible, the dipoles are scattered in space as the result of turbulent motion (turbulent diffusion) of individual volumes of the atmosphere. Consequently, the cloud continues to expand and the average concentration of dipoles in it decreases; the concentration varies in different parts of the cloud. The speeds at which different dipoles travel, both horizontally and vertically, also differ. The geometric center of the cloud as a whole is displaced relative to the intial point of expansion under the influence of wind, and loses altitude due to the force of gravity.

It has been established experimentally that in the overwhelming majority of cases the dispersion of dipoles in the horizontal plane is greater than their vertical dispersion. Therefore, a chaff cloud usually is a formation that extends in the horizontal plane, and its largest dimension coincides with the direction of the wind.

According to studies, the assumption of an equally probable orientation of dipole reflectors in the cloud is valid when dipole reflectors travel in a strongly turbulent atmosphere. If there is no appreciable turbulence in the atmosphere, certain spatial orientations predominate and the associated stratification of dipoles by descent rates predominates. The latter effect is explained as follows. When a dipole reflector is perfectly uniform through its entire length, then, as it descends under the influence of gravity in a homogeneous nonturbulent atmosphere, it aligns itself horizontally in an effort to maximize its resistance to the incident air stream. But if the weight of a dipole is unevenly distributed through its length, then, as it falls in a calm atmosphere, it aligns itself vertically with its heavy end down. Because of the fact that the dipoles that come out of the same package are not strictly identical, groups of both horizontally and vertically aligned reflectors are formed as the cloud expands.

Studies conducted on dipoles of the centimeter band disclosed that the ratio of the number of horizontally aligned dipoles to the number of vertically aligned dipoles varies in different tests and under different weather conditions from 1.5 to 8[5].

The difference in the orientation of the two groups of dipoles leads to a difference in their descent rates. The horizontally aligned reflectors form "slow" and the vertically aligned reflectors "fast" groups of dipoles. Reflectors of the intermediate spatial orientations have intermediate descent rates.

A difference in the descent rates gradually results in spatial stratification of the cloud in the vertical plane. Two dense layers of reflectors, traveling at different speeds and separated by a region in which the dipoles have a lower concentration, are formed. The distance between these layers increases monotonically with the difference between their velocities.

This pattern of the monotonic descent of the center of a chaff cloud (or of the centers of "speed groups") may not hold true in the case when the chaff is dropped from a rather high altitude. The reason for this is the fact that there are permanent air currents above our continent at altitudes of approximately 10-12 km. The "thickness" of the layer of the atmosphere in which these currents exist is approximately 1 km, and the velocity of the currents themselves varies from tens to hundreds of kilometers per hour. The air stream inside a moving layer has relatively low turbulence. However, very turbulent layers with a larger number of rising currents and eddies form at the boundaries between fast-moving air masses and adjacent, relatively motionless layers.

As they enter the upper bondary layer the dipoles are transported by the eddies that exist there. Their established spatial orientations are changed and the distribution of the spatial orientations of the reflectors approaches the equally probable distribution. The cloud expands and the average concentration of dipoles decreases. In this case the cloud may be thrown upward by rising air currents, so that the average descent rate of the cloud decreases substantially.

Such thrusts of the entire cloud or of individual parts thereof may occur repeatedly, so that in many cases the dipoles stay near the boundary layers for an inordinately long time. The length of such a cloud may increase considerably, and the concentration of the reflectors in regions near the center of the cloud decreases.

4. Spectra of Signals Reflected by Chaff Cloud

In the general case the spectrum of a signal reflected by a chaff cloud differs from the spectrum of the incident signal. This difference is attributed to the motion of the cloud as a whole relative to the victim radar and to the random motion of the dipoles in it. The curve of the spectral density $G(f)$ of a radio signal, reflected by a chaff cloud, for the case when the transmitted signal is a monochromatic wave, is shown in Figure 2.19. The central frequency, f_c of the spectrum of the reflected signal is shifted and is separated from the frequency, f_0 of the incident waves by F_d, which represents the Doppler frequency shift caused by the motion of the cloud as a whole relative to the target station.

The spectrum of the reflected signal, even in the case of monochromatic irradiation, occupies a finite band. This spreading of the spectrum is attributed to a difference in the velocities of the individual reflectors of the chaff cloud and is a function of atmospheric turbulence.

The motion of gases (or fluids), as is known, may be either laminar or turbulent. Laminar motion exhibits a certain "ordering". During such motion one layer of gas travels relative to other layers at a certain constant (or slowly changing) velocity, which is simultaneously inherent to all elements of the volume of that layer. The fact that both the velocity and the direction of individual elements of a volume of moving gas (or fluid) are decidely variable is a characteristic feature of turbulent flow. The velocity at any given point in such a volume of gas pulsates randomly around some average value; the magnitudes of these pulsations are small in comparison with the average velocity.

Under certain weather conditions, at certain chaff drop altitudes, and for certain turbulent diffusion coefficients only relatively small volumes, and consequently a relatively small number of dipole reflectors, have comparatively high velocities. The strength of the spectral components of the reflected signal, formed by turbulent diffusion of the atmosphere, also decreases monotonically as the

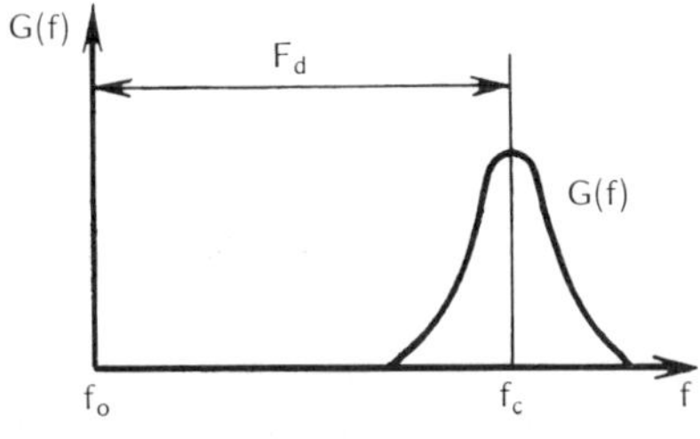

Figure 2.19

frequency of the components increases. Turbulent diffusion of the atmosphere is not the only cause of the random velocity distribution of dipoles and of the associated spectral expansion of the reflected signal. As was stated above, because of the fact that dipole reflectors are not perfectly identical, and because of different initial conditions under which chaff packages break open, the dipoles that make up a cloud have different descent rates. This entails a further broadening of the spectrum of the reflected signal. This broadening is maximum in a relatively small cloud immediately after its formation, before stationary descent rates of individual groups of dipoles are established.

Another cause of the spectral spreading of the reflected signal is a difference of wind velocities at different altitudes. If a cloud has large vertical dimensions this may result in an appreciable spectral spreading of the reflected signal.

The spectrum of a reflected signal depends to a great extent on wind velocity, and spreads increasingly as the velocity increases. However, the spectra of signals reflected by a chaff cloud remain comparatively narrow, even at substantial wind velocities and high atmospheric turbulence levels associated therewith. The effective spectral width of this jamming in the 10 cm band usually does not exceed a few tens of hertz. As λ diminishes the spectral width increases in inverse proportion to wavelength, but the spectrum remains relatively narrow within the limits of the radar operating bands.

In calculation, the spectral density, $G(f)$ of a fluctuating envelope of a reflected signal due to the motion of dipole reflectors is often approximated as a Gaussian function. For such an approximation, spectral density, $G(f)$ for $f \geqslant 0$ is written in the form [167]:

$$G(f) = G_0 \exp\left(-f^2/2\sigma_f^2\right) \tag{2.3.3}$$

where G_0 is the value of the function $G(f)$ for $f = 0$; $\sigma_f = 2\sigma_v/\lambda$ is the mean square scatter of Doppler frequencies, corresponding to one-half the width of the energy spectrum at the $0.61\,G_0$ level for a radio signal reflected by a chaff cloud; and σ_v is the mean square spread of the relative radial velocity of the radar and the dipole reflectors.

In many cases the function $G(f)$ for $f \geqslant 0$ is assumed to be [24, 116]:

$$G(f) = G_0 \exp\left(-f^2/1.6 f_{0.5}^2\right) \tag{2.3.4}$$

Here $f_{0.5}$ is one-half the width of the energy spectrum at the $0.5\,G_0$ level for the same radio signal as in Equation (2.3.3).

In connection with the comparatively small value of σ_f, the interference produced by dipole reflectors is picked up by the receiving-display systems of victim radar stations as an extremely narrowband noise signal, which explains the jamming properties of such interference.

Modern air targets usually travel at high velocities so that large Doppler frequency shifts occur in the signals reflected by them. The spectral differences between chaff and target signals can be used to filter out chaff and thus to reduce its effectiveness substantially. This effect is used extensively in pulse radar with MTI systems as well as with CW and high-prf waveforms.

However, the interference produced by dipole reflectors in their traditional application is an effective means of concealing high-speed targets (airplanes, for example) from observation by pulse radar systems without special means of counteracting such interference. The densities of chaff (chaff package drop rates) necessary for jamming such radars are relatively low. Radars using CW and high-prf waveforms should obviously be considered as satisfactorily immune to this type of jamming.

5. Chaff Corridors

Jamming with chaff may be employed to protect a target such as an airplane or group of airplanes from detection. This is exactly the way the problem is formulated in application to the jamming of early warning and target tracking radars. The problem also arises frequently in application to the jamming of angle and range tracking radar systems.

Jamming is usually performed by periodically dropping packages of chaff from an airplane or from some other aircraft, called the jamming delivery vehicle. If chaff packages are dropped frequently enough, the clouds they produce merge, forming rather wide and long zones of space, in which chaff particles are randomly "scattered." Such zones of space are called jamming corridors. The radar signals reflected by the irradiated bands light up the radar display scope at the points corresponding to the coordinates of the chaff corridors. This causes corridors to light up on the screen. The brightness of a corridor depends on the strength of the reflected signal, which under otherwise identical conditions is proportional, in turn, to the number of chaff particles that simultaneously reflect the probing radar signal. When this brightness is sufficiently great it is difficult or impossible to identify the blip of a target such as an airplane that is flying in or above the chaff band, against its background. If the output system of the station is not a visual display, but rather a computer, the computer will not be able to distinguish the target signal from the noise-like signal produced by reflections from the chaff corridor.

The jamming immunity of radars in relation to interference produced by chaff may vary and is estimated quantitatively through the jamming coefficient, k_{jr}.

Because of the fact that when an airplane flies in a chaff corridor the interference and useful signals arrive at the input of the station from the same distance, the strengths of the interference, J and signal, S are proportional to σ_t and σ_c, where σ_t is the effective scatter area of the target and σ_c is the total cross section of the chaff particles that are simultaneously illuminated by the radar signal, i.e., which fall within the resolution cell of the radar. The latter expresses the volume of that region of space in which the energy of one pulse, radiated by the radar, is distributed. In view of the above discussion, a radar is vulnerable to jamming when

$$\sigma_c = k_{jr}\, \sigma_t \tag{2.3.5}$$

The size of the resolution cell depends on the pulse duration, τ_p of the radar, the width of the main lobe of the antenna radiation pattern, θ_{el} and θ_{az} in the elevation and azimuthal planes, and distance R:

$$V_j = (1/2)\, c\tau_p\, R^2 \theta_{az}\, \theta_{el}$$

The width of the masked region, L_b, i.e., the region of space in which Equation (2.3.5) is valid and, consequently, in which the target is concealed by the jamming, is an important feature of a chaff corridor. The characteristic of the dispersion of chaff particles in the direction perpendicular to the axis of the corridor is most important in relation to chaff corridors intended for jamming early warning and target tracking radar. The reason for this is that chaff particles, because of their dispersion in the vertical direction, do not leave the resolution cell, since the antenna radiation patterns of these radars are usually wider in the elevation angle plane. If chaff packages are dropped at sufficiently high rates, when the clouds produced by individual chaff packages join together, the dispersion of chaff particles in the direction of the axis of the corridor does not alter their average concentrations, since the flow of particles from one cloud to another is counteracted by the backflow of chaff particles from the latter to the former. However, the dispersion of chaff particles perpendicular to the axis of the corridor results in a change in the width of the corridor and concentration of the chaff particles at various points in the corridor.

The width of the masked region, L_b depends not only on the width of the corridor and absolute density of the reflectors, but also on the pulse duration of the radar, the width of the main lobe of its antenna system, the positions of the corridor and radar and the cross section of the masked target.

L_b is determined on the basis of the assumption that the width, ℓ_b of a chaff corridor satisfies the inequality $\ell_b \gg R\theta_{az}$ in the entire operating range of R. Suppose a chaff corridor is laid out by a delivery vehicle flying at velocity, v_b, and with drop interval, t_b (the drop interval is measured as the time interval between the dropping of two successive packages). We select the coordinate system Oxyz such that the Oy axis coincides with the middle of the corridor and the Oz axis is pointed vertically upward. We assume that the radar radiates pulses of duration τ_p and has a sufficiently wide radiation pattern in the elevation plane, so that the entire chaff corridor falls within the pattern at all ranges.

Moving the radar parallel to the Ox axis we determine the number of chaff particles that fall within the resolution cell of the radar for different values of coordinate x (the axis of the main lobe always remains parallel to the axis of the corridor). Assuming that the number of chaff packages dropped in a segment of the path of the delivery vehicle of length 0.5 $c\tau_p$ (the length of the resolution cell) is equal to $c\tau_p/2v_b t_b$, and integrating in terms of the resolution cell of the radar, we find the number of chaff particles

$$N = \frac{\eta N_b}{2\sqrt{\pi D_x t}} \; \frac{c\tau_p}{2v_b t_b} \int_{-R\theta_{az}/2}^{R\theta_{az}/2} \exp\left(-\frac{x^2}{4D_x t}\right) dx \tag{2.3.6}$$

that fall within that space.

Here D_x is the turbulent diffusion coefficient on the Ox axis, which in the general case does not coincide with the direction of the wind, and N_b is the number of dipole reflectors per package. Therefore, when tables of turbulent diffusion coefficients are used instead of the true wind velocity, v_w, it is necessary to consider the value $v_w \cos\beta$, where β is the angle between the wind velocity vector and the Ox axis. Disregarding the change of the chaff concentration through the relatively narrow "width" of the resolution cell, we may write, instead of Equation (2.3.6), the following approximate equation:

$$N = \frac{\eta c\tau_p N_b}{4v_p t_p \sqrt{\pi D_x t}} \; R\theta_{az} \exp\left(-x^2/4D_x t\right) \tag{2.3.7}$$

The latter expression, like Equation (2.3.6), ignores the influence of the slip streams of the jamming delivery vehicle and therefore can be used only for rather large values of time t. It shows that the number of dipole reflectors in the resolution cell of the radar depends on the x coordinate, which characterizes the position of that space relative to the middle of the chaff corridor and time t. For given x and t, the number of reflectors in the resolution cell is proportional to the number of packages dropped in a segment of the path of length 0.5 $c\tau_p$, and to the "width" $R\theta_{az}$ of the resolution cell. The width of the chaff corridor increases with time and the density of the corridor in regions adjacent to its axis diminishes monotonically.

As already stated, in order to conceal a target located in a chaff corridor it is necessary to satisfy Equation (2.3.5), which in consideration of Equation (2.3.7) acquires the form

$$\frac{\bar{\sigma}\,\eta N_p c\tau_p}{4v_b t_p\,\sqrt{\pi D_x t}}\; R\theta_{az}\,\exp\left(-\frac{x^2}{4D_x t}\right) \geq k_{jr}\,\sigma_t \qquad (2.3.8)$$

If the target is relatively small and is always located inside the resolution cell of the radar, then it is possible to determine the effective width of the masked region, i.e., the width of that part of the chaff corridor within which the inequality in Equation (2.3.8) is satisfied. Placing the equal sign between the right and left hand parts of the latter expression, and solving the resulting equation relative to $x = 0.5\,L_b$, we obtain:

$$L_b = 4\left(D_x t\,\ln\frac{R\theta_{az}\,\bar{\sigma}\,\eta N_b c\tau_p}{4k_{jr}\sigma_t\,\sqrt{\pi D_x t}\,v_b t_b}\right)^{0.5} \qquad (2.3.9)$$

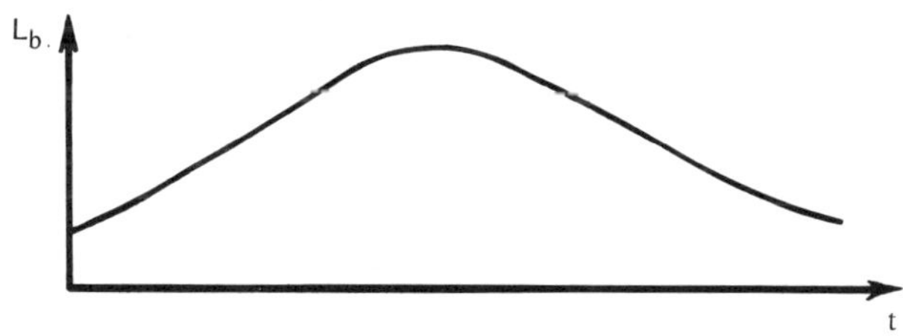

Figure 2.20

The curve shown in Figure 2.20 illustrates the dependence of L_b on t.

Differentiating Equation (2.3.9) in terms of the argument $D_x t$, and setting the resulting derivative equal to zero, we determine the value of $(D_x t)_m$ that corresponds to the maximum width of the concealed region, L_{bm}. L_{bm} itself is determined by substituting $(D_x t)_m$ into Equation (2.3.9) and it turns out to be:

$$L_{bm} = \frac{1}{\sqrt{2e\pi}}\;\frac{R\theta_{az}\,\bar{\sigma}\eta N_b c\tau_p}{k_{jr}\,\sigma_t v_b t_b} \qquad (2.3.10)$$

Hence, it follows that the maximum width of the masked region is proportional to the number of chaff packages dropped in the segment $c\tau_p/2$, to the radar cross section of the expanding package $\bar{\sigma}\eta N$ and distance to the masked target.

This width is inversely proportional to the cross section of the target and to the radar jamming coefficient. It is not a function of the turbulent diffusion coefficients of the atmosphere, which determine the time necessary for the masked region to attain its maximum width.

When $D_x t > (D_x t)_m$ the scattering of the dipole reflectors in the direction pendicular to the axis of the corridor continues. Consequently, the density of the chaff continues to decrease in the zones adjacent to the axis of the chaff corridor and the width of the masked region narrows. After a certain time interval the corridor becomes so eroded that it loses its masking properties, i.e., the width of the masked region diminishes to nothing. The lifetime of the jamming corridor can be determined by assuming $L_b = 0$ in Equation (2.3.9) and by solving that expression relative to $D_x t$. The solution yields

$$(D_x t)_{L_b = 0} = \left(\frac{R\theta_{az}\, \bar{\sigma}\eta N_b c\tau_p}{4\sqrt{\pi}\, k_{jr}\, \sigma_t v_b t_b} \right)^2$$

It is important to note that the "lifetime" of a jamming corridor is e times longer than the time interval during which the width of the masked region (assuming $\ell_b \gg R\theta_{az}$) attains its maximum value. The "lifetime" of the chaff corridor examined here is meaningful only so long as the chaff particles, which descend at an average rate of 1-1.5 m/s, remain in the air.

2.4 PASSIVE DECEPTION JAMMING

Passive deception jamming is performed with false targets, radar decoys and chaff. The distinguishing feature of this class of jamming is the fact that the electromagnetic energy of the radar signals is used to produce the jamming effect. As a rule passive deception jamming is employed to form radar patterns of targets.

False targets are imitations of actual moving targets. Various models of real targets may be used as false targets. Rockets, equipped with launch or sustainer engines, with which independent, guided or unguided flight can be sustained for long periods of time (up to several tens of minutes), are used to imitate airplanes. A rocket used as a false target is equipped with the appropriate passive repeaters to enable it to generate a signal with the same intensity and spectrum as the target it imitates.

False targets, which are used basically against target detection and tracking radars, confuse radar operators, overwhelm the computer systems of the target tracking system, increase the target pattern recognition time, and attract firepower to false targets.

Unlike false targets, radar decoys are intended for interrupting the automatic target tracking system of a radar or of the homing head of a guided missile.

Decoys are used to thwart a guided missile attack or to divert guided missiles to a distance that is safe for the masked airplane. In order for a decoy to divert the guidance (homing) system of a missile to itself its cross section must be somewhat greater than the cross section of the masked target. The cross section of decoys is increased by installing passive reradiators of electromagnetic energy (corner and lens reflectors, passive arrays, etc.) on them.

Radar decoys may be towed and dropped. Hence, they are classified as unguided and guided respectively. A towed decoy is lightweight and streamlined and, in time of danger, is released from the concealed target by means of a launching device (a winch with a thin tow line). Unguided dropped decoys are used to protect airplanes and missiles. They have no engines and are passive reradiators with a larger effective scatter area than the target which they are intended to conceal. The simplest decoy may be a Luneberg lens or a package of chaff. Chaff packages are dropped periodically from an airplane by a special automatic system at time intervals of T_b. The cross section of the expanding chaff cloud is greater than that of the target. The time interval, T_b is determined in accordance with the intrusion of the target and the nearby chaff cloud in the radar resolution cell.

The reaiming of the missile at a decoy, created by a cloud of chaff, takes place as the result of switching first of the range tracking system, and then the angle system, to automatic tracking of the chaff cloud. That there are substantial differences in velocity and acceleration from the same parameters of a real target, is a disadvantage of dropped unguided decoys. These differences can be used for the selection of decoys on the basis of the corresponding information and associated criteria. This problem is corrected in dropped controlled decoys by using engines, which permit controlled flight and provide a smoother drift of tracking gate.

Antiradar coatings and systems for controlling reradiated energy can be used during deception jamming to change the cross sections of targets [24]. Antiradar coatings are materials with special properties. These materials, when applied on reflecting surfaces, substantially reduce their radar visibility. The operating principle of the coatings is based on phenomena connected with the interference and absorption of electromagnetic waves in laminar media.

Antiradar coatings are classified as interference, absorbing and combined, in accordance with their operating principle. Interference coatings reduce the strength of reradiated energy due to the interaction of two and more coherent beams of electromagnetic waves, traveling different propagation paths between the target and the radar receiver. Coatings that reduce the cross sections of targets as a result of substantial absorption of the incident waves are called absorption coatings. Combined coatings, which derive their effect from interference and absorption of incident and reflected waves, also find practical application. Combined coatings reduce the reflection factor several tens of times in a wide frequency range (more than an octave).

Passive deception jamming can be filtered out on the basis of comparison of the attribute vector parameters of the jamming with the analogous properties of the useful signals. The target pattern recognition problem arises here. The spectrum and power of the jamming, the velocity and acceleration of the target often can be used to distinguish passive deception jamming from useful signals.

Chapter 3

Mutual Interference and Electromagnetic Compatibility of Electronic Systems

3.1 GENERAL INFORMATION ON MUTUAL INTERFERENCE AND ELECTROMAGNETIC COMPATIBILITY

Modern radio systems are used extensively in a wide range of often unrelated fields of human endeavor. The power of radiated signals is increasing, the sensitivity of receivers is being improved, and the spectra of modulating signals are expanding, in which connection the passbands of receivers are expanding.

The available radio frequency band already is overloaded with the emissions of electronic systems, operating in various radio networks and creating mutual interference, the level of which is steadily increasing. The strength of this interference sometimes is so great that receivers simply cannot function normally. Furthermore, both receivers and transmitters of electronic systems always have secondary (parasitic) frequency channels for transmission and reception, in addition to main (operating) channels, and the antennas of these systems always have side lobes in addition to the main lobes of their radiation patterns. Therefore, mutual interference can occur both in systems operating on different, often very distant frequencies, and in systems whose antenna radiation patterns have overlapping main lobes.

The operating conditions of radio receivers often deteriorate to a considerable extent as a result of interference produced by different kinds of electrical systems and installations, which inadvertently generate electro-magnetic energy. These systems include electric motors which contain arcing contact groups, the ignition systems of internal combustion engines, power transmission lines, arc welding equipment, etc. The level of extraneous interference (called industrial interference) differs from point to point on the globe and varies substantially as a function of the time of year and day. It submits only to rough, approximate forecasting, but it shows a stable tendency to increase. Already this level in most regions exceeds the level of natural atmospheric and extraterrestrial noise and is placing limitations on the design capabilities of high-sensitivity receiving channels of electronic equipment.

The operating conditions of modern radio systems give rise to the problem of their electromagnetic compatibility (EMC). The essence of this problem amounts to the determination of the design capabilities of systems with compatible properties. These design capabilities are determined primarily by the properties of parasitic receiving and transmitting channels and the spectral structure of their signals, and also by the conditions of the physical location of the systems and their separation with respect to operating frequencies and time schedule, under which interference the impairment of the operation of other systems does not occur while processes that take place in their receiving channels proceed normally.

Obviously, the successful solution to an EMC problem is possible only under conditions in which the industrial noise level at a receiving site is not too high. Therefore, the problem of controlling and finding ways to reduce the industrial noise level is extremely important. Solving an EMC problem also involves improvement of the overall noise immunity of the receiver systems of electronic equipment. The EMC problem is also of independent importance in view of a number of specific features.

The problem of ensuring electromagnetic compatibility of radio systems is not new. The need for efficient utilization of the radio frequency bands and the standardization of tolerable radio emission levels, both in main and parasitic channels, was settled long ago. All-Union standards on the frequency stability and level of spurious and out-of-band radiation of transmitting systems, permissible industrial noise levels, the spectral width of radiated signals, etc., have been adopted. International organizations and conferences have proposed corresponding recommendations. Meanwhile, the number of systems is increasing at exceedingly high rates under conditions of virtually complete occupancy of the most commonly utilized radio frequency bands. The number of functions performed by these systems, radiated power and industrial noise levels in the major regions of the earth are all steadily increasing. All this creates a qualitatively new electromagnetic situation, which requires the careful development and implementation of a whole group of measures to ensure the electromagnetic compatibility of electronic and electrical systems.

3.2 ELECTROMAGNETIC SITUATION AND EMC PARAMETERS

The electromagnetic situation in which the receiver systems of some radio installations operate is characterized by a combination of electromagnetic emissions which influence these systems and which are produced by radio frequency oscillators of other systems, by electrical systems that are not intended for the generation of electromagnetic waves, by natural sources etc. Unified methods and a unified system of parameters have not yet been developed for describing the electromagnetic situation. The electromagnetic situation is evaluated in practice both for electronic systems and for individual components of these systems.

In the former case, the site level of emissions, called "external" in relation to an examined system (the external electromagnetic situation) is usually determined. The intensity and character of emissions produced by the electronic components of the system itself are ignored. Interference that characterizes the external electromagnetic situation is usually produced by sources that are distant from a given electronic system. Such interference consists of regular signals from radio transmitters and noise-like signals, produced chiefly by the emissions of various electrical systems (power transmission lines, welding machinery, automotive ignition systems, etc.).

Regular electromagnetic emissions are evaluated by constructing special maps [76], which describe the electromagnetic situation at a given locality. Information about the frequencies and radiation levels of all the transmitters used in a given locality, their operating time modes, signal spectra, frequency stability, etc., is included in these maps. The statistical characteristics of regularly scheduled radio transmissions (for example, the transmissions of airplanes, ships and other mobile radio stations) are also indicated. These maps are used to forecast the probable interference level and to determine the right parameters and operating schedules of systems. They usually describe the electromagnetic situation in comparatively large areas.

The level of the noise-like background produced by the radiation of various electrical systems can be determined only statistically and is characteristic of comparatively small regions, surrounding the place where it is measured. It depends on the proximity of that point to power transmission lines, heavily trafficked highways, various industrial installations, etc. In many cases moving the measurement point several hundred meters results in an exceedingly large change of the noise level. The intensity of the noise-like background experiences sharp daily fluctuations and depends to a great extent on weather conditions and on the time of year.

The general site electromagnetic situation of a radio system is evaluated by plotting the results of measurement of the industrial noises that are characteristic of a given area on a map, containing information about regular emissions in that region. The internal electromagnetic situation of a radio system, i.e., the electromagnetic situation in which each specific component of the system must operate, is determined by the interference that occurs as a result of the operation of other components of the same system. These components include radio transmitters and their main and secondary radiation channels, receiver heterodynes, electrical power sources, various kinds of regulators, switches, etc.

For many radio systems, for example the radio complexes of modern aircraft, the interference produced by the systems of a complex almost completely determines the operating conditions of its receivers. Noise sources are installed in such systems side by side with systems that pick up the noise. This makes it necessary to examine numerous possible connections between oscillators and receivers, antenna feeds, etc. Considering also that systems that pick up noise are often located in the immediate vicinity of the systems that generate the noise, the difficulty of analyzing the internal electromagnetic situation of sophisticated radio electronic complexes becomes obvious.

When evaluating the electromagnetic compatibility of radio systems, it is necessary to consider a large number of their parameters. Chief among them are the following:

- the energy and spectral characteristics of the main and secondary emissions of the transmitting systems, receiver heterodynes and of transmitting systems that determine the external electromagnetic situation,
- the actual sensitivities of receivers in the main and secondary receiving channels,
- the dynamic range of receivers,
- the noise level at the receiver site, and
- the coupling coefficients of various components (including antennas) of systems that radiate and receive noise.

3.3 SECONDARY RADIATIONS OF RADIO SYSTEMS

During the design of any radio transmitting system both the carrier frequency and the primary bandwidth, corresponding to the minimum spectral width of the radiated signal, are specified. This is done to ensure the required transmission speed and quality for the class of emissions to be employed. Emissions at frequencies outside of the primary frequency band are called secondary emissions.

Secondary emissions represent a wasteful utilization of the high-frequency power of a transmitter and in many cases result in a substantial deterioration of EMC conditions by jamming the receiving channels of other systems. These emissions are usually classed as spurious and out-of-band.

Spurious emissions include a large class of secondary emissions. They are produced by a variety of nonlinear effects in transmitter components not involved in modulation. The frequencies on which spurious emissions occur may differ substantially from the carrier frequency, ω_0 of the transmitter (for example, they may be harmonics of ω_0). Therefore, secondary emissions can interfere with systems which operate on frequencies substantially different from the primary operating frequency of the offending transmitter.

The basic kinds of spurious emission are [76]:

- emissions on harmonics of the main frequency,
- emissions on subharmonics of the main frequency,
- parasitic emissions,
- combination emissions, and
- intermodulation emissions.

Out-of-band emissions are those which are radiated in the frequency bands directly adjacent to the primary frequency band of a transmitter. They are related to the modulation of the radiated signal. Both the beneficial modulation of the signal and parasitic modulation, which occurs, for example due to the instability of power voltages or other fluctuations, are taken into account. Out-of-band emissions can interfere with systems operating on frequencies close to the frequencies of the interfering transmitter.

1. Harmonic Emissions

Emissions of this kind take place on frequencies that are multiples of the carrier frequency, ω_0 of the transmitter. The mechanism responsible for the generation of harmonics of frequency ω_0 and their relative strength vary in transmitters operating in different bands and of different designs.

In oscillators using vacuum tubes with external resonant circuits, higher harmonics of the fundamental frequency occur in Class C stages.

A pulse of anode current, $i_a(t)$ of a stage biased below cut-off and conducting for a phase angle, θ_c, without saturation at high outputs, is described, as is known, by the expression

$$i_a(t) = I_{am} \frac{\cos \omega_0 - \cos \theta_c}{1 - \cos \theta_c}$$

and the amplitude of the n-th harmonic, I_{an} is connected with the amplitude of the first harmonic, I_{a1} by the relation

$$I_{an} = I_{a1} \frac{\sin n\theta_c \cos \theta_c - n \cos n\theta_c \sin \theta_c}{n(n^2 - 1)(1 - \cos \theta_c)}$$

and may comprise a significant fraction of the amplitude, I_{am} of the anode current pulse.

When a stage is biased below cut-off and driven into saturation over a phase angle, θ_s, the anode current pulse is described by the expression

$$i_a(t) = I_{am} \frac{\cos \omega_0 t - \cos \theta_c}{\cos \theta_s - \cos \theta_c}$$

and usually has an even wider frequency spectrum.

The harmonic emission level is reduced by filtering with intermediate antenna circuits. This filtering is most effective when the antenna circuit has a capacitive coupling with the inductive part of the intermediate circuit. In this case, however, the antenna of a transmitter (particularly one with high overall radiating power) may radiate noise on higher harmonics of fundamental frequency, ω_0 with considerable power.

In the SHF band, where magnetrons, traveling wave tubes (TWT) and klystrons are used as oscillators, higher harmonics occur because of the fact that the density of the electron stream is modulated in the drift space of the device, and the shape of the current differs sharply from sinusoidal, even when harmonic excitation voltage is used (for example in drift klystrons and TWT). In this case, various kinds of wave guide and coaxial filters are used to reduce the higher harmonic emission level.

2. Subharmonic Emissions

The frequencies of such emissions are an integral number of times lower than the fundamental transmitter frequency ω_0. Subharmonic emissions are inherent, for example, in transmitters in which frequency multiplication modes are used to generate highly stable frequencies from lower crystal-stabilized signals. This principle is employed especially extensively in the design of transmitters of the decimeter band. The subharmonics of the radiated frequencies in this case are higher harmonics of the frequency of the crystal oscillator. The subharmonic level is reduced by filters in the transmitter and antenna feeder system.

3. Parasitic Emissions

Parasitic emissions are spurious transmitter emissions which are not related to the oscillation of the main frequency. They can occur in oscillators using vacuum tubes with external resonant circuits because of the fact, for example, that the reactive impedance of connecting wires, coupling components, electrode leads of tubes, etc., form oscillator circuits and feedback circuits for frequencies that are often higher and occasionally lower than frequency, ω_0.

Parasitic waves, the frequencies of which correspond to the meter and even the decimeter bands, often are generated in transmitters of the medium and short wave bands. These waves are the result of the connection of the anode and grid circuits through tube electrode capacitors, closely grouped wiring elements and common power sources.

Neutralization circuits, tuned to the fundamental frequency, usually not only prevent the generation of waves with frequencies several orders of magnitude higher than the design frequency, but, conversely, themselves often form oscillating systems and feedback circuits for parasitic frequencies with elements of neutralizing circuits.

A diagram of the intermediate stage of a tube-type common cathode oscillator is illustrated in Figure 3.1a as an example. The equivalent diagram of that stage for the SHF band is illustrated in Figure 3.1b. For the SHF band, inductors L_1 and for the SHF band, inductors L_1 and L_2 have very large impedance and may be ignored. The "parasitic" anode circuit is made up of capacitor C_1, capacitor C_{a-c} between the anode and cathode of the tube, and inductors L_{w1}, L_{w2} of the connecting wires. However, the "parasitic" grid circuit consists of capacitor C_2, capacitor C_{g-c} between the grid and cathode of the tube, inductors L_{w3}, L_{w4} of the connecting wires, and cathode input inductor L_c. The equivalent diagram of the stage for the SHF band is a double loop common cathode self-excited oscillator.

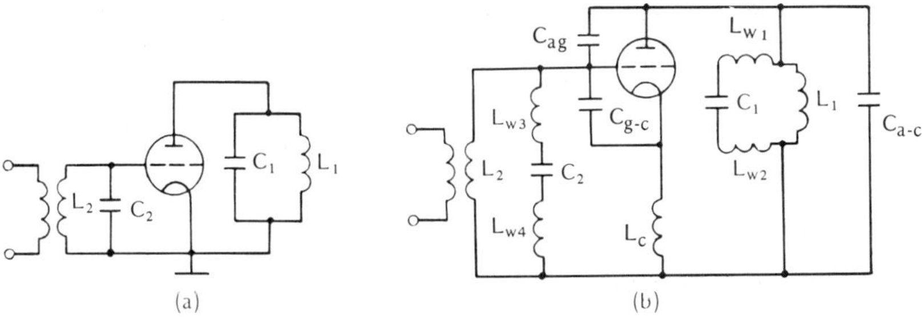

Figure 3.1

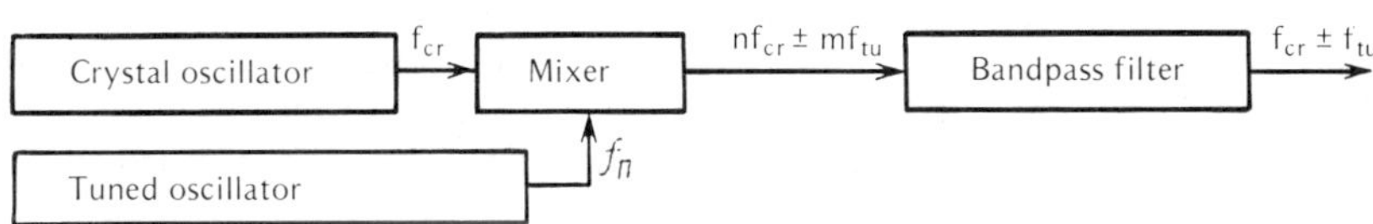

Figure 3.2

In SHF oscillators without external resonant circuits and feedback circuits (magnetrons, for example), parasitic oscillations most often occur because of the fact that the basic (type π) oscillations are interrupted for various reasons (for example, because of the instability of the power voltage) and other types of oscillations occur, the frequencies of which differ substantially from frequency ω_0 of the intended oscillations.

4. Cross-Product Emissions

Cross-product emissions occur during oscillation of the fundamental frequency as a result of the conversion of the oscillations of two or more auxiliary oscillators by nonlinear systems.

Systems of this kind are used, for example, in the design of crystal-stabilized tunable transmitters. Figure 3.2 shows the structural diagram of one such transmitter (interpolation diagram). It consists of a crystal oscillator and a fine tuning oscillator. The signals generated by these oscillators are mixed in a mixer, on whose output appear voltages with frequencies $nf_{cr} \pm mf_{tu}$. To the output of the mixer is connected a band pass filter, which passes the lower cross-product component (usually $f = f_{cr} \pm f_{tu}$). It can be shown that under certain conditions the relative output voltage instability will be higher than the relative instability of the frequency, f_{tu} of the fine tuning oscillator.

Emissions on cross product frequencies may occur in such transmitters as a result of the fact that for certain values of m and n, frequency $nf_{cr} \pm mf_{tu}$

may fall within the passband of the filter, and also because the amplitude-frequency response of this filter is different from zero outside of its pass-band.

A reduction of the cross-product emission level of transmitters of this type requires careful analysis of the frequencies and amplitudes of the useless cross-product components of the mixer output voltage and the selection of a band pass filter with the proper amplitude-frequency response.

5. Intermodulation Emissions

Such spurious emissions occur in cases when two or more transmitters operate on a common wide band antenna or when the antennas of transmitters with sufficiently high power are located in the immediate vicinity of each other.

Such situations often arise, for example, on ships and in other small objects equipped with a large number of radio systems. During simultaneous operation, the high-frequency oscillations of one transmitter act through the antenna feeder channel upon the output stages of other transmitters, which in relation to these frequencies are active two-ports with nonlinear responses. Consequently, a periodic change of a parameter of the second transmitter occurs (at frequency f_1 of the offending transmitter), which, in addition to its fundamental frequency f_2, radiates waves with frequencies $nf_1 \pm mf_2$. The latter form a noise emission spectrum. The higher the power of the interacting transmitters and the stronger the coupling between their output stages, the higher the power of intermodulation emissions. Intermodulation components with sum and difference frequencies $f_1 \pm f_2$ usually are strongest. In many cases, however, when frequencies f_1 and f_2 are close, components of intermodulation emissions with frequencies $2f_1 - f_2$ and $2f_2 - f_1$, which in view of their proximity to working frequencies f_1 and f_2 are difficult to filter out with intermediate and antenna circuits, have the highest intensity.

6. Out-of-Band Emissions

During spectral analysis and determination of the out-of-band emission level caused by modulation of a high-frequency signal of a transmitter, it is important to determine the emission bandwidth occupied by a given transmitter. In accordance with IRCC, the emission bandwidth, B_e is considered to be that part of the frequency band in which is concentrated $(100-\beta)\%$ of the average power radiated by a transmitter. Here $0.5\beta\%$ of the power is radiated on frequencies above the upper boundary of the emission bandwidth, and $0.5\beta\%$ on frequencies below the lower boundary of the emission bandwidth.

If the emission bandwidth coincides with the required frequency band, then the distribution of the power radiated by the transmitter is called "perfect" (Figure 3.3a). But if B_e exceeds the required frequency band (Figure 3.3b), then the emission is called "imperfect." In the case of imperfect emission, more than $\beta\%$ of transmitter power is radiated outside of the required frequency band. Exactly this type of power distribution is encountered, unfortunately,

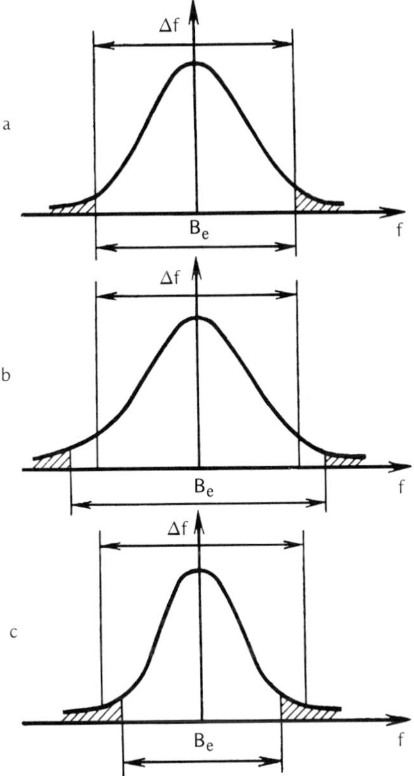

Figure 3.3

in most operating systems. Emission narrower than perfect (Figure 3.3c) is encountered very rarely and is used only in cases when the quality or speed of data transmission can be reduced.

The acceptable value of $0.5\beta\%$ is determined for each specific kind of emission.

In accordance with the general recommendations of IRCC on the reduction of the out-of-band emission level, the following measures must be taken:

- utilize the most effective kinds of modulation;
- reduce the frequency deviation in FM transmitters;
- try to reduce the steepness of the leading and trailing edges of the modulating telegraph signal for operation in the amplitude and frequency telegraphy modes;
- optimize the pulse envelope in accordance with the criterion of minimization of the side lobes of the spectrum for the prescribed main lobe width.

3.4 RECEIVER HETERODYNE EMISSIONS

The heterodynes of superheterodyne receivers, which are quiet oscillators, are capable of generating interference in nearby receivers. Such emission usually is transmitted through two main channels: through the receiver antenna or through the metallic chassis of the receiver. The latter is especially dangerous in SHF receivers with strong local oscillators, such as reflex klystrons. Such LO's, in view of the waveform of current in klystron cavities, are capable of generating a perceptible level of interference not only on the fundamental frequency, but also on higher harmonics. In addition, the interference created by LO's in sophisticated radio systems with several receivers, may also propagate through common power sources and switching systems.

Interference associated with LO emissions must be dealt with, for example, in television and radio broadcasting, where they are manifested as characteristic whistling noises, which are heard when a nearby receiver is being tuned.

The average voltage levels that are developed in a receiving antenna by the LO emission of a typical broadcasting receiver vary from a few microvolts to tens of millivolts and higher, depending on production quality (primarily the number of high-frequency amplifier stages, the selectivity of its circuits and heterodyne design), and increases as the number of UHF stages decreases and their passbands expand.

The replacement of tubes with transistors in the amplifier stages of receivers (under otherwise identical conditions) entails an increase of the power of the signals generated by the LO's and requires more reliable filtering of that emission. Such filtering can be accomplished not only by increasing the number of HF amplifier stages and improving the Q-factor of the circuits, but also by reducing the coupling between the LO and mixer to the minimum tolerable values and by including reliable filters in the HF switching and LO and mixer power circuits.

It is also helpful to design separate shielding systems for the LO and mixer. When this is done the shields must not be used as a ground because the currents that flow along the surface of the shield are capable of generating interference emissions on the LO frequency.

3.5 SECONDARY RECEIVING CHANNELS

For a number of reasons, chief among which are the frequency instabilities of the transmitter and receiver LO, the passband of the linear part of most receivers, which is determined chiefly by the total passband of their intermediate frequency amplifiers, exceeds (substantially, in many cases) the required bandwidth. This adversely affects selectivity of a receiver in relation to signals with close frequencies and complicates EMC conditions.

In addition to the frequency responses of the primary channel, it is also necessary to consider the fact that receivers (especially of the superheterodyne type) contain a large number of secondary receiving channels. The latter form as a result of the imperfect amplitude-frequency responses of the filters used in the receiver, and also as a result of the nonlinear properties of its mixer and certain amplifier stages.

Consequently, new frequency components appear in the spectrum of the signal amplified by the receiver.

Secondary receiving channels may be classified as spurious and out-of-band by analogy with secondary emission channels. Spurious receiving channels are formed as the result of the nonlinearity of the mixer and amplifier stages of a receiver and of the inadequate selectivity of the antenna circuits of HF amplifiers. These channels may be subdivided into cross-product and intermodulation channels.

Cross-product spurious receiving channels are formed as a result of the interaction signal and its harmonics with the LO voltage in the mixer stage of a receiver. As a result of such interaction, oscillations with frequencies $|mf_{int} \pm nf_h|$, where m and n are integers and f_{int} and f_h are the frequencies of the interference signal and of the LO respectively, appear on the mixer output. If any of the frequencies of this combination fall within the IF amplifier passband, they are not filtered out by its circuits and pass on to the output systems of receivers.

An interference signal that passes through any cross-product receiving response is attenuated by the circuits of all the selective elements wired in front of the mixer (the preselector of a receiver). If the preselector contains n_c identical circuits with quality, Q and the frequency of the interference signal differs from the frequency, f_s to which the receiver is tuned by Δf, then the amount of that attenuation, characterized by the attenuation factor k_{att}, may be calculated through the equation:

$$k_{att} = \left[\frac{(2\Delta fQ)^2}{f_s^2} + 1 \right]^{0.5n_c}$$

Hence, it follows that the amount of attenuation decreases as the number of circuits in the preselector increases and as the Q-factor of these circuits and frequency error, Δf increase.

The second factor that leads to the attenuation of signals propagating through cross-product receiving responses is a reduction of the gain of the mixer on frequencies corresponding to signal and LO harmonics. It may be assumed in the first approximation that the attenuation of cross-product response signals by the mixer increases with the sum of the numbers of harmonics of the signal and heterodyne m + n which form a given cross-product response.

Intermodulation receiving responses occur in cases when the receiver input is acted upon by more than one interference signal. They are caused by the interaction of two or more interfering signals with intensities sufficient for exposing the nonlinear properties of a channel, or by the interaction of several interfering signals and the receiver LO signal.

If the frequencies of the interfering signals are $f_1, f_2, \ldots, f_n$, then the interfering signal in the first of these cases reaches the receiver output only when one (or several) of the combinations of frequencies, $n_1 f_1 \pm n_2 f_2 \pm \ldots \pm n_n f_n$ fall within the IF amplifier passband, and in the second case, when even one of the frequencies $n_1 f_1 \pm n_2 f_2 \pm \ldots + n_n f_n \pm q f_h$ falls within that passband. Here $n_1, n_2, \ldots, n_n; q$ are integers and f_h is the LO frequency.

Techniques used to increase the interference immunity of receivers in relation to spurious emissions that propagate through intermodulation receiving response involve the maximum possible improvement of the selectivity properties of the preselector, reduction of gain in the HF stage of a receiver, etc.

3.6 INDUSTRIAL INTERFERENCE

Considerable attenuation has been devoted recently to analysis of phenomena that generate industrial interference and of methods of preventing them. Studies that have been conducted heretofore show that both the densities and the spectral composition of industrial interference differ substantially in different, often nearby regions. The highest interference densities occur in major cities and their suburbs.

The sources of industrial interference may be broken down into two major groups. The first group includes systems that inadvertently generate relatively regular high-frequency emissions, for example the deflection systems of cathode-ray tubes, various kinds of industrial and medical high-frequency instruments, etc. The interference emitted by such sources, both on the fundamental operating frequency and on harmonics, consists of nearly sinusoidal waves. These sources are relatively easy to identify and readily submit to simple forecasting of the effects that they produce, but they are not easy to control.

Sources of the second group, which is the largest, include various electrical installations which do not generate periodic high-frequency signals. These include power transmission lines, automotive ignition systems, high-frequency machinery for arc welding, gas discharge devices, dynamo machinery, power generators, and induction and switching machinery.

Studies have shown that the "specific weights" of the different groups of interference sources are about the same in different regions with approximately the same level of industrial development. In major industrial centers and adjacent regions, for example, the main sources of this type of interference are the ignition systems of internal combustion engines, power transmission lines and arc welding equipment.

Interference produced by ignition systems at frequencies above 30 MHz usually predominates over interference from other sources. However, at frequencies below 30 MHz, interference from power transmission lines usually predominates.

1. Ignition Noise

Interference from automotive ignition systems consists of an approximately periodic pulse signal of complex form. When large numbers of automobiles traveling at different speeds congregate at different points, test instruments record series of pulses with random amplitudes and time intervals. The duration of recorded groups of pulses varies from a few microseconds to several milliseconds, and the duration of individual pulses varies from 1 to 6 ns.

The overall intensity of the examined type of interference in a modern city is characterized by daily fluctuations of approximately 16 dB (between periods of maximum and minimum traffic density). There is a correlation between the absolute level of interference and the population of a city. The examined kind of industrial noise nearly always predominates over other types of industrial noise in 60-80m wide belts adjacent to heavily trafficked thoroughfares and highways.

2. Power Transmission Noise

High-voltage machinery and power transmission lines generate interference, which reaches maximum intensity during times of rain, snow, fog and high relative humidity.

High atmosphereic turbulence and high-level solar radiation lead to an increase of the interference level in arid regions. The interference level increases sharply when power transmission lines break down.

Interference from high-voltage transmission lines is random in character and appears in the form of nonrectangular pulses, the duration of which substantially exceeds the width of noise pulses emitted by automotive ignition systems. However, the repetition frequencies of the noise pulses from power transmission lines are considerably lower than the repetition frequencies of ignition noise pulses.

Power transmission lines produce interference because of transient processes associated with electrical discharges which occur randomly on the surfaces of wires and insulators. The pulse currents that occur during such discharges travel along the line, which may be either a coaxial wave guide (in this case the flow of energy is restricted to the inner conductor and sheath), or a single overhead line.

3. Interference from Arc Welding Machinery

Arc welding equipment generates very high-level interference. The fact alone that only a few such machines, as a rule, may be operating simultaneously witwin a given locality makes this kind of machinery the basic source of industrial noise. The spectra of the interference produced by different kinds of equipment do not coincide. Test results on a large number of different designs of machines revealed three broad resonance bands, the centers of which correspond to the frequencies 750 kHz, 3 MHz and 20 MHz. The emission spectrum of each individual machine does not necessarily include all three resonance bands.

Chapter 4

General Description of Radio Interference Control Techniques

4.1 JAMMING IMMUNITY OF RADIO SYSTEMS AND METHODS OF ANALYZING IT

1. Quantitative Description of Jamming Immunity

Modern electronic systems are most often used as component parts of different kinds of systems and complexes. Complexes are made up of functionally related semiautomatic systems, among which are the guidance systems of airplanes, missiles and spacecraft, fire control systems, etc.

Most familiar among complexes that contain radio systems are air and missile intercept and air defense complexes, strategic aviation and missile complexes for destroying surface and underwater targets, air reconnaissance complexes, etc.

The jamming immunity of an electronic complex (system) describes its ability to perform effectively while an enemy is simultaneously engaged in electronic surveillance and jamming. The jamming immunity of the radio components of complexes and systems must be evaluated quantitatively through the indices with which their jamming immunity can be determined. The main criterion of the jamming immunity of an electronic system is the probability, p_{ji} that it will perform its functions in spite of enemy electronic countermeasures. This probability is determined by the following equation:

$$p_{ji} = p_{ij}p_{nii} + (1 - p_{ij})\, p_{ni}, \tag{4.1.1}$$

where p_{ij} is the probability that intentional jamming will be used against an examined radio electronic system, p_{nii} and p_{ni} are the probabilities that the electronic system will perform its functions successfully when acted upon by intentional and natural interference in addition to useful signals, and by natural radio interference only, respectively.

The probability, p_{ij} in Equation (4.1.1) describes the security of an electronic system and the effectiveness of the enemy surveillance system, and p_{nii} and p_{ni} are the main quantitative indices of the noise immunity of an electronic system in the presence and absence of jamming, respectively.

The specific meaning of the probabilities p_{nii} and p_{ni} is determined by the function of an electronic system. For example, the p_{nii} of a detection radar that is a part of a guided missile complex is estimated as the probability of the correct detection of a single or group target for a fixed false alarm level under conditions of enemy electronic countermeasures. If p_{nii} is known for such a radar, then it is possible to determine the loss of effectiveness of the entire complex. If the radar is used for missile guidance, by beam riding for example, then p_{nii} describes the probability of angular coordinate tracking of a target within a prescribed accuracy, in spite of jamming. Special indices are also used extensively in addition to the basic noise immunity indices of electronic systems. The number of special indices, which are directly or indirectly related to the basic indices, is often large. In many cases, for instance, radio interference causes only additive random errors in coordinate measurement, guidance command transmission, etc. When such errors have a normal distribution, a knowledge of their mathematical expectations and variances, which are special noise immunity indices, is sufficient, as is known, for determining the p_{nii} and p_{ni} of the corresponding electronic systems. In addition to additive errors, however, there may also be multiplicative errors. Under such conditions it is necessary to determine the conditional mathematical expectations and variances of the errors, which are computed for each fixed value of a measured coordinate, transmitted command, etc.

When electronic systems are used as dynamic components of automatic or semi-automatic control systems, it is important for evaluating the effectiveness of such systems to know the mathematical models (dynamic equivalents) of the electronic systems, which are usually different from the model used in the absence of jamming [8, 51, 106, 107, 110].

The ratio, p_{ji}/p_{per}, where p_{per} is the probability that an electronic system will perform its functions in the absence of electronic countermeasures [192], is sometimes used in comparative estimates of the jamming immunity of electronic systems.

Also used as special jamming immunity indices of electronic systems are the effective signal to noise ratios at the outputs of linear parts of radio receivers, the false alarm and target miss probabilities, the probabilities of errors in digital data transmission, the average operating time of a tracking system before the first switching to the search mode, the probability of automatic target tracking failure, etc. A more detailed discussion of special criteria, which are used basically for analyzing anti-jamming techniques, is presented in the ensuing chapters.

2. Methods of Analyzing Jamming Immunity

General Principles

Both theoretical and experimental techniques are used to analyze the jamming immunity of electronic systems.

Experimental techniques include laboratory tests and tests under actual operating conditions. The specific features of a tested system are most completely taken into consideration during tests under actual operating conditions. Due to limited time and resources, however, it is not always possible to analyze all properties of an electronic system with respect to jamming. Major problems also arise in relation to the development of systems capable of simulating probable enemy electronic surveillance and ECM systems.

Laboratory tests often are conducted by the hybrid simulation method and are of great importance, since they provide results that are close to the performance of a tested electronic system under actual operating conditions without requiring additional outlays of time and resources.

The essence of hybrid simulation consists of partial utilization of the actual equipment, combined with an electronic computer, which simulates the operation of the elements that are represented by mathematical models. Systems that cannot be controlled with the required accuracy are used in real form. In hybrid simulation the solution of a given problem is found in real time scale and requires the development of additional equipment for mating the computer with the elements of the tested system.

A sufficiently exact theoretical analysis, the purpose of which is to evaluate the performance of an existing or planned system under ECM conditions, often encounters virtually insuperable obstacles. This necessitates the analysis of the passage of different and, as a rule, structurally complex signals and interference not only through the linear, but also through a large number of nonlinear converters with both constant and variable parameters. However, the approximate theoretical analysis, especially of the noise immunity of various systems, is practiced extensively, and many journal articles, monographs and textbooks have been published on this subject.

Difficulties that arise in the theoretical analysis of the jamming immunity of electronic systems can be attributed not only to the need to find their mathematical models, but also to the limited capabilities of the mathematical models that have been developed for the solution of sophisticated control systems. Therefore, the endeavor to obtain more accurate results from theoretical analyses necessitates the use of electronic digital computers. The latter are capable of the statistical modeling both of processes that take place in electronic systems and of radio interference to which they are subjected.

The success of jamming immunity analysis, both analytical and with a computer, depends to a great extent on the capability of developing a model of the analyzed ECM systems and of the victim system that is acceptable in terms of achieving the desired results and at the same time ignores secondary phenomena which encumber computational processes.

In a theoretical analysis, the probabilities p_{ij}, p_{nii} and p_{ni} or the dynamic equivalents of electronic systems, or the probability p_{ij} and the statistical characteristics (usually the mathematical expectations, correlation functions and dispersions) for the errors of the signals obtained are sought directly.

Principles of Determination of Probabilities p_{ij} and p_{nii}
The probability p_{ij} that jamming will be employed against an electronic system
is determined by the following equation [192]:

$$p_{ij} = p_r p_{ut} p_{jrec} \qquad\qquad (4.1.2)$$

Here p_r is the probability that the parameters of an electronic system, necessary
for jamming it, have been subjected to enemy reconnaissance, p_{ut} is the probability that the enemy will utilize jamming under conditions when the parameters
of the electronic system have been subjected to reconnaissance, p_{jrec} is the
probability that the receiver of the system will be jammed on the assumption
that its parameters have been subjected to reconnaissance and jamming will be
employed.

If, in order to jam an electronic system, it is necessary to know n_p of its independent parameters, then

$$p_r = \prod_{i=1}^{n_p} p_{ri}$$

where p_{ri} is the probability of reconnaissance of the i-th parameter. In order to
determine n_p and p_{ri} it is necessary to take into consideration the capabilities
of jamming and reconnaissance systems and the properties of the electronic system being analyzed.

The benefit that an enemy can gain by using (operating) the corresponding ECM
system depends chiefly on the bearing and range of the target electronic system
and its frequency range, in which interference signals can be generated. It is assumed here that all other parameters of an ECM system ensure the required effectiveness.

If the range to the victim system falls within tolerable limits, then [192]

$$p_{ut} = p_f p_s$$

where p_f is the probability that the operating frequency bands of the victim system and of the jamming system coincide, and p_s is the probability that the jamming antenna will be aimed at the victim system.

The probabilities p_f and p_s can be calculated only for the solution of specific
problems, the statement of which must take into consideration the characteristics
of antenna systems, operating frequency bands, the operational organization of
victim systems, etc. Some examples of the calculation of p_f and p_s under conditions of jamming are presented in [192]. We shall only mention here, by way of
illustration, that when frequency barrage jamming is used, the probability $p_f = 1$
for all electronic systems, the operating frequencies of which fall within the frequency band of the jamming transmitter. But if this condition is not satisfied,

then $p_f = 0$. Whenever the jammer and the victim system are located in space in such a way that reliable radio communications between them is assured, then $p_s = 1$.

The effect of jamming the receiver of a victim system under conditions when its parameters have been subjected to reconnaissance and the enemy employs jamming, depends on how accurately the receiver is tuned and the antenna of the ECM system aimed. It also depends on the time of utilization of the victim system, the time during which it remains within the effective range of ECM systems deployed against it, and their speed. Here, the required accuracy of tuning of an ECM system is related to the spectral width of the jamming and to the passband of the receiver of the victim system, and tolerable antenna aiming errors of an ECM system are limited by the width of its radiation pattern. In view of what has been said above we may write:

$$p_{jrec} = p_{jrec\,f}\,p_{jrec\,a}\,p_{jrec\,i}$$

Here $p_{jrec\,f}$ is the probability that the ECM system will be tuned accurately enough so that the jamming signals fall within the passband of the receiver that is used in the victim system, $p_{jrec\,a}$ is the probability that the antenna of the ECM system will be aimed accurately enough so that it will remain in radio communications with the victim system, and $p_{jrec\,i}$ is the probability that the jamming signals will reach the receiver of the victim system while it is performing its necessary functions. The probabilities $p_{jrec\,f}$ and $p_{jrec\,a}$ should be determined on the assumption that ECM will be employed, and $p_{jrec\,i}$ on the assumption that the ECM system is tuned to the prescribed frequency and its antenna is aimed at the victim system.

Errors that occur during the tuning of an ECM system are the result of measurement errors, Δf_{ts} of the operating frequency, f_{ts} of the victim system and of the error, Δf_{cms} with which the frequency, f_{cms} of the ECM system is tuned to the frequency $f_{ts} + \Delta f_{ts}$. Errors Δf_{ts} and Δf_{cms} usually can be assumed to be independent, normally distributed random values with mathematical expectations equal to zero. Under such assumptions, the total error, $\Delta f_\Sigma = \Delta f_{ts} + \Delta f_{cms}$ will also be a normally distributed random value and will have zero mathematical expectation. The value $p_{jrec\,f}$ for a system with the passband, Δf_{rec} and for a jamming signal with the spectral width, Δf_j is determined by the probability that Δf_Σ will differ from f_{ts} by $\pm 0.5\,(\Delta f_j + f_{rec})$. Therefore,

$$p_{jrec\,f} = \frac{1}{\sqrt{2\pi}\,\sigma_\Sigma} \int_{-0.5\,(\Delta f_j + \Delta f_{rec})}^{0.5\,(\Delta f_j + \Delta f_{rec})} \exp\left(\frac{\Delta f_\Sigma^2}{2\sigma_\Sigma^2}\right)\,d\Delta f_\Sigma$$

$$= \Phi\left(\frac{\Delta f_j + \Delta f_{rec}}{\sqrt{2}\,\sigma_\Sigma}\right) \tag{4.1.3}$$

where $\sigma_\Sigma^2 = \sigma_{cms}^2 + \sigma_{ts}^2$; σ_{cms}^2 and σ_{ts}^2 are the variances of the random values Δf_{cms} and Δf_{ts}, respectively; and

$$\Phi(Z) = \frac{2}{\sqrt{\pi}} \int_0^Z e^{-x^2}\, dx$$

is the tabulated probability integral.

In cases when

$$\frac{\Delta f_j + \Delta f_{rec}}{2\sigma_\Sigma} \gg 1$$

which is always the case when retransmitted ($\sigma_\Sigma \approx 0$) and frequency barrage ($\Delta f_j \gg \sigma_\Sigma$) jamming are employed, the probability $p_{jrec\ f}$ is virtually equal to unity.

Elevation and azimuth aiming errors, $\Delta\theta_{el}$ and $\Delta\theta_{az}$ of antenna A_j of the jammer depend on the accuracy with which the bearing to the victim system is determined and the accuracy of the control system of antenna A_j. But if the width of the radiation pattern of antenna A_j is substantially greater than possible maximum errors, $\Delta\theta_{el}$ and $\Delta\theta_{az}$, then $p_{jrec\ a} = 1$.

When jamming is selective in terms of direction, $p_{jrec\ a}$ can be calculated if the two-dimensional probability density, $w(\Delta\theta_{el}, \Delta\theta_{az})$, is known. It is also necessary to know the tolerable values of $\Delta\theta_{el}$ and $\Delta\theta_{az}$. In practice errors $\Delta\theta_{el}$ and $\Delta\theta_{az}$ very often can be considered as normally distributed random uncorrelated values with zero mathematical expectations. Under such conditions,

$$w(\Delta\theta_{az}, \Delta\theta_{el}) = w(\Delta\theta_{az})\, w(\Delta\theta_{el})$$

where $w(\Delta\theta_{el})$ and $w(\Delta\theta_{az})$ are the probability densities for $\Delta\theta_{el}$ and $\Delta\theta_{az}$, respectively, and

$$p_{jrec\,a} = \int_{-\Delta\theta_1}^{\Delta\theta_1} \int_{-\Delta\theta_2}^{\Delta\theta_2} w(\Delta\theta_{az})\, w(\Delta\theta_{el})\, d\Delta\theta_{az}\, d\Delta\theta_{el}$$

Here $-\Delta\theta_1$, $\Delta\theta_1$ and $-\Delta\theta_2$, $\Delta\theta_2$ are the tolerable errors $\Delta\theta_{el}$ and $\Delta\theta_{az}$.

After integrating for

$$w(\Delta\theta_{az}) = \frac{1}{\sqrt{2\pi}\,\sigma_{az}} \exp\left(-\frac{\Delta\theta_{az}^2}{2\sigma_{az}^2}\right)$$

$$w(\Delta\theta_{el}) = \frac{1}{\sqrt{2\pi}\,\sigma_{el}} \exp\left(-\frac{\Delta\theta_{el}^2}{2\sigma_{el}^2}\right)$$

where σ_{el}^2 and σ_{az}^2 are the variances of errors $\Delta\theta_{el}$ and $\Delta\theta_{az}$, we obtain

$$p_{jrec\ a} = \Phi\left(\frac{\Delta\theta_{az}}{\sqrt{2}\sigma_{az}}\right)\ \Phi\left(\frac{\Delta\theta_{el}}{\sqrt{2}\sigma_{el}}\right)$$

The time of operation and the time t_z during which the victim system remains within the range of ECM systems, and also the speed of the latter, exert the greatest influence on probability, $p_{jrec\,f}$. Speed is defined here as the time interval, t_{rrec} between the beginning of reception of the signals of the victim system by the reconnaissance receiver and the transmission of jamming signals. If the victim system operates for time t_z, then the receiver will be jammed for a period of time $t_z \geqslant t_{rrec}$.

The analytical determination of probability p_{nii}, which characterizes how well an electronic system performs its functions when receiving both useful signals and jamming signals, usually is frought with considerable difficulties. However, p_{nii} can be calculated in certain cases. To do so it is necessary to:

- take into consideration the beginning and duration of jamming,
- know the parameters of the jamming signal and their change in time and space,
- analyze the passage of the useful signals and jamming through the victim system in order to determine its resulting characteristics, and
- calculate probability p_{nii} on the basis of the tactical characteristics thus obtained.

Mathematical Models of Electronic Systems
To evaluate the noise immunity of electronic systems and complexes, it is necessary to build mathematical models for electronic coordinate estimators, command guidance links (of command transmission systems), and other electronic systems involved in data transmission in the guidance systems of airplanes, missiles, space vehicles, etc.

These kinds of electronic systems usually are referred to as radio links. The most common radio links can be conventionally divided into two groups. The first group includes the electronic systems that generate commands of the "on-off" type. The mathematical model of a single-channel system of the first group is a two-way switch, which is closed in accordance with the probability distribution law that characterizes the operation of the radio link when it is subjected to a particular kind of jamming. The second group includes systems that satisfy the functional dependence of the output signals on the measured coordinates or transmitted data: radar coordinate estimators, certain command guidance radio links, etc. A radio link of the second group contains an unsmoothed converter, which extracts from the received radio signal an information parameter (for example, it generates a voltage that describes the distance between an airplane and its target), and circuits that amplify and smooth that parameter, which usually have comparatively long time constants.

An unsmoothed converter, also called a discriminator, depending on the type of radio link, has a variety of circuitry designs. In an angle estimator, for example, it is a direction finder, in a Doppler frequency estimator, it is a receiver with a frequency detector, in a command radio link, it is a receiver combined with a decoder (decoding and demodulating systems), etc.

In the absence of jamming, the converter properties of a discriminator are characterized by the discrimination characteristic, which determines the dependence of the output signal of the discriminator on the transmitted message (the measured coordinate, its derivative, transmitted guidance command, etc.). The properties of a discriminator often change under the influence of jamming, and consequently, so does its mathematical model. The change of the properties of a discriminator is related to the presence in it of parametric and nonlinear elements. Amplitude fluctuations of a signal received by a conical-scanning angle tracker, as is known [86], lead to the following dependence of the output voltage, u_{df} of the angle tracker on error angle θ in one (the vertical, for example) tracking plane:

$$u_{df} = k_{df} \left[1 + m_j(t) \right] \, \theta + (2 k_{df}/k_m) \, m_j(t) \cos \Omega_{sc} t \tag{4.1.4}$$

Here k_m is the steepness of the modulation characteristic of the antenna, $m_j(t)$ is the jamming modulation index, k_{df} is the slope of the tracker in the absence of jamming, and Ω_{sc} is the scanning frequency.

As can be seen by the above equation, which was written under the assumption that components with the scanning frequency and its harmonics are completely eliminated by the output filter, the slope $k_{dfl} = k_{df} \left[1 + m_j(t) \right]$ of the tracker changes along with $m_j(t)$ when it receives amplitude-flucutating signals and is, therefore, a random function of time. At the same time, an additive component of voltage u_{df}, equal to $(2k_{df}/k_m) \cos \Omega_{sc} t$, also appears.

Equation (4.1.4) describes the mathematical model of a conical scanning tracker when it receives amplitude-fluctuating signals. In addition to a random change of the steepness of the discrimination characteristic and the appearance of additive error, jamming also results in a distinct narrowing of the band in which the discriminator output signal is a linear function of the transmitted parameter.

The equations for discriminators are derived in explicit form as a result of the solution of the problem of the passage of signals and jamming through all components of the system. However, writing such equations directly usually involves unsurmountable difficulties. Especially serious problems arise in connection with the analysis of strong jamming.

It is usually possible to determine theoretically and experimentally only the statistical characteristics of the discriminator voltage, u_d for fixed values of a transmitted signal or measured coordinate, x. These characteristics can be used to construct the mathematical model of a discriminator in accordance with a prescribed criterion of statistical equivalence. In practice, it is usually required to match the mathematical expectations, M_d, $M_{s\,eq}$ and the spectral densities $G_d(\omega)$, $G_{s\,eq}(\omega)$ (or correlation functions) for the output signals, u_d and $u_{s\,eq}$, which appear on the outputs of the real discriminator and its mathematical model, also called statistical equivalent.

The equation for the statistical equivalent which establishes the relation between $u_{s\,eq}$ and x is arbitrary because it is possible to determine many functions, $u_{s\,eq}(x)$ that yield $M_{s\,eq}$ and $G_{s\,eq}(\omega)$ which coincide with M_d and $G_d(\omega)$, respectively. In practice, however, it is advantageous to use an n-th power polynomial as the equation of the statistical equivalent. Under these conditions the identity of the real discriminator and its statistical equivalent is assured by determining n and the coefficients of the polynomial. Then, if the coefficients are random, it is necessary to find their mathematical expectations and their spectral and joint spectral densities.

The use of a polynomial as the equation for the statistical equivalent is justified by the fact that a linear relation exists between $u_{s\,eq}$ and the coefficients of the polynomial, which simplifies the computation. Furthermore, polynomials readily yield to mapping with standard computers, which may be used for analyzing different kinds of telecontrol systems.

A method of determining statistical equivalents in accordance with the above criterion is examined in detail in references [51, 106, 107, 110]. Therefore, we shall only give two examples as illustrations.

The first example refers to an angle tracker, for which the mathematical expectations, M_{df} and spectral density, $G_{df}(\omega)$ are determined theoretically or experimentally for voltage, u_{df}, generated by the tracker for any fixed value of angle θ between the line to the target and the axis of the antenna.

Let mathematical expectation, M_{df} and spectral density, $G_{df}(\omega)$ of the voltage generated by the tracker be

$$M_{df} = k_{df}\theta \tag{4.1.5}$$

$$G_{df}(\omega) = G_0(\omega) + G_1(\omega)\theta + G_2(\omega)\theta^2 \tag{4.1.6}$$

Since, in the example under consideration, M_{df} changes in proportion to θ and $G_{df}(\omega)$ is a function of θ and θ^2, it is natural to assume that a real tracker is a linear estimator for angle θ, and the equation of its statistical equivalent acquires the form

$$u_{s\,eq} = \xi_0(t) + \xi_1(t)\,\theta \tag{4.1.7}$$

Here $\xi_0(t)$ and $\xi_1(t)$ are the coefficients of the equation, and may be constants and deterministic or random functions of time.

Mathematical expectations, $M_{\xi 0}$ and $M_{\xi 1}$, spectral densities, $G_{\xi 0}(\omega)$ and $G_{\xi 1}(\omega)$, and joint spectral densities, $G_{01}(\omega)$ and $G_{10}(\omega)$ of coefficients, $\xi_0(t)$ and $\xi_1(t)$, should be selected such as to satisfy the following equations:

$$M_{s\,eq} = M_{df} \tag{4.1.8}$$

$$G_{s\,eq}(\omega) = G_{df}(\omega) \tag{4.1.9}$$

where $M_{s\,eq}$ and $G_{s\,eq}(\omega)$ are the mathematical expectation and spectral density, respectively, of voltage $u_{s\,eq}$ for a given value of θ.

From Equation (4.1.7), on the assumption that $\xi_0(t)$ and $\xi_1(t)$ are random functions of time, and angle θ is fixed, it follows that

$$M_{s\,eq} = M_{\xi 0} + M_{\xi 1}\,\theta \tag{4.1.10}$$

$$G_{s\,eq}(\omega) = G_{\xi 0}(\omega) + G_{\xi 1}(\omega)\,\theta^2 + [G_{01}(\omega) + G_{10}(\omega)]\,\theta \tag{4.1.11}$$

Comparison of Equations (4.1.5) and (4.1.10) shows that Equation (4.1.8) is valid when $M_{\xi 0} = 0$ and $M_{\xi 1} = k_{df}$. In order that $G_{s\,eq}(\omega) = G_{df}(\omega)$, it is necessary that

$$G_{\xi 0}(\omega) = G_0(\omega); \; G_{\xi 1}(\omega) = G_2(\omega); \; G_{01}(\omega) + G_{10}(\omega) = G_1(\omega)$$

Consequently, the equation for the statistical equivalent of the tracker may be written as

$$u_{s\,eq} = [k_{df} + \xi_1^0(t)]\,\theta + \xi_0(t) \tag{4.1.12}$$

Here $\xi_1^0(t) = \xi_1(t) - M_{\xi 1}$ is the centered random function $\xi_1(t)$.

As the second example for illustrating the method of determining the statistical equivalent; and at the same time for demonstrating the identity of the results obtained from the exact solution of the problem and by using the statistical equivalent, we shall examine a system consisting of a high-frequency linear amplifier and an amplitude detector, loaded with parallel-connected resistor, R, and capacitor, C. Let the input of the high-frequency amplifier receive an additive mixture of a sinusoidal amplitude-modulated signal and wideband stationary fluctuations, independent of each other, and let the useful signal be voltage, $\eta(t)$, generated on the load.

We assume that $(2\pi/\omega_0) \ll RC \ll \tau_c$, where ω_0 is the angular frequency of the signal, which coincides with the center frequency of the amplitude-frequency response of the high-frequency amplifier, and τ_c is the correlation time of the envelopes of the signal and fluctuations. Proceeding from this assumption, and using a quadratic detector for which the responses of the nonlinear element are

$$g(u) = \beta_g \, u^2 \quad \text{for } u > 0$$

$$g(u) = 0 \qquad \text{for } u \leqslant 0$$

we find [172],

$$\eta = \frac{1}{4} \beta_g \, RV^2(t) \tag{4.1.13}$$

In Equation (4.1.13), which is valid for $\beta_g \, RV(t) \leqslant 0.1$,

$$V^2(t) = [U_s(t) + A_s(t)]^2 + A_s^2(t) \tag{4.1.14}$$

is the square of the envelope of the voltage produced by the high-frequency linear amplifier. $U_s(t)$ is the envelope of the received signal, and $A_s(t)$ and $A_s(t)$ are the envelopes of the cosine and sine terms, which produce fluctuations $\xi(t) = A_s(t) \times \cos \omega_0 t + A_s \sin \omega t_0$ on the output of that high-frequency amplifier.

On the other hand, the equations that determine the mathematical expectation, M_V and correlation function, $R_V(\tau)$ for the square, $V(t)$ of the envelope of the mixture of the sinusoidal signal with a constant amplitude, U_s and narrow-band noise, $\xi(t)$, are universally known. These equations are [172]:

$$M_V = 2\sigma^2 + U_s^2 \tag{4.1.15}$$

$$R_V(\tau) = 4\left[\sigma^2 \rho^2(\tau) + U_s^2 \rho(\tau)\right] \sigma^2 \tag{4.1.16}$$

where σ^2 and $\rho(\tau)$ are the dispersion and envelope of the correlation coefficient for fluctuations $\xi(t)$.

From the standpoint of the problem under examination here, when the envelope $U_s(t)$ of the received signal can be viewed as a slowly changing function of time, M_V and $R_V(\tau)$ are the conditional mathematical expectation and correlation function for the square of the envelope, calculated on the assumption that $U_s(t) = U_s$. Then the conditional mathematical expectation, $M_\eta(U_s)$ and correlation function, $R_\eta(U_s, \tau)$ of process $\eta(\tau)$ will be:

$$M_\eta(U_s) = \frac{1}{4} \beta_g \, R \, (2\sigma^2 + U_s^2) \tag{4.1.17}$$

$$R_\eta(U_s, \tau) = \frac{1}{4} \beta_g^2 \, R^2 \sigma^2 \rho(\tau) \left[\sigma^2 \rho(\tau) + U_s^2\right] \tag{4.1.18}$$

Assuming the functions $M_\eta(U_s)$ and $R_\eta(U_s, \tau)$ to be given, we find the equation of the statistical equivalent that connects $\eta_{s\,eq}$ and U_s, where $\eta_{s\,eq}$ is the output signal of the statistical equivalent, equal to

$$\eta_{s\,eq} = \xi_0(t) + \xi_1(t)U_s + \xi_2(t)U_s^2 \qquad (4.1.19)$$

The mathematical expectations M_{ξ_0}, M_{ξ_1} and M_{ξ_2} and the correlation functions $R_{\xi_0}(\tau)$, $R_{\xi_1}(\tau)$ and $R_{\xi_2}(\tau)$ for the coefficients $\xi_0(t)$, $\xi_1(t)$ and $\xi_2(t)$, in accordance with the criterion of the equality of the mathematical expectations and correlation functions for the output signals of a real system and of its statistical equivalent, should be

$$M_{\xi_0} = 0.5\beta_g R\sigma^2 \,;\; M_{\xi_1} = 0;\; M_{\xi_2} = 0.25\beta_g$$

$$R_{e_0}(\tau) = 0.25\beta_g^2 R^2 \sigma^4 \rho^2(\tau)$$

$$R_{\xi_1}(\tau) = 0.25\beta_g^2 R^2 \sigma^2 \rho(\tau); \; R_{\xi_0}(\tau) = 0$$

Comparing Equations (4.1.13) and (4.1.19) in consideration of Equation (4.1.14), we readily see that $\eta_{s\,eq} = \eta$, if we assume that

$$\xi_0(t) = 0.25\beta_g\, R\left[A_c^2(t) + A_s^2(t)\right]$$

$$\xi_1(t) = 0.5\beta_g\, RA_c(t) \text{ and}$$

$$\xi_2(t) = 0.25\beta_g\, R$$

The above examples dramatically demonstrate the feasibility of deriving equations for statistical equivalents on the basis of given conditional mathematical expectations and correlation functions (or spectral densities) for the output signals of a single-channel electronic system. If the electronic system is a multichannel system, then to find the statistical equivalent for each of the channels, it is necessary to consider also channel couplings that occur as a result of jamming. However, the essence of the method of determining the statistical equivalents for multichannel electronic systems is the same as for single-channel systems [106].

4.2 SECURITY ENHANCEMENT TECHNIQUES

The security of an electronic system is its capacity to resist enemy reconnaissance of its radio signals and thus to resist jamming.

Frequently, structure, time, amplitude and space methods are used to increase security.

The essence of the frequency methods consists in tuning transceivers during operation and frequency camouflaging of the radiated radio signals. In addition, it is desirable in conical-scanning angle-tracking systems to use a secret (constant or changing in time) scanning frequency. The frequency method is used to impede an enemy's efforts to discover the operating frequency of an electronic system with a prescribed probability or to measure it with the necessary accuracy.

A comparatively detailed analysis of the capabilities of search and non-search systems that perform the detection of radio signals and measurement of their carrier frequencies is presented in reference [24], and therefore, systems of this type are not discussed here.

The use of a secret scanning frequency often prevents an enemy from employing jamming that is selective in terms of that frequency. The larger the scanning frequency range, the less effective jamming becomes. Frequency camouflaging of radiated radio signals amounts to the simultaneous utilization of N_{tr} ($N_{tr}>1$) transmitters in the same electronic system. All N_{tr} transmitters operate on different carrier frequencies, spaced far apart from each other, but they transmit the same modulation signals. Only 1, 2, ..., N_{tr} signals can be processed in the receiver. Under this condition, and when the operating frequencies of the N_{tr} transmitters of the target electronic system are spaced very far apart, N_{tr} barrage jamming transmitters and, of course, N_{tr} frequency-selective jamming transmitters may be used. The use of N_{tr} jamming transmitters is precluded in a considerable number of cases because of weight, size, power, consumption and other limitations. The solution of the EMC problem is further complicated by the use of multiple frequency transmission.

The structural method of enhancing security is based on the use of quasi-random signals. The essence of this method consists in generating a noise-like signal or code pulse groups (compound signals), the processing law of which is unknown to the enemy. Typical examples of such code groups are M-trains [27], Barker codes [172] and other types of compound signals which it is possible to use the compression principle. In the simplest cases structural security is achieved by adding to the pulse train that represents transmitted messages (for example missile guidance commands) pulses which are perceived by the enemy as having random repetition frequencies and at the same time are known to the intended receiving system. The enemy has trouble determining the kind of modulation, codes, etc., which reduces his effective jamming capability.

Quasirandom signals may be used in all kinds of radio systems. The time method of improving security is based on a reduction of the duration and repetition frequency of transmitted radio signals. If radio signals are not transmitted during the time of preparation of system for operation and during its operation, which is characteristic of passive homing, passive radio direction finding, and other systems, then a system with only a receiver is secure for practical purposes. Radio transmission time is reduced substantially by using combined (complex) systems for measuring coordinates, and which contain electronic and non-electronic (for example inertial) data sensors. Nonelectronic data sensors, calibrated to the signals of electronic measuring instruments, are capable of measuring coordinates for a comparatively long time with an acceptable accuracy. A detained examination of complex systems and analysis of their basic properties are presented in Chapter 9 of this book.

The amplitude method of improving security calls for a reduction of the transmitter power of the victim system in order that the input signal level of the enemy receiver will be below its actual sensitivity. This method is impractical, because it detracts from the noise immunity of the victim system.

Spatial security is determined by the radiation patterns of the antenna systems of a victim system. The better spatial security is, the narrower the antenna radiation pattern and the lower the side lobe level.

Nontracking and tracking antennas may be employed in the electronic systems of radio telephone and radio telegraph, command guidance links, data transmission and other nonradar systems. Narrowing of the radiation patterns of nontracking antennas is limited by the requirements of uninterrupted radio communications between transmitting and receiving stations at all possible locations. In this case, the necessary shape and width of the radiation pattern are determined in accordance with the familiar radio link calculation methods.

If the antennas of nonradar electronic systems are of the tracking type, then the narrowing of their radiation patterns is limited by the errors of the tracking systems and by the tolerable sizes of the antennas.

The required radiation patterns of antennas used in tracking radar also depend to a considerable degree on the required measurement accuracy and tolerable dimensions. The choice of the shape and width of antenna radiation patterns for radars of a different type depends on the specific problems to be solved and is made in accordance with recommendations of radar theory and engineering.

4.3 NOISE IMMUNITY ENHANCEMENT TECHNIQUES

1. General Information

The problem of protecting electronic systems from natural, mutual and intentional radio interference is by no means conclusively solved. However, there are numerous ways of controlling individual kinds and groups of interference.

Industrial noise and mutual interference usually can be controlled to a considerable extent. This is done by separating the carrier frequencies, by properly positioning the antennas of mutually interfering radio transmitters, by coding signals, by using noise suppression systems in electrical machinery, etc.

Natural interference, for example, receiver noise and signal fluctuations which occur as a result of reflection from radar targets and from propagation, almost always exist. Site noise and echos from local objects are also often permanent.

Modern electronic systems are designed to be able to operate normally under conditions of continuous interference. The design of such systems is an exceedingly difficult task. However, this problem already has been solved satisfactorily, and considerable progress is being made in optimum radio reception theory and practice.

The development of an optimum or nearly optimum receiver in realtion to receiver noise and jamming interference, which is similar to receiver noise in terms of structure, is the essential task of every designer. Therefore, a special section of this chapter is devoted to signal reception optimization problems. At the same time, electronic systems are designed so as to be protected from not only the most probable, but also the most dangerous kinds of intentional jamming. This is done even in cases when the employed safety techniques are complicated and expensive. All electronic systems are also protected from kinds of jamming that are easy and inexpensive to counteract.

The control of natural, mutual and intentional interference is based essentially on the difference of the structure and laws of change of the parameters that are inherent to useful signals and interference. This is accomplished by preventing the overloading of receivers, by cancellation of radio interference, by primary, secondary and functional selection, adaptation and complete utilization of information, and by using interference signals in systems containing electronic components that are vulnerable to jamming.

2. Prevention of Radio Receiver Overload

When a receiver is subjected to the action of a useful signal along with strong interference, its amplifiers operate in the nonlinear mode. This results in distortions of the envelope of the useful signal. When the level of the interference reaches the point where the amplifiers switch periodically from saturation to cutoff, a phenomenon called receiver overloading occurs. Overloading is dangerous during the reception of continuous AM signals and pulses and of unmodulated radio pulses.

To prevent receiver overloading, and thus to protect the most important parameters of the useful signal and to make efficient utilization of noise supression systems in HF and IF amplifiers and in video circuits, it is necessary to adjust the dynamic range of the voltage, u_r that acts upon the input of a group of amplifiers or of each amplifier of a receiver, and which represents a mixture of the useful signal and noise. AGC systems and amplifiers with linear logarithmic responses are employed for this purpose. Fast AGC is most desirable, because jamming may consist of strong pulses with a low duty factor (intermittent jamming), and an AGC system with a large time constant cannot abruptly change the gain of a receiver. Fast automatic gain control (FAGC) systems are not always acceptable in practice. They cannot be used, for example, in conical scanning trackers.

The subject of the prevention of radio receiver overloading using various systems and the results obtained are examined at length in Chapter 5.

3. Jamming Cancellation

The concept of the cancellation of jamming was first propounded by the Soviet scientist, Academician N. D. Papaleksi in his book *Radio Interference and Its Control*, published in 1942. The problem of jamming cancellation was stated and solved in that book as follows. In addition to the main receiver, which reacts to the signal-interference mixture, an additional cancellation receiver is used, the antenna of which picks up jamming only. The intensities and phases of the jamming in the cancellation and main receivers are set identical and opposite, respectively. As a result, as confirmed in reference [122], the jamming at the output of the main receiver is cancelled and the useful signal remains undistorted.

These are the classical statement and solution of the problem of jamming cancellation and correspond completely to the case when the main and cancellation receivers perform linear conversions of real signals and jamming. The above problem, presented mathematically, is posed as follows. On the input of the main receiver is the additive mixture

$$u_{in}(t) = u_{sm}(t) + u_{jm}(t)$$

of useful signal $u_{sm}(t)$ and jamming $u_{jm(t)}(t)$, and on the input of the cancelling receiver appears only noise $u_{jc}(t)$, which is functionally related to $u_{jm}(t)$.

If statements O_0 and O_c, which describe processes that take place in linear voltage converters $u_{in}(t)$ and $u_{jc}(t)$, respectively, are correctly selected, then it can be proved that

$$\Delta u(t) = O_0[u_{in}(t)] - O_c[u_{jc}(t)] = O_0[u_{sm}(t)]. \tag{4.3.1}$$

Here $O_0[u_{in}(t)] = O_0[u_{sm}(t)] + O_0[u_{jm}(t)]$ is the useful signal and jamming at the output of the main channel and $O_c[u_{jc}(t)]$ is the output noise of the cancelling receiver. Since statement O_0 is known, reconstruction of signal $u_{sm}(t)$ poses no difficulty.

A useful signal and jamming may appear both simultaneously and at different times. The latter case is characteristic of pulse systems, subjected to pulse jamming, generated in time intervals during which there is no useful signal. Under these conditions jamming can be cancelled both by the classical method, called amplitude-phase or coherent, when the main and cancelling receivers are linear converters, and by cancelling the jamming after preliminary shaping of their envelopes, called amplitude or incoherent cancellation.

These jamming cancellation methods, of which there are presently numerous modifications, are used extensively for cancelling jamming in the side lobes of the main receiving antennas. The important requirement here is that an electronic system have two receivers. One receives only jamming and the other both jamming and the useful signal. However, jamming may be cancelled with only one antenna and one receiver. The necessary condition here is that the useful signal and the jamming be pulses with repetition period T_p, which appear in that time interval at different times. In addition, the intensities of the useful signal and jamming should change in time and remain constant, respectively. This condition is characterized mathematically by the fact that the input signal

$$u_{in}(t) = u_s(t) \quad \text{for } \tau_0 + kT_p \leqslant t \leqslant \tau_0 + \tau_p + kT_p$$

$$u_{in}(t) = u_j(t) \quad \text{for } \tau_{jm} + kT_p \leqslant t \leqslant \tau_{jm} + \tau_j + kT_p$$

$$u_{in}(t) = 0 \quad \text{for other t.} \tag{4.3.2}$$

Here τ_0 is the time of appearance of the first pulse, characterizing the useful signal, τ_p is the duration of the useful pulse, τ_{jm} is the time of appearance of the first jamming pulse, τ_j is the duration of the jamming pulse, k is an integer, which acquires the values 0, 1, 2, ... , $u_s(t)$ is a function that describes modulation of the useful signal in time, and the amplitude of that signal changes in time T_p; $u_j(t)$ is a function that describes the modulation of the jamming in time, and, in contrast to $u_s(t)$, the amplitude $U_j(t)$ of pulse $u_j(t)$ remains the same during time T_p.

If the receiver that picks up jamming contains a system that delays the voltage $u_{in}(t)$ for T_p and generates the signal

$$u_{in}(t - T_p) = u_s(t - T_p) + u_j(t - T_p) \tag{4.3.3}$$

then, by feeding $u_{in}(t)$ and $u_{in}(t - T_p)$ into a computer it is possible to determine the difference voltage

$$\Delta u(t) = u_{in}(t) - u_{in}(t - T_p) = u_s(t) - u_s(t - T_p) \tag{4.3.4}$$

which characterizes the jam-free useful signal.

The above-examined jamming cancellation method, based on the fact that jamming is made up of periodic, nonoverlapping pulses, is called the interperiod method.

Jamming also can be cancelled by means of decorrelation. The essence of this method consists in the following. Suppose there is a mixture

$$u_{in}(t) = u_{sm}(t) + u_{jm}(t)$$

of useful signal $u_{sm}(t)$ and interference $u_{jm}(t)$ at the output of the main receiver and reference voltage, $u_r(t)$, generated by an auxiliary receiver or by a radar transmitter. For definition, we shall assume in the ensuing discussion that the main receiver is radar receiver and reacts to continuous signals, and $u_r(t)$ describes the undistorted references voltage generated by the radar transmitter.

Having at our disposal $u_r(t)$ we can generate the *a priori* known voltage, $u_r'(t)$, which differs, for example, in terms of phase from $u_r(t)$. As the result of the joint transformation of $u_r(t)$ and $u_{in}(t)$, and also of $u_r'(t)$ and $u_{in}(t)$, we obtain two expressions:

$$u_1(t) = u_{s_1}(t) + u_{j_1}(t), \quad u_2(t) = u_{s_2}(t) + u_{j_2}(t),$$

where $u_{j_2}(t)$ and $u_{j_2}(t)$ are uncorrelated interference signals and $u_{s_1}(t)$ and $u_{s_2}(t)$ are functionally related useful signals.

Addition or subtraction of $u_1(t)$ and $u_2(t)$ (depending on the specific problem to be solved) yields the voltages $u_\Sigma(t) = u_1(t) + u_2(t)$ or $u_\Delta(t) = u_1(t) - u_2(t)$. The jamming is $u_\Sigma(t)$ and $u_\Delta(t)$ weaker in comparison with the useful signal component than in the original mixture, $u_{in}(t)$. A detailed discussion of the jamming cancellation method and of the resulting basic properties is presented in the next chapter of this book.

To summarize what we have said above, we should point out that the following three basic jamming cancellation methods are employed at the present time:

- jamming cancellation using auxiliary cancelling and receivers,
- interperiod jamming cancellation, and
- decorrelation jamming cancellation.

These methods are used to counteract a comparatively large number of different kinds of jamming.

4. Primary Selection

Primary selection is defined as the extraction of a useful signal from a mixture with jamming in various parts of a radio receiver using only those parameters of the useful signal which depend on the design principle of the electronic system. Accordingly, there are spatial, polarization, frequency, phase, time, amplitude, structural and combined primary selection.

Spatial selection is performed by the antenna. The narrower its pattern and the lower the side lobe level, the better spatial selection becomes. This type of selection is used to combat multiple point jamming, i.e., jamming produced by several sources scattered in space. However, the inevitable presence of side lobes prevents the complete elimination of the influence of such jamming.

Polarization selection is based on a difference of the polarization of received signals and jamming. It is used to counteract both natural and manmade jamming, particularly for various kinds of radar angle tracking systems. Polarization selection of manmade jamming may be passive and active. The former is accomplished by matching the polarizations of the received signal and antenna, and the latter with a polarization filter. It is an effective means of combatting cross-polarization jamming, which is considered to be universal for jamming systems.

A polarization filter is a network of closely arranged parallel wires or metal plates. This network, when placed in the aperture of antennas, reflects radio waves whose polarization plane is parallel to the axes of the wires (or plates) and is transparent to waves with orthogonal polarization. The function of a polarization filter may also be performed by the antenna reflector, if it is made transparent to jamming signals. There are also other kinds of polarization filters.

Primary frequency selection is based on the spectral difference of the useful signal and jamming. It is accomplished by tuning and by making the receiver passband as narrow as possible. In conical scanning angle tracking systems, the scanning frequency, $F_{sc} = \Omega_{sc} / 2\pi$ is also tuned for the purpose of eliminating or substantially reducing the influence of selective and barrage jamming in relation to F_{sc}. Such jamming may be a continuous or a pulsed carrier frequency signal, amplitude-modulated by a sinusoidal voltage with frequency F_{sc} or noise, the spectrum of which is near F_{sc}. The receiver passband is narrowed by using high-frequency systems for stabilizing the frequency of the radiated signals and by correcting the LO frequency of the receiver, and also by using systems that track the Doppler frequency and its rate of change in time.

Primary frequency selection is most effective against active and passive deception jamming. It also helps to eliminate completely or reduce substantially the influence of reflections from the ground and local objects.

Phase primary selection is based on a difference between the phase-frequency characteristics of received useful signals and jamming. This kind of selection is accomplished with phase automatic frequency control systems which completely suppress jamming that is orthogonal with the reference signal in terms of phase, and substantially reduce the receiver output level of wide band noise. The weakening of the influence of wide band noise by phase selection is attributed to the fact that this kind of noise contains components which are in phase and in quadrature with the reference signal in terms of phase.

Primary time selection is based on the possibility of distinguishing between pulse signals and jamming in terms of duration and the times of their appearance, and also in terms of the pulse repetition frequency. This kind of selection is performed with automatic pulse selectors, which select pulses in accordance with their duration, their position in time, and repetition frequency.

An automatic pulse duration selector passes signals whose times of appearance fall within a prescribed range. Because of various kinds of instability, this range is somewhat larger than the duration of the itilized pulse signals, which impedes the complete suppression of jamming.

The selection of pulses by repetition frequency, F_p is performed with a system that contains an AND circuit and a delay line. The latter delays input video signals for time $T_p = 1/F_p$. The delayed and undelayed pulses are supplied to the AND circuit, which passes only those pulses with repetition period T_p and is open in relation to jamming pulses if their repetition frequency is not F_p. Automatic pulse repetition frequency selectors theoretically can operate in both the video- and radio-frequency bands. Automatic pulse duration and repetition frequency selectors are suitable for all kinds of electronic systems, the transmitters of which radiate periodic pulse signals with constant durations.

Primary time pulse position selection is performed with automatic range trackers, the signals of which are used to gate receivers for the duration of received useful pulses. Such automatic range trackers provide protection both against jamming which is synchronous with the useful pulses in terms of repetition period but different from the latter in terms of their arrival times, and against jamming from echos reflected by distributed targets. Situations are possible in which an airplane, traveling away from a radar, drops chaff in the rear hemisphere. The purpose of jamming of this type is to trick the automatic range finder, and along with it the angle tracker to track automatically not the airplane, but the chaff cloud. But if the axis of the tracking split-gates of a radar range tracker coincides not with the center of gravity of the received pulse signal, but with its trailing edge, then it is possible to range the farthest reflecting object of the group that makes up the distributed target, and thus to determine automatically the coordinates of the target (airplane), and not of the chaff cloud [9]. At the same time it is possible to protect an automatic range tracker and a radar as a whole against range gate pull-off jamming.

The separation of useful signals from jamming on the basis of their different intensities at the input of a receiver or some part of it is called primary amplitude selection. Jamming that is weaker than the useful pulse signal at the receiver input, and which does not coincide with it in terms of arrival time, is easiest to filter out. A bottom clipper may be used for this purpose. Such a clipper is also often useful in systems with continuous useful radio signals. Transmitter power should be turned up as high as possible to prevent the bottom clipper from distorting the transmitted information.

Pulse level selectors which do not pass jamming pulses greater in amplitude than the useful signal may also be used in pulse electronic systems in addition to bottom clipper that removes only jamming pulses, and a NOT circuit. The output pulses of the clipper and the mixture of the useful signal and jamming are supplied to that circuit. When two signals reach the NOT circuit at the same time its output effect is equal to zero, so that it passes only the complete signals. Jamming cannot be completely eliminated under actual conditions. However, its influence can be reduced substantially. Amplitude selection can also be accomplished by the storage method and by angle gating.

The essence of the storage method, which is suitable for all kinds of electronic systems, lies in that a decision about the presence of a signal is made not immediately after its arrival in the receiver, but only after some comparatively long period of time, T_{st}. T_{st} is selected so that it will be possible to discover the statistical properties of jamming without appreciably altering the monitored parameters (angular coordinates, aircraft guidance commands, etc.).

In pulse systems a signal is stored by adders, and in continuous systems by integrators. Adders and integrators reduce the effectiveness of wide band jamming. The reason for this is that the stored signals are coherent and noise is added in terms of energy. The storage of a pulse signal for a time equal to n repetition periods improved the signal to noise ratio by a factor of n, in comparison with the input signal to noise ratio of an adder (or integrator); the equivalent n for an integrator is the value T_{st}/τ_c, where τ_c is the jamming correlation time. For proof of what we have just said the reader is referred to Chapter 7 of this book and to reference [185].

In addition to temporary storage discussed above, it is also possible to use frequency and code diversity. Whereas time signals are transmitted successively in time, frequency diversity is characterized by the transmission of a given message simultaneously on several carriers or subcarriers. In code diversity each message is represented by the corresponding code combination of pulses, which is repeated n times simultaneously or in succession.

Angle gating is a means of increasing the resolving power of an electronic system in terms of angular coordinates by virtue of a special kind of signal processing in the receiver. Therefore, additional opportunities are available to control jamming. The essence of the angle gating method is explained below by way of example of the protection of an angle-tracking system from two-point blinking jamming. The tracking system is assumed to be turned off during the time of operation of a jamming transmitter, located on a target that need not be tracked or destroyed with a homing missile. The fact that a jamming transmitter is in operation is established as follows.

In the absence of blinking jamming the angular coordinates are determined comparatively accurately and the error signal in the tracking system is close to zero. It remains so for practical purposes even during the operation of a jamming transmitter that coincides in space with the target. As soon as a jamming transmitter located outside of the target begins to operate, the error signal increases sharply. This is established by an amplitude selector and is used to turn off the tracking system. As soon as the jamming transmitter ceases operation, the error signal decreases and the tracking system locks on again. There are also other possible angle gating techniques.

Structural selection is based on the structural difference of jamming and useful signals. The structure of the latter depends on the kinds of modulation employed. For example, radar pulse signals with a linear frequency-modulated carrier are used for compression in the receiver. In accordance with the compression principle, a pulse of long duration with an *a priori* known law of change of its carrier frequency, is transformed into a narrow pulse. Since individual half-waves of the useful signal are related by a rigid functional dependence and the jamming (for example, noise) is random, the accumulation of the latter during the compression process is relatively minor. At the same time the amplitude of the narrow pulse increases substantially in comparison with the amplitude of the wide pulse.

Structural selection can be performed to the fullest extent only with pattern (signal) recognition systems. Such systems are in the initial stage of development at the present time.

Combined primary selection consists of various combinations of the above-examined kinds of selection and may be frequency-time, space-time, space-frequency, etc. Combined primary selection is used quite frequently in practice.

5. Secondary Selection

Secondary selection involves monitoring the corresponding signal parameters which are formed during special coding on the transmitter end to increase the noise immunity of the system. This means that in order to perform secondary selection, it is necessary to increase the energy of the radiated signals. There are frequency, phase, time, amplitude and structural varieties of secondary selection.

Frequency secondary selection is accomplished by modulating the carrier signal with extra subcarriers. For example, a continuous illumination signal for a semi-active guidance head can be frequency-modulated with the ranging signal [147], and frequency division of channels by means of special sub-carriers with different frequencies is often utilized in multichannel radio communications systems and command guidance radio links.

Secondary frequency selection in receivers is performed by demodulators and filters which are directly connected to final controls or are components of tracking systems. Secondary frequency selection reduces the noise level and, in homing systems with continuous illumination signals, also prevents locking into the automatic tracking of jamming signals transmitted by targets located beyond the effective range of a missile.

Phase, time, and amplitude secondary selection refer to sinusoidal and pulse subcarriers, and not to carrier waves. These kinds of secondary selection are similar to the like kinds of primary selection in terms of their capabilities and techniques. With regard to amplitude secondary selection, it should be borne in mind that subcarrier angle gating is not used in this case. Special reference signals must be transmitted for secondary pulse position selection.

Not only attendant parameters of signals, but also auxiliary signals are used for secondary structural selection, based on analysis of receiver video signals. At the present time there is secondary structural selection without feedback and with feedback. In the former case, the signals radiated by the transmitter of the system to be protected are code pulse groups with a structure that is known on the receiving end. Redundant number and various kinds of non-number codes, as well as quasirandom signals, may be used for this purpose. Widely known among redundant binary codes are detection codes, codes with simultaneous detection and error correction, and also non-numerical codes, a typical representative of which is a time code, which consists of a group of pulses with *a priori* known pauses between them. If the structure of a received signal after conversion to video pulses differs from the signal generated in the transmitter, then the presence of jamming is established in selection without feedback and voltage is not supplied to the final control system.

Selection without feedback can be used in any electronic system and is useful for combatting pulsed concealment jamming. In structural selection with feedback both the structure of the radio receiver output signals is monitored and data (guidance command) transmission errors are corrected. This is done by using two-way radio communications systems, also called feedback systems.

There are interrogation feedback and comparison feedback [66,75]. If, in the case of interrogation feedback, the receiver, by analyzing its input signal establishes that it does not reflect any of the possible messages, then an interrogation signal is sent to the transmitter. After receiving the signal the transmitter retransmits the previous signal through the direct communications circuit. This process continues until interrogation transmission ceases.

When comparison feedback is used, the receiver in the direct communications line (main receiver) informs the transmitter of incoming signals by sending through the direct communications circuit signals called receipts. In the transmitter the information provided by the receipts is compared with the information that was transmitted. If the compared data do not agree the transmitter sends the main receiver a command to stop using the previous message and to retransmit the necessary message. If the receipt corresponds to the transmitted information no additional signals are transmitted through the direct communications circuit and the receiver identifies the message that was received before the last receipt. In the simplest systems signals transmitted from the output of the main receiver are used as receipts.

Structural selection with feedback can be used only in electronic communications systems (command transmission links, data transmission systems, radio telephone stations, nonautonomous radio navigation systems, etc.) for transmitting only digital data. The design of radar stations in this manner is precluded, since the sources that generate radio signals for them are enemy targets or other types of radar targets. Structural selection with feedback is used to counteract pulsed concealment jamming.

6. Functional Selection

Functional selection, also called functional signal processing, increases the resolving power of radar angular coordinate estimators by processing signals from several independent receivers. The essence of this kind of selection is explained below by way of example of the direction finding of one of two noncolocated sources of radio emissions u_1 and u_2 on the assumption that they are covered by the main lobe of the antenna of an amplitude monopulse tracker [231].

A typical monopulse angle measuring system, as is known, has four antennas. At the output of each of them appears a signal, containing information about the angles of elevation and azimuth of sources u_1 and u_2. By processing independently the output signals of all four antennas and of the receivers associated with them it is possible in the final analysis to derive a system of equations with four unknowns. In the example at hand, these equations establish the relation between the combined amplitudes of the signals produced by the corresponding receivers and the angular coordinates of sources u_1 and u_2. The joint solution of the equation system thus derived is used to resolve sources u_1 and u_2 and to determine their angular coordinates. In the general case, when there are N_{so} sources of radio emission, it is necessary to have $2N_{so}$ independent receiving channels.

Accordingly, by virtue of functional selection, an electronic angle-measuring system is protected from jamming from transmitters located outside the target. This kind of selection is useful in radar and radio navigation systems with angle-measuring systems, and in radio communications with one carrier frequency for several simultaneously operating transmitters.

7. Adaptation, Complete Utilization of Information, Utilization of Jamming

Adaptation (to external conditions) involves a change in the structure and parameters of an electronic system to enable it to perform best its functions under the conditions of any kind of jamming. The problem of the adaptation of electronic systems under conditions of jamming is by no means completely solved, but studies on this subject are being expanded. Certain recommendations already have been made on the basis of theoretical analyses, aimed at improving individual components of radar coordinate estimators, command guidance links, radio communications systems, etc.

For example, the well known automatic noise control system performs adaptation of a receiver by maintaining its output false alarm level constant in the presence of wide band noise. Another example of an adaptive system is the antenna of a radar angle measuring system with an adaptive pattern. It is also possible to control automatically the sampling time of transmitted information by level [29] and the transconductances of gauges and command guidance links [110] in accordance with the jamming level.

The principles of the complete utilization of information are incorporated in mobile radar and radio navigation instruments. In such instruments information about the desired coordinates is obtained as a result of processing the signals generated by electronic and nonelectronic transducers. Nonelectronic transducers track the vehicle that carries an electronic instrument. By making complete utilization of information, it is possible to further reduce the passband of an electronic estimator without increasing the dynamic error level. This is possible by virtue of the fact that fluctuations of a parameter in a radio signal to be measured, which are caused by the motion of the object, are monitored by nonelectronic transducers. Examples of nonelectronic transducers, capable of operating in conjunction with electronic instruments, are accelerometers, gyroscopic sensors and aircraft air speed measuring instruments.

During operation electronic and nonelectronic transducers can perform mutual correction, which makes it possible not only to narrow the passband of an electronic coordinate estimator, but also to turn it off periodically. In this case signals generated by nonelectronic transducers are used as useful signals. In order to be able to perform the required switching operations, the combined measuring instrument of an electronic system operating with jamming must include a jamming analyzer.

Complete utilization of information is a comparatively universal method. It can be used to counteract both cancealment and deception jamming.

Jamming signals are utilized to benefit the victim system usually with passive target angular trackers (for example, missile homing heads). Passive direction finding is also employed at military air, naval and ground radar installations and in active and semiactive missile homing heads. Passive radio fuses have also been developed.

4.4 SIGNAL PROCESSING OPTIMIZATION

1. General Information

The traditional fields of optimum signal processing theory are:

- signal detection and recognition,
- parameter estimation,
- filtering, and
- signal resolution.

In electronic measuring instruments, the detection of a signal, i.e., the establishment of the fact of its existence, usually always precedes measurement. The reason for this is that only measurement results obtained with an *a priori* guaranteed accuracy are useful. For this reason, automatic measuring instruments and automatic command guidance radio links contain detectors which activate measurement circuits or recording equipment when a certain signal to noise ratio is reached. In binary communications systems, the detector is the receiver, which establishes the presence or absence of binary symbols in a received signal.

Detection theory is related in terms of the mathematical tools and circuitry employed to signal recognition (discrimination), in which it is necessary to determine which, from among all possible signals, is present in a received message.

Estimation of parameters is one of the classical problems of mathematical statistics. It consists of determining (estimating), on the basis of a finite number of samples of some random value, the distribution of which is presumed known, the parameters of that distribution. In electronic applications this problem consists of determining certain characteristics (parameters) of a received signal. The value to be measured is assumed to be coded in one of the parameters of a signal, received along with noise. In optimum analysis this parameter should be measured (estimated) in the best manner in terms of whatever quality criterion is used.

As follows from the general formulation of the parameter estimation problem, its solution requires a knowledge of the distribution of the received signal-noise mixture. Another basic assumption which is made for estimation of a parameter is that the parameter remains constant during the time of measurement or, as is often said, during the observation time. This assumption places great limitations on the practical importance of parameter estimation in application radar and radio navigation moving target coordinate estimators.

Optimum filtering is devoid of this assumption. Optimum filtering is used to separate in the best possible way, from a mixture of a useful signal and noise, the useful signal, which in the general case is a sample of some random process. If the extracted signal is constant but unknown, then results similar to those yielded by parameter estimation theory are obtained. Therefore, filtering is sometimes called process estimation [100].

In statistics, optimum resolution amounts to establishing the fact of the presence or absence of a given signal in the presence of interference signals with similar parameters. Consequently, the problem to be solved amounts to the detection of a signal under conditions when signals that are similar in structure are received along with the one that is to be detected.

In view of everything that has been said above, we should like to point out that a general idea of optimum signal processing problems can be gained by simply examining optimum detection and filtering problems.

2. Signal Detection

To clarify the presentation we introduce the basic concepts and characteristics of optimum detection, illustrating them by way of the simplest example. We assume that a useful signal, characterized by a constant value, u_s, (Figure 4.1a) may or may not exist in detection (processing) time, T. A segment of a sine wave with a constant amplitude could be used as the desired signal, but this does not alter the essence of the detection procedure examined below.

Noise (Figure 4.1b) with which a signal is mixed (if it exists) hampers reliable detection. Consequently, it is necessary to establish the fact of the existence of a signal on the basis of sample functions that contain a mixture of the signal and noise (Figure 4.1c) or just noise.

Thus, the detection procedure amounts to the processing of sample functions of the random function

$$Y(t) = U_s + U_n(t). \qquad\qquad (4.4.1)$$

The signal to be detected may or may not exist in each of these sample functions.

We denote random functions and random values with capital letters: $Y(t)$, $U(t)$, etc., and the set of their possible values, and consequently, the arguments (independent variables) of the corresponding distribution laws, with small letters, for example, y, u_n, etc.

A sufficiently complete statistical description of the random function $Y(t)$ is its multivariate distribution density. It is introduced as follows. We examine the values of a random function at discrete moments in time, $t_1, t_2, \ldots, t_m$, which

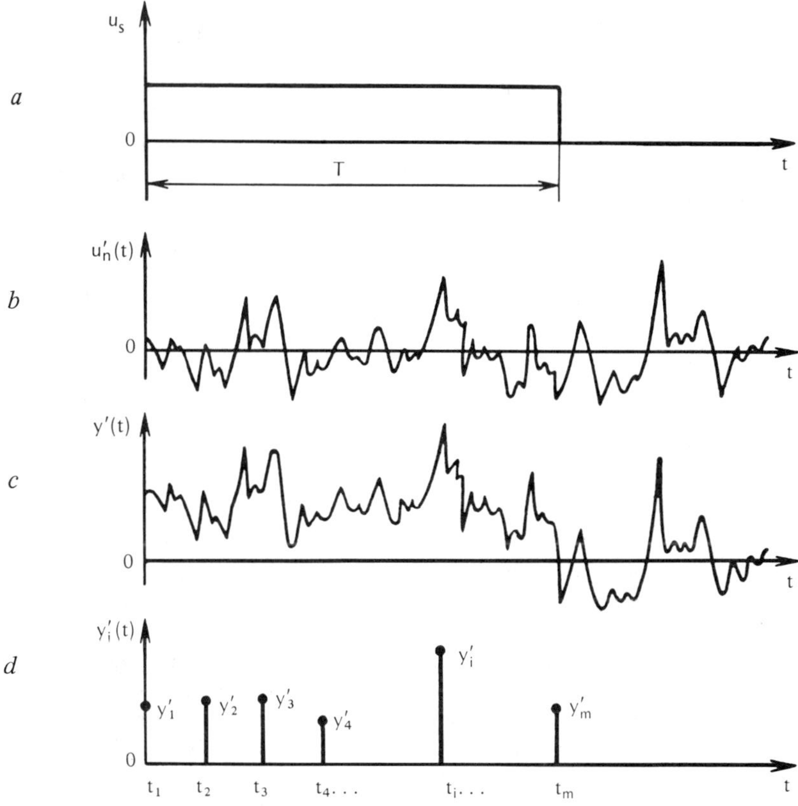

Figure 4.1

means that we replace the random function with a random sequence. In the entire set of possible sample functions, the values of random function, $Y(t)$ at moments of time, t_1, t_2, ..., t_m are a sequence of random values $Y_1 \equiv Y(t_1)$, $Y_2 \equiv Y(t_2)$, ..., $Y_m \equiv Y(t_m)$. This sequence is sometimes called a multivariate random value or random vector.

The combined distribution density of these random values, $w_m(y_1, ..., ..., y_m/u_s)$, where $y_1 = y(t_1)$, ..., $y_m = y(t_m)$, is taken in practice as the statistical characteristic of random function, $Y(t)$. This multivariate distribution density depends on the presence or absence of a signal. In this respect it is conditional, which is indicated by including the symbol u_s in the argument of the distribution density. The details of a random process, which are disclosed when it is described by an m-variate distribution density, become clearer as m increases.

The set of values y_1' ..., y_m' of random values, Y_1, ..., Y_m, obtained from each specific sample function (Figure 4.1d), is called a sampling. In many cases, the set of random values Y_1, ..., Y_m is called a general sampling.

When multivariate distributions are used for detection, meaningful results and feasible detectors can be obtained only under certain conditions which must be satisfied by the random sequence being processed. In most cases, these conditions are not troublesome for practical applications. One such constraint may be the requirement that random values $Y_1, ..., Y_m$ must be normal. Another often utilized constraint is that these random values be independent. This condition can be satisfied by selecting the right reading times, $t_1, t_2, ..., t_m$.

We shall further assume that such independence exists. Then, in the presence of a signal

$$w_m(y_1, ..., y_m/u_s) = \prod_{i=1}^{m} w_1(y_i/u_s),$$
(4.4.2)

and in the absence of a signal

$$w_m(y_1, ..., y_m/0) = \prod_{i=1}^{m} w_1(y_i/0).$$
(4.4.3)

Consequently, multivariate distribution densities are expressed sufficiently simply through univariate densities.

In statistical radio engineering, which utilizes the tool of mathematical statistics, conclusions concerning the presence of a signal and its parameters are made on the basis of sample functions that are run and of the samplings corresponding to them. Although these sample functions do contain all the information about the phenomena in which we are interested, it is often impossible to obtain this information directly from sample functions or a sampling. They must be subjected to processing (analysis). An important aspect of this analysis is the determination of certain average characteristics of a sample. Processing a sample on the basis of the probability function and probability ratio is extremely productive and in many cases optimum.

In mathematical statistics [22], the probability function is obtained from the multivariate distribution density (equation 4.42) of random values by replacing in it the independent variables $y_1, ..., y_m$ with the values of a sample $y'_1, ..., y'_m$, obtained by running each specific sample function.

In detection problems, the probability functions for the existence and non-existence of a signal will be, respectively,

$$l'(u_c) = w_m(y'_1, ... y'_m/u_s) = \prod_{i=1}^{m} w_1(y'_i/u_s),$$
(4.4.4)

$$l'(0) = w_m(y'_1, ..., y'_m/0) = \prod_{i=1}^{m} w_1(y'_i/0).$$
(4.4.5)

In the literature ususally no distinctions are made in the symbole of the arguments
of distributions in equations (4.4.2) and (4.4.3) and the numerical data of each
specific sample shown in equations (4.4.4) and (4.4.5), which sometimes
causes confusion. In the ensuing discussion, therefore, the sample data, representing a set of random numbers and functions thereof, which are also random
numbers in the sample set, will be written with primes.

The probability function for each specific sample tells which of the two events
$u_s \neq 0$ or $u_s = 0$ is more probable. When formulating signal detection procedures
on the basis of the above-examined statistical characteristics of sample data, it is
more convenient to compare not the values $l'(u_s)$ and $l'(0)$, but their ratio

$$l_r' = l'(u_s)/l'(0) \tag{4.4.6}$$

called the probability ratio, with threshold h. This probability ratio is also a random value in the sample set.

For reasons which will become clear in the ensuing discussion, preference is
given to comparison not of the value l', but of its natural logarithm with the
threshold, i.e.,

$$v' = \ln l_r' \gtrless C, \tag{4.4.7}$$

where $C = \ln h$. Since the logarithmic function is nondiminishing and l' is non-
negative, comparison of l' with threshold h and v' with threshold C are equivalent procedures.

The detection process consists of the following. The logarithm of the probability
ratio is calculated and compared with threshold C for each sample function. If it
turns out that $v' > C$, then it is decided that a signal exists in the sample function,
and when $v' < C$ a signal is assumed not to exist.

At comparatively low signal to noise ratios, and also because of the random
nature of the sample functions that are taken, the logarithm of the probability
ratio is a random value, and the inequality $v' > C$ may be valid in the absence
of a signal.

In this case, the detector makes a wrong decision concerning the presence of a
signal. Errors of this kind are called false alarms. And, conversely, if $v' < C$ in
the presence of a signal, then a wrong decision is made, which is called a miss.

When the threshold is low, misses virtually do not occur, but the percentage of
false alarms increases sharply. Increasing the threshold increases the number of
misses and reduces the number of false alarms. One may feel intuitively that
an optimum threshold exists. This is actulaly the case, but it depends on a number
of conditions, in particular on the criterion on which is based the design of the
optimum detector. The choice of some optimal criterion for a system, including
signal detection systems is, to a considerable extent, subjective, i.e., the criterion

is not derived from theory, but is designated arbitrarily in consideration of the performance features of a specific optimized system. The soundness and worth of a quality criterion for a system are checked in practice.

For example, it has been established that to optimize radar detectors is is advantageous to use the Neyman-Pearson criterion, but for communications systems the perfect observer criterion is better. When the Neyman-Pearson criterion is used the false alarm level is assigned and it is required that the probability of detection be maximum. The perfect observer criterion requires that the total error, caused by both false alarms and misses, be minimal.

Once a criterion has been selected, the optimum threshold, C is determined on the basis of the requirements of the given criterion and the structure of the optimum detector is established.

The logarithm of the probability ratio, v', determined by equation (4.4.7), is the sample value of some random value V. The kind of distribution density of this random value depends on whether or not a signal exists in a given sample function

We denote the distribution density, V, through $w(v/u_s)$ in the presence of a signal, and through $w(v/0)$ in the absence of a signal. In accordance with the above definitions, the false alarm probability is expressed as

$$p_{fa} = \int_c^\infty w(v/0)dv, \qquad (4.4.8)$$

and the signal miss probably as

$$p_{ms} = \int_{-\infty}^c w(v/u_s)\, dv. \qquad (4.4.9)$$

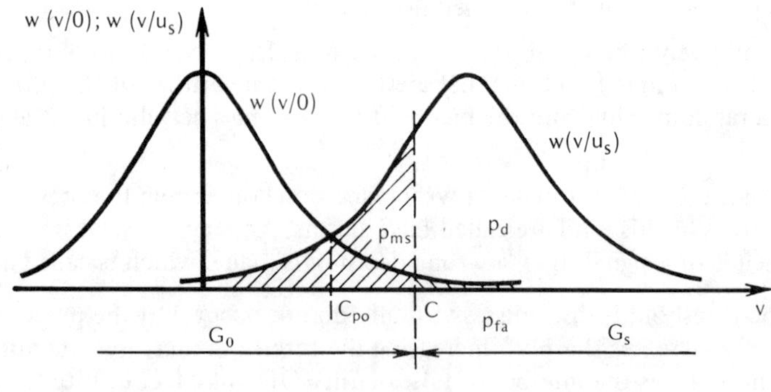

Figure 4.2

These results give a graphic geometric picture (Figure 4.2). Here the distribution densities of random value V (of the logarithm of the probability ratio) are shown for the existence and nonexistence of a signal, respectively. False alarm probability, p_{fa} represents the area under the curve $w(v/0)$ to the right of threshold C(ray G_s), and the signal miss probability p_{ms} represents the area under the curve $w(v/u_s)$ to the left of the threshold (ray G_0).

Obviously, the correct detection probability, P_d is

$$p_d = 1 - p_{ms} = \int_C^\infty w(v/u_s)\,dv. \tag{4.4.10}$$

This probability is defined as the area under the curve $w(v/u_s)$ to the right of threshold C.

As follows from the figure, the false alarm probability decreases as the threshold level increases, but at the same time the correct detection probability decreases. The opposite is true when the threshold decreases.

The combined distribution densities $w_m(y_1, ..., y_m/u_s)$ and $w_m(y_1, ..., y_m/0)$ of the sequence of random values $Y_1, ..., Y_m$ that yield the analyzed samples $y_1', ..., y_m'$, may also be used for calculating probabilities p_{fa}, p_{ms} and p_d. This is done by the rules of detection formulated previously. The space of all possible samples (the domain of random vector Y) is broken down into two nonintersecting ranges, G_s and G_0. The existence of a given specific sample in range G_s is equivalent to the case when random V acquires the value of v' that falls within ray g_s of the v axis (Figure 4.2). If the sample falls within range G_0, then v' will be on ray G_0. Hence it follows that

$$p_{fa} = \int_{G_s} 0_0 ... \int w_m(y_1, ..., y_m/0)\,dy_1 ... dy_m, \tag{4.4.11}$$

$$p_{ms} = \int_{G_o} 0_0 ... \int w_m(y_1, ..., y_m/u_s)\,dy_1, ..., dy_m, \tag{4.4.12}$$

i.e., when this technique is used to determine p_{fa} and p_{ms} it is necessary to compute m-multiple integrals, for which reason equations (4.4.8) and (4.4.9) are often employed in practice.

The above analysis shows that by calculating the probability ratio it was possible to transform an m-dimensional (and at the limit an infinitely dimensional) space of samples (space of observations) into a one-dimensional space. Such transformations are used extensively in mathematical statistics and comprise the essence of the analysis of experimental data for the purpose of drawing certain conclusions therefrom.

If the transformation is performed so that no information contained in the original sample is lost, them it is called sufficient, and the random value obtained as a result of it is called a sufficient statistic. The probability ratio is a sufficient statistic.

The optimality of the Neyman-Pearson criterion consists in the fact that when it is disclosed only as a result of comparison with other processing procedures, which do not yield sufficient statistics. Such a comparison shows that the Neyman-

Pearson procedure yields the highest correct detection probability for a given false alarm level. The threshold $C_{N\text{-}P}$ for the Neyman-Pearson criterion is found by solving the equation

$$P_{fa0} = \int_{C_{N\text{-}P}}^{\infty} w(v/0)\,dv, \qquad (4.4.13)$$

in which the distribution density, $w(v/0)$, and tolerable false alarm probability, P_{fa0}, are given.

To determine threshold level, C_{po} using the perfect observer criterion, it is necessary to compute the complete error probability

$$P_{er} = p_0 \int_{c}^{\infty} w(v/0)\,dv + p_s \int_{-\infty}^{c} w(v/u_s)\,dv \qquad (4.4.14)$$

where p_0 and p_s are the *a priori* (i.e., assigned before the beginning of the analysis of a sample) probabilities of the nonexistence and existence of a signal, respectively.

To find threshold C_{po}, which minimizes p_{er}, it is necessary to set the derivative of the right hand side of equation (4.4.14) in terms of C equal to zero. The result of this is the equation

$$\frac{p_0 w(C_{po}/0)}{p_0 w(C_{po}/u_s)} = 1. \qquad (4.4.15)$$

The solution of this equation relative to the threshold yields the value of C_{po} which corresponds to the perfect observer criterion.

The perfect observer criterion theoretically is better than the Neyman-Pearson criterion, since it takes into consideration past experience, reflected in the *a priori* probabilities p_0 and p_s. In practice, however, it is very difficult to find such relations in which p_0 and p_s can be assigned ahead of time with sufficient accuracy. Therefore, they are often set equal to $p_0 = p_s = 0.5$. Then the solution for determining C_{po} will be

$$\frac{w(C_{po}/0)}{w(C_{po}/u_s)} = 1 \qquad (4.4.16)$$

This special case of the perfect observer criterion is often called the maximum probability criterion.

Equation (4.4.16) means that the threshold should correspond to the point of intersection of the curves $w(v/0)$ and $w(v/u_s)$ in Figure 4.2. Therefore, false alarms and misses will occur with different probabilities. When the Neyman-Pearson criterion is used, the threshold is usually set such that the false alarm probability will be substantially smaller than the miss probability. Therein lies the chief difference between the two criteria.

A common feature of these criteria is the fact that the detection procedures employed for each of them are formulated on the basis of calculation of the probability ratio. This circumstance is attributed to the fact that they are included as subclasses in a more general criterion, called the Bayes criterion, or the minimum mean risk criterion.

Bayes detection, developed in statistical decision theory, consists of assigning, in addition to the sample and *a priori* probabilities p_0 and p_s, certain costs or penalties which are associated with false alarms and misses. These data are used to calculate the mean risk involving making a decision as to the existence or non-existence of a signal. The threshold, C is set so that the mean risk is minimal. If costs caused by false alarms and misses are assumed to be identical, then the Bayes criterion becomes the perfect observer criterion.

Since it is extremelu difficult to assign sound costs for real situations, the practical worth of the Bayes criterion is negligible. However, the optimality of all detection procedures based on calculation of the probability ratio, can be better substantiated in the theoretical sense.

To draw the structural plans of detectors that use the above criteria, and to obtain information about the performance of these detectors, it is necessary to assign the specific kind of distribution densities $w_m(y_1, ..., ..., y_m/u_s)$ and $w_m(y_1, ..., y_m/0)$ of the sequence of random values $Y_1, ..., Y_m$, from which the sample $y_1', ..., y_m'$ is taken.

To change to continuous processing, the multivariate conditional distribution densities $w_m(y_1, ..., y_m/u_s)$ and $w_m(y_1, ..., y_m/0)$ are transformed (when possible) to the functionals $F[y(t)/u_s(t)]$ and $F[y(t)/0]$, respectively, and the algorithm of the probability ratio is written as

$$v' = \ln \frac{F[y'(t)/u_s(t)]}{F[y'(t)/0]} \tag{4.4.17}$$

If a signal is detected in white noise, with spectral density $G(w) = N_0$, then calculations by equation (4.4.17) yield [100]

$$v' = \frac{2}{N_0} \int_0^T y'(t)u(t)dt - \frac{1}{N_0} \int_0^T u_s^2(t)dt. \tag{4.4.18}$$

Here $\int_0^T u_s^2(t)\,dt = E$ is the energy of the signal transmitted at time T. In problems of the detection of a known signal, E is assumed to be given. The value of v', obtained from each sample function is compared with the threshold

$$C_{N-P} = \sqrt{2q_1}\, x_{fa} - (q_1/2) \tag{4.4.19}$$

for the Neyman-Pearson criterion and

$$C_{po} = 0 \tag{4.4.20}$$

for the perfect observer criterion. Here $q_1 = (2E/N_0)$ is the ratio of double the signal energy to the spectral density of the noise; $x_{fa} = \arg \Phi(x)$ is the argument of the probability integral, calculated for the given false alarm probability.

The structural diagram of an optimum detector, illustrated in Figure 4.3, was based on equation (4.4.18).

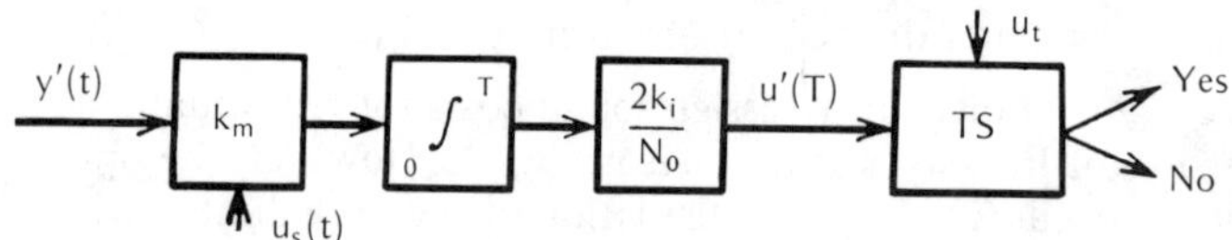

Figure 4.3

The basic operations performed by a detector of this type consist of the following. The received signal-noise mixture, or just the noise alone, is multiplied in a system with gain k_m by a copy of signal $u_s(t)$, which is stored in the receiver. The k_m of the multiplier [1/V] is introduced only for dimensional matching purposes, and its magnitude is unimportant. Therefore, it is often assumed to be equal to one. The multiplier output voltage is supplied to an integrator, where it is integrated during time T, and from there it passes through a circuit with gain $2k_i/N_0$ to threshold system TS. Multiplier $2/N_0$ is included for normalization, and the coefficient k_i [V^2], like k_m, matches the dimensions of the signal processing channel.

At the end in integration, the signal

$$u_o(T) = \frac{2k_{mi}}{N_0} \int_0^T y'(t)\, u_s(t)\, dt, \tag{4.4.21}$$

appears at the output of the $2k_i/N_0$ circuit, and this signal is compared in the threshold with voltage u_t for the purpose of making a decision as the to the existence or nonexistence of a signal in the received sample. Then the integrator is reset and the detection cycle starts all over again. The coefficient $k_{mi} = k_m k_i$ [V].

The threshold voltages u_{po} and $u_{t\,N-P}$ for the perfect observer and Neyman-Pearson criteria are found from (4.4.18) - (4.4.20) and are

$$u_{po} = k_{mi}\, q_1/2, \tag{4.4.22}$$

$$u_{t\,N\text{-}P} = k_{mi}\sqrt{2q_1}\, x_{fa} \tag{4.4.23}$$

The integrator output voltage $u_o(T)$, measured at moment of time $t = T$ is a sampling of some random value, U, which has normal distribution. Its mathematical expectation, M_{u_0} and variance, $\sigma_{u_0}^2$ in the absence of a signal are, respectively,

$$M_{u_0} = 0 \quad\text{and}\quad \sigma_{u_0}^2 = k_{mi}^2 q_1; \tag{4.4.24}$$

in the presence of a signal

$$M_{u_0} = K_{mi}^2 q_1 \quad\text{and}\quad \sigma_{fa}^2 = k_{mi}^2\, q_1. \tag{4.4.25}$$

The false alarm (p_{fa}) and correct detection (p_d) probabilities are calculated through the equations:

$$p_{fa} = \int_{u_t}^{\infty} \frac{1}{\sqrt{2\pi k_{mi}^2 q_1}}\, \exp\left[-\frac{u^2}{2k_{mi}^2 q_1} \right]\, du$$

$$= \frac{1}{2}\left[1 - \Phi\left(\frac{u_t}{k_{mi}\sqrt{2q_1}} \right) \right]. \tag{4.4.26}$$

$$p_d = \int_{u_t}^{\infty} \frac{1}{\sqrt{2\pi k_{mi}^2 q_1}}\, \exp\left[-\frac{(u - k_{mi}q_1)^2}{2k_{mi}^2 q_1} \right] du$$

$$= \frac{1}{2}\left[1 - \Phi\left(\frac{u_t - k_{mi}q_1}{k_{mi}\sqrt{2q_1}} \right) \right]. \tag{4.4.27}$$

Here $\Phi(x) = \dfrac{2}{\sqrt{\pi}} \displaystyle\int_0^{x} e^{-t^2}\, dt.$

For calculations with equations (4.4.26) and (4.4.27), threshold level u_t is determined by equations (4.4.22) and (4.4.23), depending on which detection criterion is used.

Tables and graphs have been prepared [71] to facilitate calculations by equations (4.4.22), (4.4.23), (4.4.26) and (4.4.27).

The circuit in Figure 4.3 represents an optimum correlation detector, also called a correlation receiver. It can be shown that this circuit is equivalent in

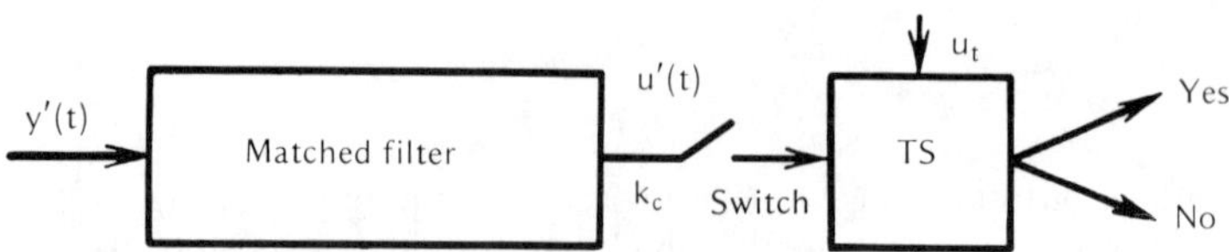

Figure 4.4

in terms of detection performance to one with a matched filter (Figure 4.4).
The matched filter is determined by weighing function $h(\tau)$ or by the response
frequency $H(j\omega)$ and $h(\tau)$ is the mirror image of the signal relative to the
center point $t = T/2$. Switch Sw is closed at signal completion time T.

The choice of detector as a correlation receiver or matched filter is dictated
only by design considerations.

Detection systems that have been developed often fall far short of these optimum
detectors in terms of performance. There are several reasons for this, which
may be divided conditionally into two groups.

The first group consists of those related to a change of assumptions used in the
synthesis of an optimum detector, relative to the detected signals and interference, due to the following circumstances: interference cannot be reduced to
white noise; the phase of the received signal is not known on the receiving
end; a fluctuating signal is received; the position of the received signal on the
time axis is not known, etc.

The second group is related to a decision not to use elements of the optimum
system which are complicated in practice.

The deterioration of the performance associated wtih the above-listed factors is
customarily characterized as a detector sensitivity loss.

Nonwhite Gaussian noise is characterized by a zero mean and by correlation
function $R_n(t_1, t_2)$. This kind of noise is also called correlated or colored.
In order to find the alogrithm of operation of an optimum signal detector for
the case when the signal is mixed with correlated noise, it is necessary to perform the same operations as in the case of white noise, i.e., to compute the
logarithm of the probability ratio and to compare it with a threshold, the value
of which depends on the criterion employed. Here the difference from signal
detection in white noise consists only of major problems that arise in connection
with calculating the probability ratio. These problems are related to the fact
that in noise a general sampling is a system of correlation random values,
$Y_1, ..., Y_m$, the combined distribution density of which can no longer be represented as the product of the distribution densities of each of these values.

These are two widely known approaches to the computation of the probability
ratio, to which correspond two structural diagrams of the optimum detector. In
the first method, the probability ratio is calculated directly on the basis of the

multivariate distribution densities of correlated random values, $Y_1, ..., Y_m$, with and without a signal [184,6]. In the second method, the random function $Y(t)$ is expanded in the interval 0-T into an orthogonal series, which is usually called a Karhunen-Loeve series. This expansion is convenient to use because the coefficients of the series form a system of uncorrelated random values, and if the processes being analyzed are normal, then these coefficients are also statistically independent. Therefore, the previously examined method of constructing optimum detectors for a signal in white noise can be used. Taking independent readings of a correlated normal process is sometimes called whitening the colored noise [26].

The basic results of these two approaches to the synthesis of optimum detectors are examined below. The greatest difficulty that arises in calculations of the multivariate probability density of statistically independent random values consists of finding the matrix $Q_n = R_n^{-1}$, which is inverse in relation to correlation matrix, R_n. In the case of the continuous processing of received sample functions, matrix inversion amounts to the solution of the integral equation [184,6]

$$\int_0^T R_n(t_1, t) Q_n(t, t_2) dt = \delta(t_1 - t_2),\qquad(4.4.28)$$

where $Q_n(t_1, t_2)$ is the continuous analog of the inverse correlation matrix. By analogy with inverse matrix Q_n, the function $Q_n(t_1, t_2)$ is sometimes called an inverse correlation function.

The finite limits of integration pose the basic difficulties in the solution of equation (4.4.28). If Equation (4.4.28) is solved and $Q_n(t_1, t_2)$ is determined, then the logarithm of the probability ratio is written as

$$v' = \int_0^T\int y'(t_1) u_s(t_2) Q_n(t_1, t_2) dt_1 dt_2\qquad(4.4.29)$$

$$- \frac{1}{2} \int_0^T\int u_l(t_1) u_l(t_2) Q_n(t_1, t_2) dt_1 dt_2$$

If the noise is white, i.e., $R_n(t_1, t_2) = N_0 \delta(t_1 - t_2)/2$, then from Equation (4.4.28) we find $Q_n(t_1, t_2) = 2\delta(t_1 - t_2)/N_0$. Substituting this value of the inverse correlation function into Equation (4.4.29), we obtain Equation (4.4.18), which we derived previously.

To simplify the construction of the structural plan of the detector, we introduce the function $\Psi(t)$, defining it as

$$\Psi(t_1) = \int_0^T u_s(t_2) Q_n(t_1, t_2) \, dt_2\qquad(4.4.30)$$

Then

$$v' = \int_0^T y'(t_1)\, \Psi(t_1)\, dt_1 - \frac{1}{2} \int_0^T u_s(t_1)\, \Psi(t_1)\, dt_1 \tag{4.4.31}$$

If function $\Psi(t)$ is assumed to be some general reference signal, then the analogy in Equation (4.4.18) and (4.4.31) can be examined and the structural plan of the detector can be drawn up in the form illustrated in Figure 4.5. Here the voltage $u_\Psi(t) = k_\Psi\, \Psi(t)$ is multiplied by sample $y'(t)$, and the product is integrated in time interval T.

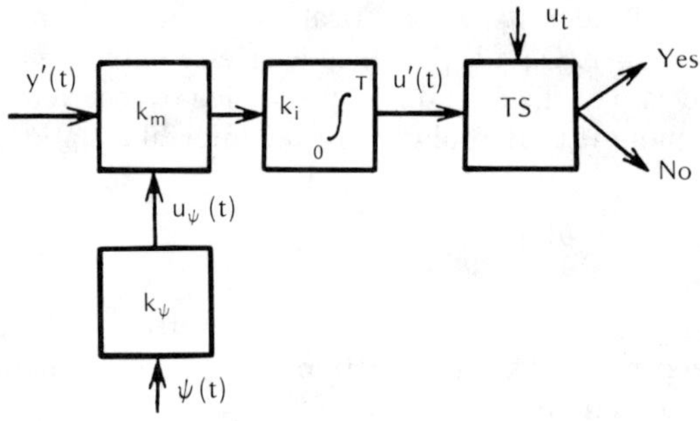

Figure 4.5

The integrator output voltage, $u'(T)$ at moment of time T,

$$u'(T) = k_y k_i' \int_0^T y'(t) u_\psi(t)\, dt \tag{4.4.32}$$

is compared in the threshold system with threshold level, u_t, which is determined by Equations (4.4.22) and (4.4.23) if it is assumed that

$$q_1 = \int_0^T u_s(t)\, \Psi(t)\, dt \tag{4.4.33}$$

The detector performance indices are, as before, the false alarm and detection probabilities, which are calculated by Equations (4.4.26) and (4.4.27). Equation (4.4.33) shows that the detection reliability now depends on the kind of signal. We recall that when a signal is detected in white noise, the value q_1 is determined only by the energy of the signal and by the spectral density of the noise, and the form of signal has no influence on it.

The function $\Psi(t)$ may be calculated directly on the basis of the correlation function of noise $R_n(t_1, t_2)$ without converting to inverse correlation function $Q_n(t_1, t_2)$. To do this, Equation (4.4.28) must be multiplied on the right and on the left by $u_s(t_2)$, the resulting relation must be integrated from 0 to T, and the integration variable replaced by

$$\int_0^T R_n(t, \lambda)\, \Psi(\lambda)\, d\lambda = u_s(t). \tag{4.4.34}$$

The optimum detector of a signal in correlated noise may also be designed on the matched filter principle. The weighting function of such a filter is calculated in accordance with the general signal, $\Psi(t)$, which is determined by Equation (4.4.34). Therefore, integral Equation (4.4.34) must be solved in any case for the construction of an optimum detector.

It has been shown [184] that is integration is done in infinite limits, or if the spectral density of interference is described by a fractional rational function of frequency w, then the solution of Equation (4.4.34) will be obtained in closed form.

The structure of a matched filter for correlated noise is such that the filter has more attenuation for those received spectral components of which correspond to the frequencies of greater intensity in the noise spectrum.

The process whereby a signal is detected in correlated noise, based on the conversion to statistically independent samples, amounts in the case of continuous processing to the inclusion of a so-called whitening filter in the detector circuitry. A structural diagram of such a detector is shown in Figure 4.6. The output process of a whitening filter (WF), $\widetilde{y}'(t) = \widetilde{u}_s(t) + \widetilde{u}_n(t)$ is a mixture of the converted signal and white noise.

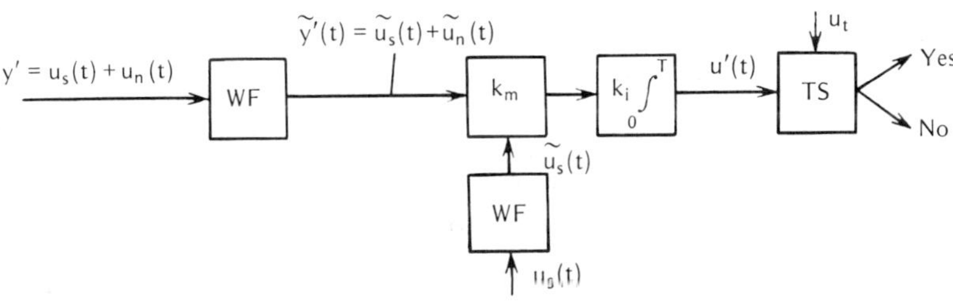

Figure 4.6

To maintain the required ratios between the converted sample and the reference signal at both inputs of multiplier k_m, a similar filter is also included in the multiplier channel. The rest of the system is identical to the optimum detector for a signal in white noise.

To find the parameters of the whitening filter we assume that only a noise process $u_n(t)$ with a zero mean and correlation function $R_n(t_1, t_2)$ acts upon its input. The weight function $h_{wf}(t_1, t_2)$ of the filter should be such that noise $u_n(t)$ at its output has a correlation function equal to $N_0 \delta(t_1 - t_2)/2$, i.e.,

$$\frac{N_0}{2} \delta(t_1 - t_2) = \int_0^T h_{wf}(t_1, \lambda) h_{wf}(u, t_2) R_n(\lambda, u) \, d\lambda du. \qquad (4.4.35)$$

Thus, to determine the structure of a whitening filter it is necessary to solve integral Equation (4.4.35). This solution depends exclusively on the correlation function of the input noise. If the duration of the process sample, T is longer than the filter storage time, then the range of integration may be expanded to infinity. In this case, when the noise is a stationary process, Equation (4.4.35) is readily solved with the aid of the Fourier transform

$$N_0 = |H_{wf}(j\omega)|^2 G_c(\omega).$$

Hence, we find an expression for the combined transfer function of the filter $H_{wf}(j\omega)$ through the one-sided spectral density $G(\omega)$ of the correlated noise at its input and spectral density N_0 of the white noise at its output

$$|H_{wf}(j\omega)|^2 = N_0/G_c(\omega). \qquad (4.4.36)$$

The problem examined previously, in which only the amplitude and initial phase of the received signal were known, is not encountered in practice and the condition introduced above is a convenient mathematical abstraction, used for finding the limiting detection performance. The conditions under which signals actually are received are vastly more complex. The first approximation to such conditions corresponds to the case when the frequency of the useful signal and its position on the time axis are known precisely at the receiver to the accuracy of the period of high-frequency waves, and the initial phase and amplitude are unknown.

In application to radar problems this situation characterizes the detection of a signal reflected by a target for a constant and *a priori* known distance between the target and receiver. It is further assumed that the frequency of the radar transmitter is absolutely stable and the influence of instability is excluded by storing the frequency of the radiated signal until the arrival of the reflected pulse.

If some parameter of the signal is not precisely known and only its statistical characteristics are given, the optimum reception theory suggests two different approaches for this case. In the first one the unknown parameter must be measured, i.e., its optimum estimate must be found, and the detector receives a signal, which, instead of the unknown parameter, contains an estimate of that parameter. This procedure leads to rather complicated designs with simultaneous detection and measurement [23, 164, 98]. However, if the influence of unknown parameters on detection performance is negligible, such design complication is unjustified. In this case preference is given to the other approach, in which the probability ratio must be averaged over the unknown parameters in order to exclude them from the structure of the optimum detector. This approach is based on the (not entirely valid) assumption that the unknown parameters do not carry information about the detected signal. Such parameters often are called uninformative and even parasitic [52], from which follows the need for averaging. The second approach is considered more traditional in relation to the synthesis of optimum detectors.

The next step of approximation to actual detector operating conditions is to make an assumption concerning the unknown carrier frequency of the signal and its unknown position on the time axis. The frequency of the signal is not known because of the frequency instability of the transmitter and the Doppler frequency shift, caused by the relative motion of the transmitter and receiver. Because of the absence of data on the distance between the radar and a target, and also between the two terminals in the communications system, the position of the signal on the time axis becomes unknown.

In theory, the problem amounts to what is known as complex or multialternative detection. The optimum detector in this case is designed as a multichannel system. The possible range of delay of the signal is broken down into intervals, each of which corresponds to one element of target range resolution. An optimum detector is built for each such interval. It is noteworthy that such a multichannel detector performs a detection and measurement procedure, since the appearance of a signal in any channel makes it possible to establish the time delay of the signal, and consequently, the range to the target in accordance with the number of the channel. A multichannel frequency division system is designed in like fashion if the frequency of the signal is unknown.

Optimum signal detection theory, based on analysis of probability ratios, presumes the probability distributions of the received samples to be unknown. The probability distribution determines the structure of the detector, and a knowledge of the parameters of that relationship makes it possible to calculate the threshold necessary for attaining the required detection reliability.

In mathematical statistics, methods in which the distributions of analyzed processes must be known in order to draw statistical conclusions are called parametric methods. Parametric methods, in spite of their extensive application in statistical radio engineering, may encounter fundamental difficulties, as in the case of insufficient statistical data in the description of the input processes of an electronic system, or when such data change in time in an unpredictable manner. The simplest, but very characteristic situation of this kind is an increase in the receiver output noise level as a result either of an increase in its gain or of the level of wide band noise. If the parameters of the detector are left unchanged, the probability of false alarm increases.

In order to stabilize the false alarm level, an extra receiving channel is included in these types of parametric detector, in which the noise level is estimated. In radar systems such a channel may be provided by additionally gating the receiver at a distance (time interval) where a target signal is known *a priori* not to exist. The measured noise level is used only to adjust the threshold or to normalize the noise. Some false alarm stabilization algorithms that work by changing the threshold are described in reference [82] and [179]. Theoretical substantiation of the normalization of noise in an optimum detector with an unknown intensity leads to a rule called Student's t-test [12]. This rule is used approximately in automatic noise control systems.

The basic deficiency of these false alarm stabilization systems is that the noise intensity estimate obtained in such systems differs from the true intensity by the amount of measurement error, to which parametric detectors are extremely sensitive. It was shown in reference [62], for example, that a 10% measurement error of the average noise level causes the false alarm probability to change by approximately one order of magnitude. This fact, and also the sensitivity of such detectors to a change of the noise distribution law, prompted the development of nonparametric detectors, for the design of which very little information about the distribution of the analyzed samples is required.

Nonparametric theory yields algorithms (on the basis of which statistical conclusions are drawn), which are invariant with the distribution. In practice, however, in application to signal detection, this subject does not arise very often. Nonparametric detection is usually defined as an algorithm which makes some detection quality characteristic independent of the distribution. This characteristic often the false alarm level. In nonparametric detectors, consequently, the false alarm level is stabilized as reception conditions change. This property is gained at the cost of a loss of optimality. However, the performance of such detectors may be made sufficiently close to optimum [12].

The simplest nonparametric detector is a sign detector [12, 52]. This detector is designed on the basis of the following assumptions relative to the statistical properties of samples. If a signal does not exist and sample y', ..., y'_m consists only of noise components, then the random values Y_1, ..., Y_m are assumed to have a symmetrical distribution density, i.e., $w(y) = w(-y)$. If the sample contains a signal, then symmetry is violated. In other words, positive and negative noise deviations are assumed to be equally probable and the fact that a signal exists in the sample violates this principle.

The algorithm of a sign detector is constructed as follows. The sample function to be analyzed is quantized to two levels, 0 and 1, and at zero threshold, i.e., random value U, with the sampled value

$$u'(y_i') = \begin{cases} 1 \text{ for } y_i > 0, \\ 0 \text{ for } y_i \leqslant 0. \end{cases}$$

is formed. Then the sume of these sampled values is found and is compared with threshold C_z

$$\sum_{l=1}^{m} u'y_i \gtrless C_z.$$

(4.4.37)

Threshold C_z is calculated on the basis of the required false alarm probability.

One modification of the sign detector is the so-called phase autocorrelator [179], a functional diagram of which is shown in Figure 4.7. Wide band and narrow band filters (WF and NF) are tuned to the frequency of the signal. The passband B_2 of the narrow band filter is matched with signal duration T, i.e., $B_2 = 1/T$. The following condition is satisfied for the ratio of the bands B_1 and B_2 of filters WF and NF: $B_2/B_1 \gg 1$.

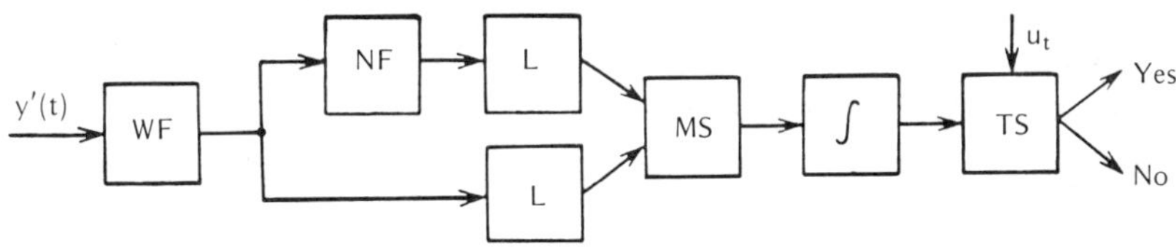

Figure 4.7

The filter output voltages are supplied to limiters, L, and then to a matching stage (MS), which forms pulses of standard amplitude, the duration of which is proportional to the time of coincidence of the positive polarities of the voltages from the limiters. Then come the integrator and threshold system (TS). A signal is detected when the integrator output voltage exceeds threshold level u_t. An improved version of the sign detector is examined in the article [188].

3. Filtering

In radio communications the problem of optimum filtering consists of the extraction of a transmitted message from a received mixture of signal and noise in the best possible way. In radar and radio nevigation, signals which describe the change in time of target coordinates relative to a radar or of a mobile facility equipped with a radio navigation system, relative to certain reference points, are subjected to such a procedure. As they apply to automatic measuring systems, such signals are often called input or useful signals. Such concepts as message and useful signal are similar in that they may be viewed as separate sample functions of some random process.

There are various models for assigning a random process. In radio communications, for example, a voice message is identified with the sample function x(t) of a Gaussian process. Even though this is not completely valid, experiments have shown that is a design is based on the assumption that a filtered process is normal, then the resulting system operates quite satisfactorily, in spite of certain deviations of the input process from Gaussian [26].

Samples of some stationary random process [149] also are used as a statistical model of the motion of a radar target. A polynomial model, for which measured coordinate x(t) is represented as the polynomial

$$x(t) = \sum_{i=0}^{n} a_i t^i \qquad\qquad (4.4.38)$$

is also examined [93]. The coefficients, a_i, of the polynomial are left constant for each target fix, but they change randomly from one fix to another. The statistical characteristics of the coefficients are assumed to be known. The laws of change of coordinates are also assigned similarly in radio navigation measurements.

The carrier wave is modulated by the message (the process being measured). Therefore, the received signal $u_s(t, x(t))$ is a deterministic function of time and of random process $x(t)$. The signal is accompanied by noise (disturbances) $u_n(t)$. The filtering problem consists of extracting from the received mixture

$$u_y(t) = u_s(t, x\,(t)) + u_n(t) \qquad\qquad (4.4.39)$$

the useful message $x(t)$.

The extractor is designed on the basis of the statistical properties of the process $x(t)$ and of noise $u_n(t)$, the kind of coding of the carrier wave by the message and the filtering quality criterion employed. Filtering is classed as linear or nonlinear, depending on whether the carrier wave is coded linearly or nonlinearly by process $x(t)$. An example of linear coding is amplitude modulation, and phase and frequency modulation are examples of nonlinear filtering.

The most complete results are obtained in linear filter theory. These results are of great importance for nonlinear filtering as well. In fact, systems of optimum linear and nonlinear filtering of radio signals, in most cases of practical importance, may be broken down into two parts: an unsmoothed discriminator (demodulator) and frequency-selective filter circuits. Message $x(t)$ itself is "extracted" from signal $u_s(t, x(t))$ in the discriminator, and the message is extracted from the noise in linear bandpass optimum filter circuits [52, 8].

Situations are also encountered in which an optimum filter does not have a discriminator and consists only of frequency-selective circuits. Examples of such situations are: the use of an optimum filter for reprocessing radar data, filtering in nonelectronic sensors and combined processing of data received from several electronic and nonelectronic sensors.

Therefore, the synthesis of frequency-selective circuits, which provide the optimum extraction of message $x(t)$ from its mixture with noise, is of fundamental importance in filter theory.

The optimum linear filter problem is formulated as follows. Process $x(t)$ to be filtered forms, along with noise (interference) $\xi(t)$, the additive mixture

$$y(t) = x(t) + \xi(t) \qquad\qquad (4.4.40)$$

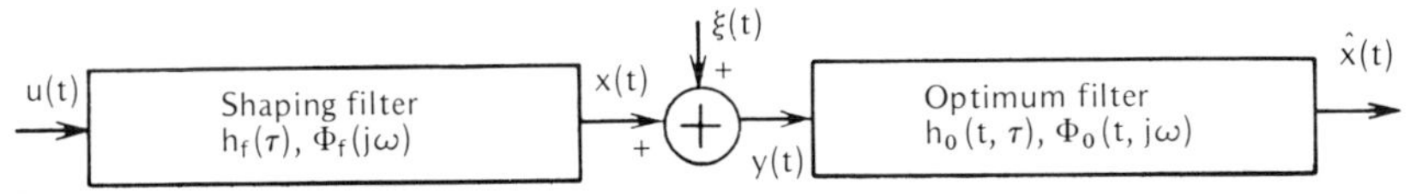

Figure 4.8

Both the individual terms and the sum as a whole represent samples of certain random processes. According to the condition used on p. 119 samples the components of Equation (4.4.40) should be written with the prime (denoting a sample).

In this section, however, the distribution laws do not figure in, and consequently, there is no risk of confusing the sample values and the arguments of the distribution functions. To simplify the notation, therefore, we shall omit the prime.

In filtering either message x(t) itself, or some process x*(t), related to x(t) by a given functional transform, is subjected to reproduction. An example of such a functional transform, for example, is the differentiation or integration of x(t). In this case, the derivative or integral of x(t) is reproduced at the filter output. The reproduction of x(t) may take place at the instant of arrival of the data (a problem strictly of filtering), or at time t_0 after the arrival of the data (a problem of interpolation or smoothing). The future behavior of process x(t) at time t_0 after the arrival of the data (a problem of extrapolation or prediction) may also be predicted.

The optimality criterion of the processing procedure is the minimum mean square error. If process x(t) itself is reproduced, then the condition

$$M(\Delta x^2) = M([x(t) - \hat{x}(t)]^2) = \min,$$

must be satisfied. Here $\hat{x}(t)$ is the filter output process, and the symbol $M(\cdot)$ denotes the statistical averaging operation.

In linear filter theory the best results are obtained for a process x(t), which is stationary and in particular, for the important practical case when the spectral density of the process is described by a fractional-rational function of frequency. The shaping filter technique gives a graphically clear idea of how such processes take place. The essence of the method consists of generating process x(t) at the output of a linear filter, to whose input is fed white noise.*

The procedure by which message x(t) is generated and extracted from a mixture with noise ξ(t) with the optimum filter is illustrated in Figure 4.8. We shall assume for the sake of simplicity that noise ξ(t) is white and is not correlated with x(t). Allowing for the difference of the noise from white does not introduce anything fundamentally new in the optimum filter circuit procedure, but it does make the synthesis procedure more unwieldy.

---

* The shaping filter is a means of describing the spectral density of the expected signal or message, and is used to derive the optimum filter. Ed.

A shaping filter is characterized by weighting function $h_f(\tau)$, or by frequency response $\Phi_f(j\omega)$. For an optimum filter we have, respectively, weighting function $h_0(t, \tau)$ (nonstationary in the general case), or frequency response $\Phi_0(t, j\omega)$, the forms of which should be determined during the synthesis process.

Noise processes $u(t)$ and $\xi(t)$ are described by the correlation functions

$$R_u(\tau) = \delta(\tau) \quad \text{and} \quad R_\xi(t) = \frac{N_\xi}{2}\delta(t)$$

or by spectral densities $G_u(\omega) = N_u$ and $G_\xi(\omega) = N_\xi$, respectively.

The correlation function $R_x(\tau)$ and spectral density $G_x(\omega)$ of process $x(t)$ are connected with shaping filter characteristics $h_f(\tau)$ and $\Phi_f(j\omega)$ by the following relations:

$$R_x(\tau) = \frac{N_u}{2} \int_0^\infty h_f(\lambda)h_f(\lambda + \tau)\,d\lambda, \tag{4.4.41}$$

$$G_x(\omega) = N_u \mid \Phi_f(j\omega) \mid^2. \tag{4.4.42}$$

The problems of determining the structure and parameters of an optimum filter under these conditions was solved by Wiener. He established that weighting function $h_0(t, \tau)$ should satisfy the following integral equation:

$$\int_0^t [R_x(\tau; \lambda) + R_\xi(\tau, \lambda)]\, h_0(t, \lambda)\,d\lambda = R_x(t, \tau). \tag{4.4.43}$$

or, in consideration of white noise,

$$\frac{N_\xi}{2} h_0(t, \tau) + \int_0^t R_x(\tau, \lambda)\, h_0(t, \lambda)\,d\lambda = R_x(t, \tau). \tag{4.4.44}$$

According to Equation (4.4.44) the structure of an optimum filter depends on the kind of correlation function $R_x(\tau)$ of the filter process, and that, in turn, is determined by the weighting function $h_f(\tau)$ of the shaping filter in Equation (4.4.41). However, the relation between the weighting functions is quite complicated, and $h_0(t, \tau)$ cannot be determined directly on the basis of $h_f(\tau)$ without solving integral Equations (4.4.43) or (4.4.44).

Considerable difficulties arise in the solution of Equation (4.4.43), especially when a nonstationary mode is examined, i.e., when the time the optimum filter is turned on is taken into consideration and the complex gain of the shaping filter is the ratio of high-order polynomials. Some examples of the calculation of the weighting function of an optimum filter for different correlation functions

$R_x(\tau)$ are examined in reference [8]. It is a little easier to determine the characteristics of a filter in the steady state mode, when the upper limits of the integrals in Equations (4.4.43) and (4.4.44) are assumed to be infinite. In this case Equations (4.4.43) and (4.4.44) are solved by the Fourier transform method, and the result of the solution will be the complex frequency characteristic, $\Phi_0(j\omega)$ of the optimum filter [149].

After determining $h_0(t, \tau)$ or $\Phi_0(t, j\omega)$, it is necessary to reproduce (model) the filter either as the algorithm of a computer, or as some circuit consisting of resisters, capacitors, inductors, tracking systems, etc. In modern engineering, when many electronic systems have outputs to computers, it is preferable to assign a filter in the form of an algorithm. Therefore, it is necessary to convert from $h_0(t, \tau)$ to differential equations that describe processes that take place in an optimum filter. Although such a conversion is, in principle, always possible [125], it involves unwieldy computations.

These problems, related to the solution of Equations (4.4.43) and (4.4.44), and to the modeling of the Wiener filter, are partly responsible for its limited practical application. They also stimulated the search for new approaches to the solution of the optimum filtering problem. The idea of assigning a shaping filter as a system of first-order differential equations turned out to be very productive and led to the discovery of an extremely simple relationship between the structures of shaping and optimum filters. The synthesis procedure proposed by Kalman and Bucey [73] is based on this concept. Filters of this design are called Kalman filters.

To explain the essence of this problem we shall introduce the basic concepts of the theory of Kalman filters by way of examination of the simplest example. We assume that process $x(t)$ to be filtered is generated by passing white noise through a low-frequency filter with the complex frequency characteristic

$$\Phi_f(j\omega) = \frac{\omega_0^2}{(j\omega)^2 + 2d\omega_0 j\omega + \omega_0^2} \tag{4.4.45}$$

where ω_0 is the eigen-frequency of the filter and d is the attenuation coefficient.

To the frequency characteristic in Equation (4.4.45) corresponds the following weighting function:

$$h_f(\tau) = \frac{\omega_0}{\sqrt{1 - d^2}} \exp(-d\omega_0 \tau) \sin \omega_0 \tau \sqrt{1 - d^2} \tag{4.4.46}$$

where u is the instantaneous value of the white noise supplied to the shaping filter. Here and henceforth the argument t of functions of time will be omitted to simplify the notation.

Instead of Equation (4.4.47) we write a system of two first-order equations, using the symbols $x_1 = x$, $x_2 = ds/dt$,

$$\left. \begin{aligned} \frac{dx_1}{dt} &= x_2 \\[2em] \frac{dx_2}{dt} &= -\omega_0^2 \, x_1 - 2d\omega_0 x_2 + \omega_0^2 u. \end{aligned} \right\} \tag{4.4.48}$$

Equation system (4.4.48) is written in vector form as

$$\begin{bmatrix} \dfrac{dx_1}{dt} \\[1.5em] \dfrac{dx_2}{dt} \end{bmatrix} = \begin{bmatrix} 0 & 1 \\[0.5em] -\omega_0^2 & -2d\omega_0 \end{bmatrix} \begin{bmatrix} x_1 \\[0.5em] x_2 \end{bmatrix} + \begin{bmatrix} 0 & 0 \\[0.5em] 0 & \omega_0^2 \end{bmatrix} \begin{bmatrix} 0 \\[0.5em] u \end{bmatrix}$$

or

$$\frac{dx}{dt} = \mathbf{F}x + \mathbf{C}u \tag{4.4.49}$$

As it applies to the example at hand, $\mathbf{x}$ is a vector column with elements x_1 and x_2; matrices $\mathbf{F}$ and $\mathbf{C}$ are, respectively,

$$\mathbf{F} = \begin{bmatrix} 0 & 1 \\[0.5em] -\omega_0^2 & -2d\omega_0 \end{bmatrix}; \quad \mathbf{C} = \begin{bmatrix} 0 & 0 \\[0.5em] 0 & \omega_0^2 \end{bmatrix}$$

and, finally, noise vector $\mathbf{u}$ consists of elements, $0, u$.

A model of the formation of a mixture of useful signal, x, and noise, ξ, with a shaping filter, assigned as differential equations, is shown in Figure 4.9. The elements of vector $\mathbf{x}$ describe the state of the model, and therefore they are called variables of state. Integrator output signals are usually taken as variables of state [56].

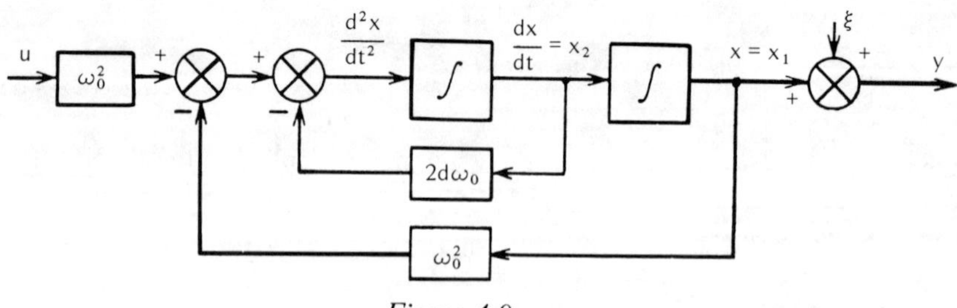

Figure 4.9

Equation (4.4.49) describes a shaping filter of arbitrarily high order. Furthermore, matrices F and C in the general case may be nonstationary, i.e., they may consist of elements that depend on time. In this case, message x will also be nonstationary.

The relation

$$y = Hx + \xi \tag{4.4.50}$$

also must be added to Equation (4.4.49) in order to construct a general model of a filtered process-noise mixture. This relation is sometimes called the equation of observation, since matrix H shows what variable state must be filtered (observed) in an optimum system. Thus, if, in the example at hand, $H = [1\ 0]$, then process $x = x_1$ will be filtered. When $H = [0\ 1]$, the derivative $x = x_2$ of that process will be filtered. Finally, when $H = \begin{bmatrix} 1 & 0 \\ 0 & 1 \end{bmatrix}$ both the process itself and its derivative will be filtered. In the latter case, the noise vector ξ should consist of two elements, i.e., two noise sources or one source with two outputs must be used.

A general diagram of the production of vector y is illustrated in Figure 4.10 (on the left hand side).

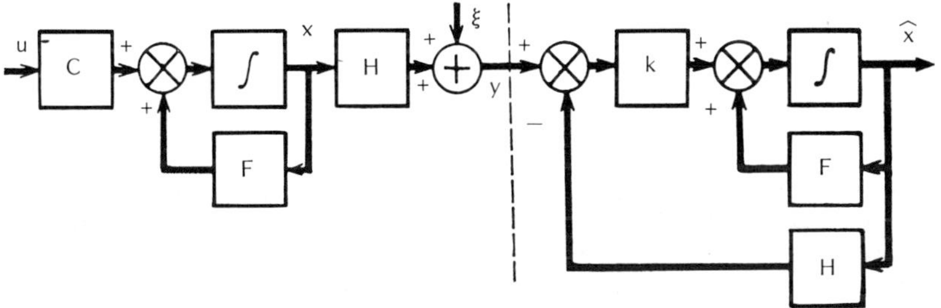

Figure 4.10

The optimum filter that reproduces process x with the minimum mean square error is described by the following vector equation [4, 26, 73]:

$$\frac{d\hat{x}}{dt} = F\hat{x} + k\,(y - H\hat{x}), \tag{4.4.51}$$

with the initial conditions $\hat{x}(0) = \hat{x}_0$, which characterize *a priori* data about process $\hat{x}$ at the filter output at moment of time $t = 0$. If there are no such data, then $\hat{x}(0) = 0$ is used. In Equation (4.4.51), k is the matrix gain of the optimum filter.

A structural diagram of an optimum filter is shown in Figure 4.10 (right hand side). The process mixture y and filter process x are fed to a comparison system.

The difference, determined as the result of comparison, determines the difference of newly arriving data from those at the output of the synthesized filter. This difference, with weighting coefficient $\mathbf{k}$, goes to the smoothing part of the filter, which is identical to the shaping filter. Therefore, determination of the structure of an optimum filter poses no problem. We recall that the relation between the weighting functions of shaping and optimum filters in Equations (4.4.41) and (4.4.44) in the Wiener problem was not so simple.

The main problem that arises in the design of a Kalman filter is determination of matrix gain $\mathbf{k}$. It is given by the following equation system [4, 26, 73]:

$$\mathbf{k} = \sigma_x^2 \, \mathbf{H}^T \left(\frac{\mathbf{N}_\xi}{2}\right)^{-1}$$

$$\frac{d\sigma_x^2}{dt} = \mathbf{F}\sigma_x^2 + \sigma_x^2 \mathbf{F}^T - \sigma_x^2 \mathbf{H}^T \left(\frac{\mathbf{N}_\xi}{2}\right)^{-1} \mathbf{H}\sigma_x^2 + \mathbf{C}\frac{\mathbf{N}_u}{2}\mathbf{C}^T. \qquad (4.4.53)$$

Here $\sigma_x^2 = M[x-\hat{x})(x-\hat{x})^T]$ is the symmetrical covariance matrix, which determines the accuracy of filtering, and $\mathbf{N}_\xi/2$ and $\mathbf{N}_u/2$ describe the correlation matrices of white noise, ξ and u, i.e.,

$$\mathbf{R}_\xi(t, \tau) = M[\xi\xi^t] = \frac{\mathbf{N}_\xi}{2}\delta(t - \tau),$$

$$\mathbf{R}_u(t, \tau) = M[u\,u^T] = \frac{\mathbf{N}_u}{2}\delta(t - \tau).$$

The letter "T" signifies transposition of the matrix. The expressions for the correlation functions of noise take into consideration the fact that these noises may be nonstationary, and consequently matrix elements $\mathbf{N}_\xi$ and $\mathbf{N}_u$ may be functions of time. Equation of variances (4.4.53) is a matrix nonlinear Riccati equation.

$$\mathbf{F}\sigma_x^2 + \sigma_x^2 \mathbf{F}^T - \sigma_x^2 \mathbf{H}^T \left(\frac{\mathbf{N}_\xi}{2}\right)^{-1} \mathbf{H}\sigma_x^2 + \mathbf{C}\frac{\mathbf{N}_u}{2}\mathbf{C}^T = 0. \qquad (4.4.54)$$

is solved instead of differential Equation (4.4.53).

The covariance matrix, and consequently gain $\mathbf{k}$, do not depend on the incoming data contained in the process mixture, y. Therefore, they are calculated beforehand, before the very beginning of the filtering procedure. In order to do the calculation, it is necessary to know the parameters of the shaping filter and the characteristics of noises u and ξ.

To illustrate the Kalman filter synthesis procedure, we introduce an example which we examined earlier. From Equation (4.4.52) we find the gain of the filter

$$
k = \begin{bmatrix} k_{11} & k_{12} \\ k_{21} & k_{22} \end{bmatrix} = \begin{bmatrix} \sigma^2_{x\,11} & \sigma^2_{x\,12} \\ \sigma^2_{x\,21} & \sigma^2_{x\,22} \end{bmatrix} \begin{bmatrix} 1 \\ 0 \end{bmatrix} \frac{2}{N_\xi} = \begin{bmatrix} \dfrac{2\sigma^2_{x\,11}}{N_\xi} \\[2ex] \dfrac{2\sigma^2_{x\,21}}{N_\xi} \end{bmatrix} \tag{4.4.55}
$$

Hence,

$$
k_{11} = \frac{2\sigma^2_{x\,11}}{N_\xi} \;;\; k_{21} = \frac{2\sigma^2_{x\,21}}{N_\xi} \;;\; k_{12} = k_{22} = 0
$$

In view of Equations (4.4.55) and (4.4.49), Equation (4.4.51) of the optimum filter is written as

$$
\begin{bmatrix} \dfrac{dx_1}{dt} \\[2ex] \dfrac{d\hat{x}_1}{dt} \end{bmatrix} = \begin{bmatrix} 0 & 1 \\ -\omega_0^2 & -2d\omega_0 \end{bmatrix} \begin{bmatrix} \hat{x}_1 \\ \hat{x}_2 \end{bmatrix} + \begin{bmatrix} k_{11} \\ k_{21} \end{bmatrix} (y - \hat{x}_1),
$$

$$
\begin{bmatrix} \dfrac{d\hat{x}_1}{dt} \\[2ex] \dfrac{d\hat{x}_2}{dt} \end{bmatrix} = \begin{bmatrix} \hat{x}_2 \\ -\omega_0^2 \hat{x}_1 - 2d\omega_0 \hat{x}_2 \end{bmatrix} + \begin{bmatrix} k_{11}(y - \hat{x}_1) \\ k_{21}(y - \hat{x}_1) \end{bmatrix}. \tag{4.4.56}
$$

To matrix Equation (4.4.56) corresponds the system of two scalar equations

$$
\left. \begin{aligned} \frac{d\hat{x}_1}{dt} &= \hat{x}_2 + k_{11}(y - \hat{x}_1), \\[2ex] \frac{d\hat{x}_2}{dt} &= -\omega_0^2 \hat{x}_1 - 2d\omega_0 \hat{x}_2 + k_{21}(y - \hat{x}_1) \end{aligned} \right\} \tag{4.4.57}
$$

The structural diagram of the filter illustrated in Figure 4.11 was drawn on the basis of these equations. The part of the diagram enclosed within the dotted line is a replica of the structure of a shaping filter. In the initial statement of the problem we were required to extract the optimum value $x_1 = \hat{x}$ of process x from the useful signal-noise mixture. Optimum estimate $\hat{x}_2$ of the derivative of this process is formed in the synthesized filter as a byproduct. This is a rather general property of a Kalman filter: the filter gives the optimum estimates of all variables of state, regardless of what process is directly measured.

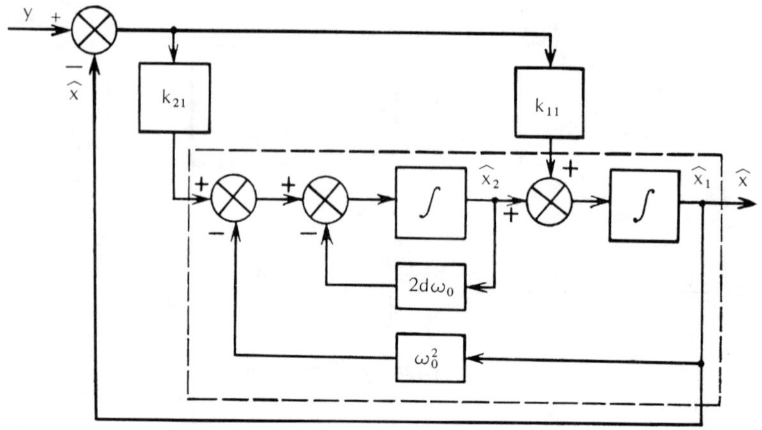

Figure 4.11

To the frequency-selective circuits of the filter, which play the basic role in useful signal extraction, are supplied the filtered variables of state $\hat{x}_1$, $\hat{x}_2$ and new incoming data about process x, contained in mixture y being processed. Variables $\hat{x}_1$ and $\hat{x}_2$ characterize *a priori* information (contained in the initial conditions) and the results of previous measurements. New incoming data update these results in accordance with the actual state of filtered process x. As follows from Equation (4.4.55), weighting coefficients k_{11} and k_{21}, with which new incoming data are supplied to the frequency-selective circuits, are inversely proportional to the spectral density of noise, N_ξ, that accompanies the useful signal. By virtue of the structure of these coefficients, the significance of the existing and of incoming data is redistributed in accordance with N_ξ. For instance, if the noise level increases the proportion of new data in the formation of the optimum value of the filtered process decreases. At the limit, when $N_\xi \to \infty$, the filter gets by without new data and relies only on *a priori* information. If $N_\xi \to 0$, then the role of new incoming data increases and at the limit, when $N_\xi = 0$, coefficients k_{11} and k_{21} become infinitely large. This means that filtering is not necessary and the input process, i.e., $\hat{x}_1 = y$, should be used as the optimum value of the extracted process.

Variances $\sigma_{x_{11}}^2$ and $\sigma_{x_{21}}^2$, which determine gains k_{11} and k_{21}, are found by solving Equation (4.4.53).

It was mentioned earlier that in addition to assigning a useful process (message) as a random process with a fractional-rational spectral density in relation to frequency, a polynomial model of an input process also is used. Such a model often is used for synthesizing radar and radio navigation instruments for measuring the coordinates of moving targets. In this case coefficients a_0, a_1 and a_2 of the polynomial in Equation (4.4.38) characterize the initial coordinate, velocity and acceleration of a target, respectively.

The shaping filter for a polynomial model of an input process is illustrated in Figure 4.12. The coefficients of the polynomial in this model are random values, which represent the initial conditions at the outputs of the corresponding integrators. For this model, it is a simple matter to write the equation of state

$$\frac{dx}{dt} = \mathbf{F}x \qquad (4.4.58)$$

with initial conditions $x_1(0) = a_0$, $x_2(0) = a_1$, $x_3(0) = a_2$ and matrix

$$\mathbf{F} = \begin{bmatrix} 0 & 1 & 0 \\ 0 & 0 & 1 \\ 0 & 0 & 0 \end{bmatrix}$$

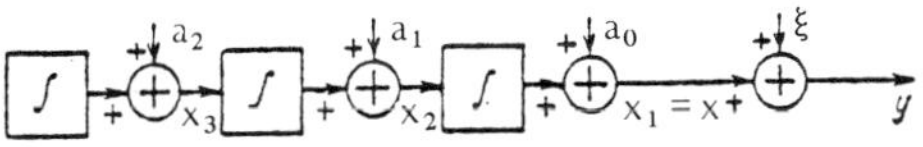

Figure 4.12

The chief advantage of the linear filter systhesis method, developed by Kalman and Bucey, is that it provides a solution of the problem of optimum filtering directly in the form of Equations (4.4.51) through (4.4.54), for which it is comparatively easy to build a model of a filter using an analog or digital computer. Therefore Kalman filters enjoy extensive practical application, particularly in radar and radio navigation systems [16, 49], in spite of the short period of time that has passed since the development of their theoretical principles (1961).

Another important advantage is that a nonstationary problem, in which the elements of matrix $\mathbf{F}$ and the levels of noises u and ξ change in time, can be solved. Theory also permits generalization for the case of nonwhite noise, ξ [20].

Using the fundamental principles of Kalman filter theory it is easy both to synthesize a multivariate optimum filter in which the results of measurements of a given process, x, with several instruments are processed, and to determine the gain obtained from using such a filter [40]. It turns out that the gain depends on the statistical properties of process x. For processes x that are most commonly used in analyses (Markov and Wiener processes, "black" noise), the gain in terms of mean square error does not exceed $\sqrt{n}$, where n is the number of measurements, and for a Markov process it is less than $\sqrt[4]{n}$.

In problems of nonlinear filtering of radio signals, the process to be filtered may be represented as the output voltage of some model (Figure 4.13). In contrast to the model of a signal examined above, a modulator is used here, in which the carrier $U_0 \sin \omega t$ is modulated by message x.

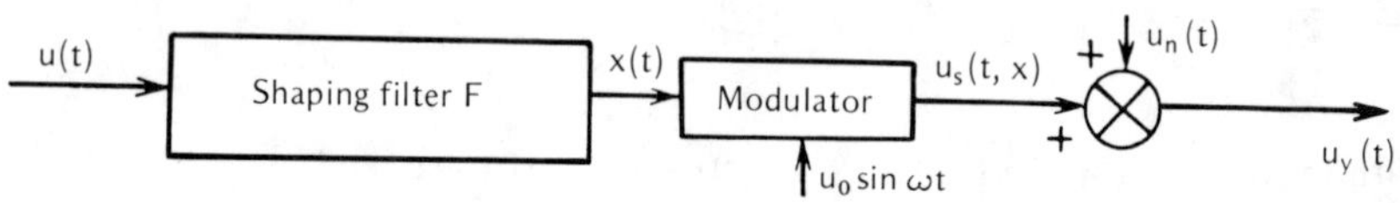

Figure 4.13

The filtering process $u_y(t)$ is a mixture of useful signal $u_s(t, x)$ and noise $u_n(t)$:

$$u_y(t) = u_s(t, x) + u_n(t). \tag{4.4.59}$$

Useful signal $u_s(t, x)$ is a known function of time and of random process x, the statistical properties of which are determined by the kind of shaping filter employed. If message x is related to signal $u_s(t, x)$ by a nonlinear function, then the problem of nonlinear filtering arises.

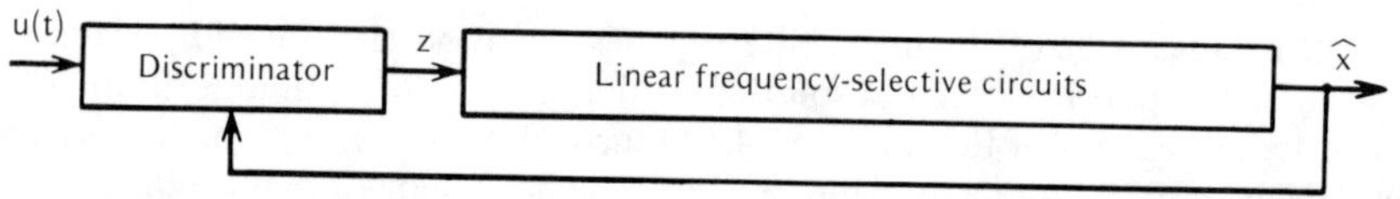

Figure 4.14

In many cases of practical importance an optimum nonlinear filter may be represented as illustrated in Figure 4.14 [52]. Process z, which is linearly connected to x, is generated at the output of an unsmoothed discriminator. Therefore subsequent filtering takes place in linear frequency-selective circuits in accordance with linear optimum filtering theory as given above. This filter gives the minimum mean square error at any given moment of time.

The operation by which process z is formed in optimum nonlinear filtering is described by the following expression [52]:

$$z = - \left| \frac{\partial Q(t, u_y, x)}{\partial x} \right|_{x=\hat{x}}$$

(4.4.60)

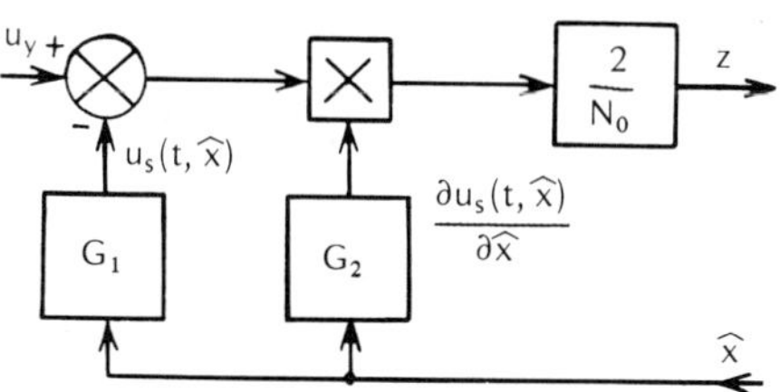

Figure 4.15

The function $Q(t, u_y, x)$ entering in Equation (4.4.60), for the case when signal $u_s(t, x)$ is deterministic, i.e., is completely known, with the exception of process x, and additive noise $u_n(t)$ are white noise with spectral density $G(w) = N_0$, expressed by the formula

$$Q(t, u_y, x) = \frac{1}{N_0} \left[u_y(t) - u_s(t, x) \right]^2$$

(4.4.61)

Then

$$z = \frac{2}{N_0} \frac{\partial u_s(t, \hat{x})}{\partial \hat{x}} \left[u_y(t) - u_s(t, \hat{x}) \right]$$

(4.4.62)

Here, for the sake of brevity, we introduce the definition

$$\frac{\partial u_s(t, \hat{x})}{\partial \hat{x}} = \frac{\partial u_s(t, x)}{\partial x} \bigg|_{x = \hat{x}}.$$

Figure 4.15 shows a structural diagram of a discriminator, built on the bases of Equation (4.4.62). The main parts of the discriminator are a computer, multiplier (synchronous detector), and two generators G_1, G_2, which generate the functions $u_s(t, x)$ and $\partial u_s(t, \hat{x})/\partial \hat{x}$. To the inputs of the generators is supplied filtered message $\hat{x}$, which is generated at the output of the frequency-selective circuits of the optimum filter. Generators G_1 and G_2 are essentially nonlinear elements of the discriminator.

As an example let us examine the passage through the discriminator of a mixture consisting of a phase-modulated useful signal and noise

$$u_y(t) = U_0 \sin(\omega t + k_x x) + u_n(t), \tag{4.4.63}$$

where amplitude U_0 and frequency ω of the signal are known and coefficient k_x is used for dimensional matching of the phase and message x. It follows from Equation (4.4.63) that message x enters nonlinearly into the useful signal.

Susbtituting Equation (4.4.63) into Equation (4.4.62) and performing simple conversions, we obtain

$$z = \frac{k_x U_0^2}{N_0} \sin k_x (x - \hat{x}) + \frac{2k_x U_0}{N_0} u_n(t) \cos(\omega t + k_x \hat{x}) \tag{4.4.64}$$

plus terms at the second harmonic.

Since the frequency-selective circuits that come after the discriminator do not pass harmonics of the carrier, these terms are ignored.

In optimum filtering the values of $\hat{x}$ will be close to x, and consequently, the sine in Equation (4.4.64) may be replaced with its argument.

We introduce the definitions

$$x' = \frac{k_x^2 U_0^2}{N_0} x; \quad \hat{x}' = \frac{k_x^2 U_0^2}{N_0} \hat{x};$$

$$\xi' = \frac{2k_x U_0}{N_0} u_n(t) \cos(\omega t + k_x \hat{x}); \quad y' = x' + \xi'.$$

Then Equation (4.4.64) is written as $z = y' - \hat{x}'$. Hence, in relation to filtered process y' and message estimate $\hat{x}'$, the discriminator of an optimum non-linear filter performs an operation similar to that performed by the comparator, connected to the input of a linear optimum filter. Therefore, the smoothing circuits of a filter are found by linear filtering.

Chapter 5 Protection of Receivers from Overloads and Cancellation of Radio Interference

5.1 PROTECTION OF RADIO RECEIVERS FROM OVERLOADS

Radio receivers intended for receiving pulse and amplitude-modulated (AM) signals may be overloaded by strong radio interference. An overloaded receiver does not react to a change in the amplitude of the input signal, and consequently, it cannot reproduce the transmitted message. Overloading occurs when the operating mode of electronic amplifiers becomes sharply nonlinear and close to the switching mode: they change periodically from saturation to cutoff. This causes the differential gain to decrease, sometimes to negative values.

Any part of a receiver, including the intermediate frequency (IF) amplifier, amplitude detector, or video amplifier, may become overloaded. However, the last IF stage is the first to be overloaded, and this will be discusses below.

Every AM or pulse receiver is so designed that overloading will not occur in the anticipated dynamic range of input signal levels (i.e., under normal operating conditions). This is done by using automatic gain control (AGC) systems.

Before the onset of overloading the amplitude response of the linear part of a receiver, i.e., the dependence of the amplitude U_o of the IF output voltage on the amplitude U_i of the mixer input signal or antenna output signal, increases monotonically. After the upper boundary $U_{i\,max}$ of the dynamic range is reached amplitude U_o ceases to increase and it either remains constant or decreases (Figure 5.1, curves 1 and 2, respectively).

The conditions of reception of the useful signal, in a background of high-level radio interference, deteriorate sharply due to this constraint. We will explain this by way of example of a pulse radio receiver. If the input of the receiver is acted upon by radio interference (for example a noise modulated carrier) in addition to a signal, then, as the result of their interaction, the envelope of the resulting signal will fluctuate (because of the fact that the phase difference of the carrier interference and radio pulses is random). For a normal receiver, the resulting amplitude of the output signal will fluctuate during the reception of signal pulses, which makes it possible to detect the signal in the noise background (curves 1 in Figure 5.2). In the case of high-level interference, when overloading begins to occur, these fluctuations disappear and the signal is "cut off" (straight line 2 in Figure 5.2). When a receiver is acted upon by noise, the magnitude of which fluctuates sharply, overloads may occur during short periods of time, after which the receiver, by virtue of the operation of its AGC system, regains its capacity to reproduce the input signal envelope. Such overloading may be called temporary.

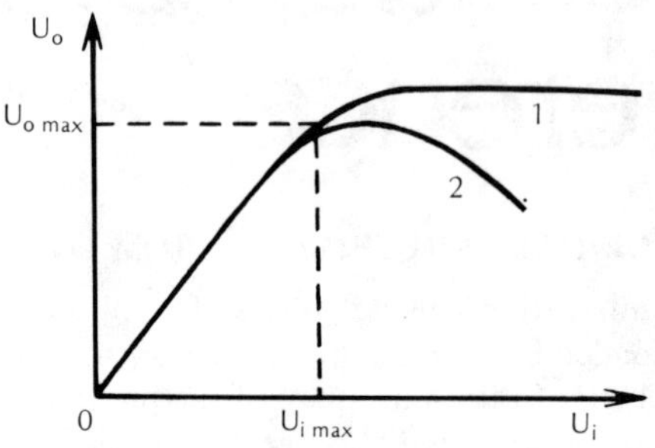

Figure 5.1

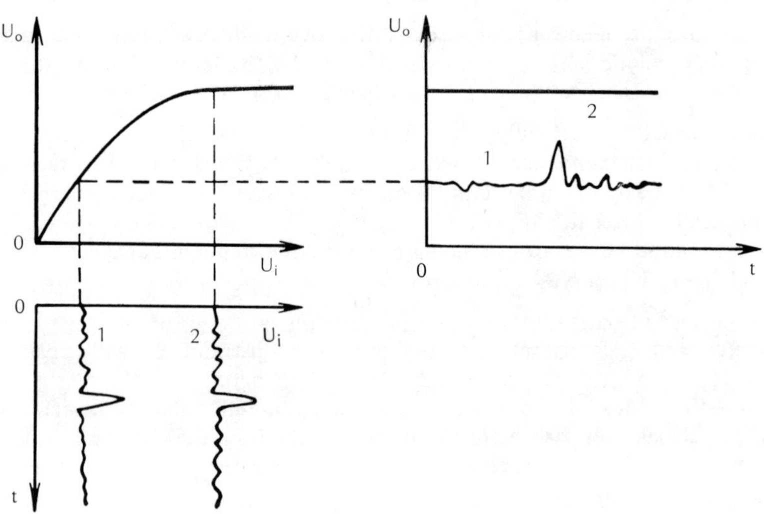

Figure 5.2

Logarithmic amplifiers and AGC systems with special built-in measures to lessen or eliminate overloading are used along with ordinary AGC systems to protect receivers from overloading.

In certain types of electronic systems, whether or not logarithmic amplifiers can be used depends on the operating principle of the systems. Automatic mono-pulse angle tracking systems, described extensively in the literature [101, 134], are an example.

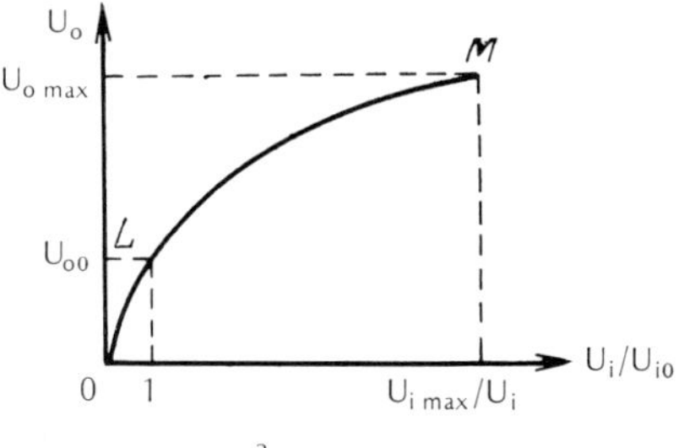

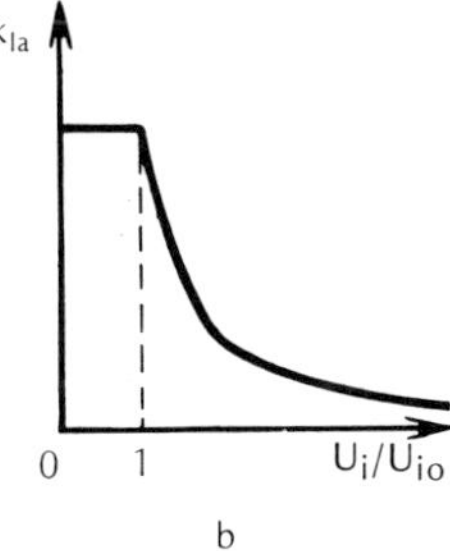

Figure 5.3

1. Logarithmic Amplifiers

Logarithmic amplifiers for IF signals have great dynamic range. Beginning at some (minimum) input voltage, the amplitude of the output voltage is approximately proportional to the logarithm of the relative change of the input signal level. The amplitude response for weak signals is virtually linear (segment OL in Figure 5.3a); this is followed by logarithmic segment LM. The output voltage is

$$U_o = \begin{cases} k_o U_i, & U_i < U_{io} \\ A \log_a \dfrac{U_i}{U_{io}} + B, & U_{io} > U_{io} \end{cases} \tag{5.1.1}$$

Here U_{io} is the input voltage corresponding to the transition from the linear part to the logarithmic part, and A and B are constant coefficients. Base a of the logarithm is selected in advance in accordance with the required amplifier characteristics.

To determine A and B, we assume that the transition from the linear part to the logarithmic part takes place without an abrupt change of the derivative, i.e., the derivative of the logarithmic part at transition point L coincides with the gain $k_0 = U_{oo}/U_{io}$ of the linear part. This yields

$$\left(\frac{dU_o}{dU_i}\right)_L = \left(\frac{A}{b}\frac{1}{U_i}\right)_L = \frac{A}{bU_{io}} = k_o$$

where $b = \ln a = 1/\log_a e$, and e is the base of natural logarithms.

Hence, we obtain the coefficient $A = b\,k_0\,U_{i_0}$.

When $U_i = U_{i_0}$ the equality

$$U_{o_0} = B = k_0 U_{i_0}$$

is valid. Therefore, we obtain for the logarithmic segment

$$U_o = k_0 U_{i_0} \left(b \log_a \frac{U_i}{U_{i_0}} + 1 \right) = k_0 U_{i_0}\, b \log_a a^{1/b} \times \frac{U_i}{U_{i_0}}$$

$$= k_0 U_{i_0}\, b \log_a e\, \frac{U_i}{U_{i_0}} \qquad \text{for } U_i/U_{i_0} > 1. \tag{5.1.2}$$

For the selected natural logarithms $b = 1$, and for such an amplifier

$$U_o = k_0 U_{i_0} \left(\ln \frac{U_i}{U_{i_0}} + 1 \right) = k_0 U_{i_0} \ln \frac{eU_i}{U_{i_0}} = U_{o_0} \ln \frac{eU_i}{U_{i_0}} \tag{5.1.3}$$

Transition point L on the logarithnic section is usually selected such that U_{i_0} will fall below the receiver noise level by about 20 dB. Consequently, virtually the entire range of input signals falls within the logarithmic section, which provides great input dynamic range.

The gain of a log amplifier for any base of logarithms is expressed by the formula

$$k_{la} = \frac{dU_o}{dU_i} = k_0 U_{i0} \frac{1}{U_i} = \frac{U_{o0}}{U_i} \tag{5.1.4}$$

and diminishes in inverse proportion to the input signal amplitude (Figure 5.3b). The output dynamic range is:

$$D_o = U_{o\,max}\,/U_{o_0} \; ; \quad D_o(dB) = 20 \lg D_o \tag{5.1.5}$$

Here $U_{o\,max}$ is the maximum output signal level, at which the actual amplitude response remains close to logarithmic.

The input dynamic range is

$$D_i = U_{i\,max}/U_{i0} ; \quad D_i(dB) = 20 \lg D_i$$

Clearly

$$D_o = b \log_a e\, D_i$$

To achieve great input, dynamic range logarithms with a low base are selected. The required input dynamic range (100 dB and higher) can usually be achieved by using an amplifier of several stages.

When a log amplifier is acted upon by jamming, signals are suppressed by the jamming. When the input noise level U_{ji} is high, the output noise voltage is

$$U_{jo} \approx U_o \, b \, \log_a e \, \frac{U_{ji}}{U_{i_0}}$$

$$U_o = k_o U_{i_0}$$

At the same time the amplitude of the signal (the increment of the amplitude of the output voltage, attributed to the influence of the signal) is

$$U_{so} = \Delta U_o = k_{la} U_{ms} = (U_{o_0}/U_{ji}) \, U_{ms}.$$

The last relations take into consideration the fact that $U_{ji} \gg U_{ms}$. Consequently, the output signal-to-noise ratio is

$$\frac{U_{jo}}{U_{so}} = \frac{U_{ji}}{U_{ms}} \, b \, \log_a e \, \frac{U_{ji}}{U_{o_0}}$$

Since $U_{ji} \gg U_{i_0}$, the output noise-to-signal ratio is always higher than the input ratio, and it increases with the noise level. The physical explanation for this is that the gain in terms of the signal is determined by the input noise and diminishes as the noise level increases.

Let us examine the features of the passage of an AM signal through a log amplifier. We assume that the receiver input is an AM signal with carrier amplitude U_{m_0} and modulation coefficient, m. The amplitudes of the maximum and minimum output voltages are, respectively (we assume $k_o = 1$),

$$U_{o\,max} = U_{o_0} \left[b \, \log_a e \, \frac{U_{m_0}(1+m)}{U_{i_0}} \right]$$

$$U_{o\,max} = U_{o_0} \left[b \, \log_a e \, \frac{U_{m_0}(1-m)}{U_{i_0}} \right]$$

The amplitude of the output voltage is

$$\Delta U_o = U_{o_0} \, b \, \log_a \frac{1+m}{1-m} \tag{5.1.6}$$

or for $a = e$

$$\Delta U_o = U_{oo}\, b \ln \frac{1 + m}{1 - m} \tag{5.1.7}$$

and the modulation coefficient is

$$M_o = \frac{\Delta U_o}{U_{oo}\, b \log_a e \dfrac{U_{mo}}{U_{io}}} = \frac{\log_a \dfrac{1+m}{1-m}}{\log_a \dfrac{e U_{mo}}{U_{io}}} \tag{5.1.8}$$

We notice that the amplitude of the output signals ΔU_o does not depend on the amplitude, U_{mo}, of the input voltage, but is determined by the logarithm of the ratio $(1 + m)/(1 - m)$. If $m \leqslant 0.5$, then

$$\Delta U_o \approx (2 - 2.1)m\, U_{io}. \tag{5.1.9}$$

As follows from Equation (5.1.8), the modulation coefficient decreases as a signal passes through the amplifier.

Analysis of the noise response is difficult, since the amplifier is a nonlinear system. The variance σ_{out}^2, and mathematical expectation, M_U, of the output voltage envelope were found [100,166] on the assumption that its input receives narrowband normal noise with variance σ_n^2 (the envelope has a Rayleigh distribution) and with uniform phase within $\pm \pi$. Approximate analysis leads to the following expressions:

$$\sigma_o^2 = \frac{\pi^2}{24}\, b^2\, k_o^2\, U_{io}^2 \tag{5.1.10}$$

$$M_U = \frac{b \log_a 2\sigma_n^2 - C}{2} \approx 0.5 b \log_a 2\sigma_n^2 \tag{5.1.11}$$

where $C = 0.577$ is the Euler constant.

Hence, it is clear that the variance of the output voltage is determined in the first approximation only by the characteristics of the linear part and by transition point, L between the linear and logarithmic parts.

The dependence of σ_o^2 on the variance of the input noise derives, obviously, from assumptions made relative to the character of the input signal and idealization of the amplifier response. For real amplifiers, the dependence of σ_o^2 on σ_n^2 does exist, but it is weak. Analysis of Equation (5.1.10) leads to the conclusion that the transition point L should be set as low as possible, in particular, below the receiver noise level.

The mathematical expectation of the output noise envelope increases nearly in proportion to the logarithm of the input noise dispersion.

There are several ways to design logarithmic amplifiers [34-36, 92]. The most commonly used technique is cascaded detection and subsequent summing. Logarithmic amplifiers of this type have series- and parallel-connected amplifiers. A functional diagram of a log amplifier with series-connected stages for pulsed signals is illustrated in Figure 5.4. The system has n identical amplifier stages 1, 2, ..., n, and the same number of limiters L_1, L_2, ..., L_n (with identical limiting levels) and detectors D_1, D_2, ..., D_n. The outputs of detectors $D_1 - D_{n-1}$ are connected to delay lines ($TD_1 - TD_{n-1}$), which are designed to compensate time lags as pulses pass through the amplifier stages. Each amplifier, limiter, detector and delay line comprise a unique cell. The output voltage is generated as a result of summing the voltages taken from each of these cells. It is

$$U_o = \sum_{i=1}^{n} U_{oi} \qquad (5.1.12)$$

where U_{oi} is the output voltage of the i-th cell.

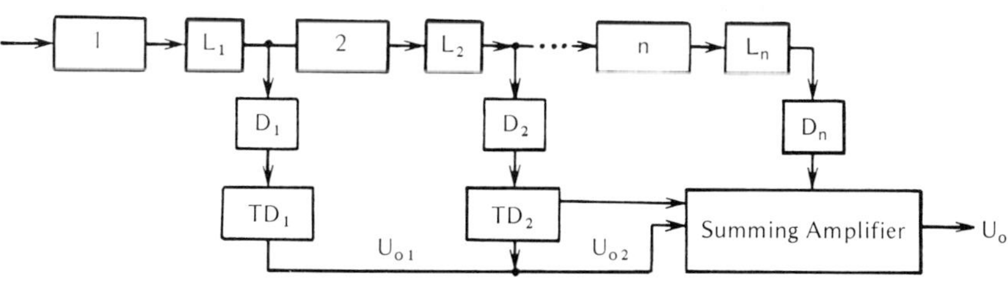

Figure 5.4

When the input signal is small, all amplifier stages operate in the linear mode and the output voltage of the n-th cell is

$$U_{on} = k^n k_d U_i$$

where k is the gain of one stage in consideration of the limiter, and k_d is the gain of the detector. As U_i increases to some value, U_{in} limiting begins to take place in the last cell, after which the output voltage from the n-th cell remains constant and equal to $k_d k^n U_{l_0}$. As voltage U_i continues to increase, the limiting level is reached in the (n − 1)-st cell, after which constant voltages will be taken from the n-th and (n − 1)-st cells. The input voltage at which this level is reached is

$U_{i_2} = kU_{i_0}$. Consequently, as U_i continues to increase, constant voltages are taken from the n-th and (n − 1)-st cells, and these voltages are, respectively,

$$U_{on} = k^n k_d U_{i_0}$$

$$U_{o(n-1)} = k^{n-1} k_d U_{i_2} = k^n k_d U_{i_0}$$

Then the limiting level is reached in the (n − 2)-nd cell. This happens when $U_i = U_{i_3} = k^2 U_{i_0}$, whereupon constant (and identical) voltages $k^n k_d U_{i_0}$ will be taken from three cells, etc.

The character of the dependence of U_o on U_i (the amplitude response of the amplifier) is illustrated in Figure 5.5. This function is represented by a broken line, consisting of individual segments, each of which corresponds to a certain number of cells in which the limiting level is reached. It is easy to write the expression for the output voltage corresponding to the different segments of the amplitude response.

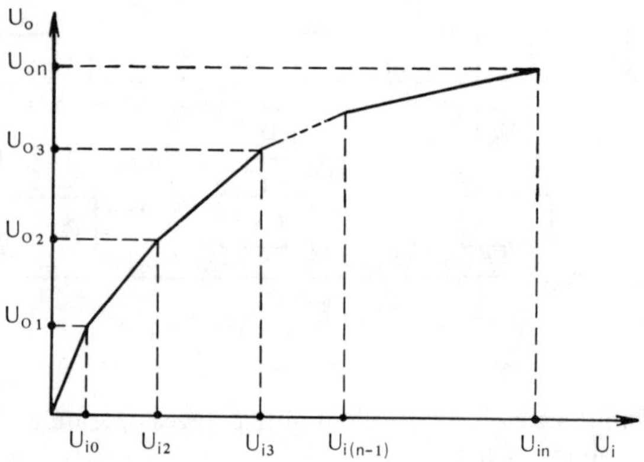

Figure 5.5

For the first segment

$$U_o = U_i k_d k^n \left(1 + \frac{1}{k} + \frac{1}{k^2} + \ldots + \frac{1}{k^{n-1}} \right) \approx U_i k_d k^n \left(1 + \frac{1}{k} \right) .$$

The voltage at the end of that segment (i.e., when $U_i = U_{i_0}$) is

$$U_{oi} = U_{o_0} = U_{i_0} k_d (k + k^2 + \ldots + k^n) \approx U_{i_0} k_d k^n \left(1 + \frac{1}{k} \right) .$$

For the second segment

$$U_o = [k^n U_{i_0} + k^{n-1} U_i (1 + \frac{1}{k} + \ldots + \frac{1}{k^{n-2}})] k_d$$

$$\approx [k^n U_{i_0} + U_i k^{n-1} (1 + \frac{1}{k})] k_d .$$

The voltage at the end of the 2nd segment is determined by substituting $U_{i_2} = kU_{i_0}$ in place of U_i. Then

$$U_{o_2} = [k^n U_{i_0} + U_{i_0} k^n (1 + \frac{1}{k} + \ldots + \frac{1}{k^{n-2}})] k_d$$

$$\approx k^n U_{i_0} (2 + \frac{1}{k}) k_d .$$

Proceeding with the analysis, we find for the n-th segment

$$U_o = [(n - 1) k^n U_{i_0} + U_i k] k_d .$$

At the end of the n-th segment, when $U_i = U_{in} = k^{n-1} U_{i_0}$, we have

$$U_{in} = [(n - 1)k^n U_{i_0} + U_{i_0} k^n] k_d = nk^n U_{i_0} .$$

When the input voltage exceeds U_{in} the output voltage ceases to increase.

The function thus obtained is approximately logarithmic. Actually, if the relationship between U_o and U_i were logarithmic, then, with the logarithmic scale plotted on the abscissa axis and linear scale on the ordinate axis, the graph of $U_o(U_i/U_{i_0})$ would be a straight line. For the amplifier under examination, we plot the value U_i/U_{i_0} in logarithmic scale on the abscissa axis. Then, for the i-th segment

$$U_i = k^{i-1} U_{i_0} \quad \text{and} \quad \log_a \frac{U_i}{U_{i_0}} = (i - 1) \log_a k.$$

Consequently, starting at i = 1, the points on the abscissa axis will be laid out uniformly with step $\log_a k$. The ordinates corresponding to the inflection points will then fall on straight line segment OM (Figure 5.6), since the difference of the ordinates of two segments

$$\Delta U_{oi} \approx [k^n U_{i_0} (i + \frac{1}{k}) - k^n U_{i_0} (i - 1 + \frac{1}{k})] k_d = k^n U_{i_0} k_d .$$

is constant.

In the spaces between the inflection points, obviously, the function differs from logarithmic (Figure 5.6). However, if the number of amplifiers is sufficiently large this difference may be ignored.

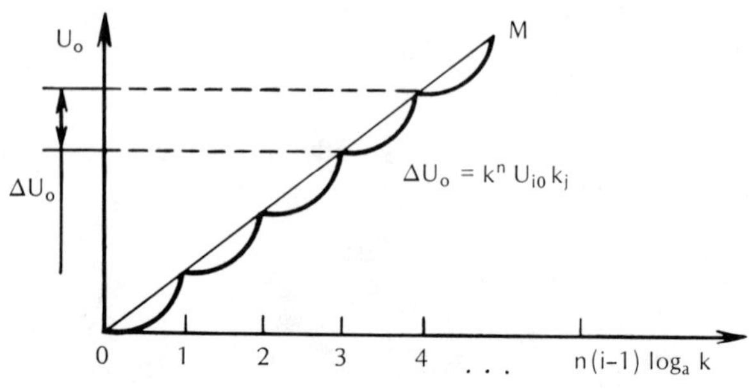

Figure 5.6

There are also other ways to design logarithmic amplifiers. A more general class of nonlinear amplifiers (functional amplifiers with a wide dynamic range) is examined in great detail in reference [36].

2. Basic Dynamic Feature of AGC Systems: Response to Low-Level Fluctuation Interference

The influence of an AGC system on the response to interference usually is ignored in analysis of radio interference in a receiver. In order to take this influence into consideration, it is necessary to examine the dynamic characteristics of the AGC system.

A functional diagram of an AGC system is illustrated in Figure 5.7. The output voltage of the linear part of the receiver (REC) goes to the AGC detector, which simultaneously receives delay voltage E_d. Then come the amplifier and filter (F). Output voltage u_r from the filter output goes into controlled stages, where the gain $k(u_r)$ of the receiver is changed.

Between the envelopes $U_o(t)$ and $U_i(t)$ of the output and input voltages of the receiver exists the approximate relation

$$U_o(t) = k(u_r) U_i(t). \tag{5.1.13}$$

Voltage u_r is found from the relation

$$u_r = F(D)(U_o - E_d)k_a, \tag{5.1.14}$$

which is valid for $U_o \geqslant E_d$. Here $F(D)$ is the transfer function of the filter and k_a is the gain of the AGC detector and amplifier.

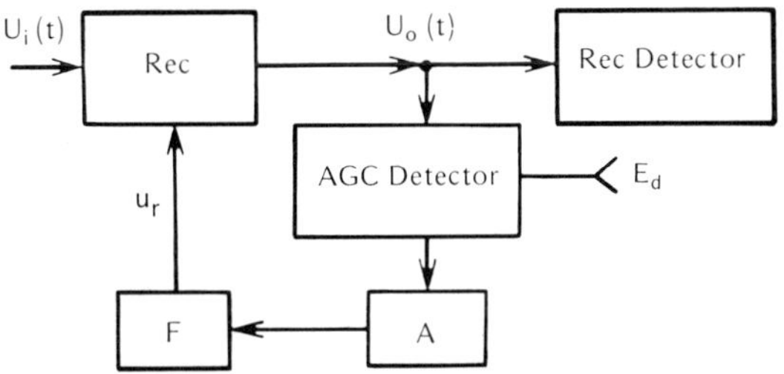

Figure 5.7

It is assumed in Equation (5.1.13) that IF amplifier has the same passband and the control voltage is changed so slowly by the filter that transient processes in REC, caused by the action of voltage u_r, may be ignored. This assumption is valid in practice, with the exception, perhaps, of cases of control with transistors, when it is sometimes necessary to consider the input time constant of the transistors.

Equations (5.1.13) and (5.1.14) are nonlinear and this equation system can be solved only for certain special cases. As shown in the literature, results that are adequate for practice are obtained in the case of linear approximation of the function $k(u_r)$. We assume that the receiver input is an amplitude-modulated voltage and stationary additive noise. The envelope of the resulting voltage is assumed to be quasistationary with regular and random components. The mathematical expectation of the input voltage is assumed to be a slowly changing function of time or a constant from test to test. We linearize the function $k(u_r)$ at the point $u_r = U_{r0}$ corresponding to the response to a receiver input voltage U_{i0}, equal to the mathematical expectation. Then

$$k = k_0 - \alpha u_r, \tag{5.1.15}$$

and

$$k_{av} = k_0 - \alpha \overline{U}_r. \tag{5.1.16}$$

Here $\overline{U}_r$ is the control voltage at the point relative to which linearization is performed (Figure 5.8), k_0 is the gain corresponding to the intersection of the ordinate axis and straight line of Equation (5.1.16), k_{av} is gain k for $u_r = \overline{U}_r$, and $\alpha = \tan \psi$ is the angular tangent coefficient.

A method described in reference [8] and further developed in [200, 201, 45, 46] is a productive way of analyzing an AGC system under these conditions. An approximate analysis of an AGC system can be carried out by an easier method.

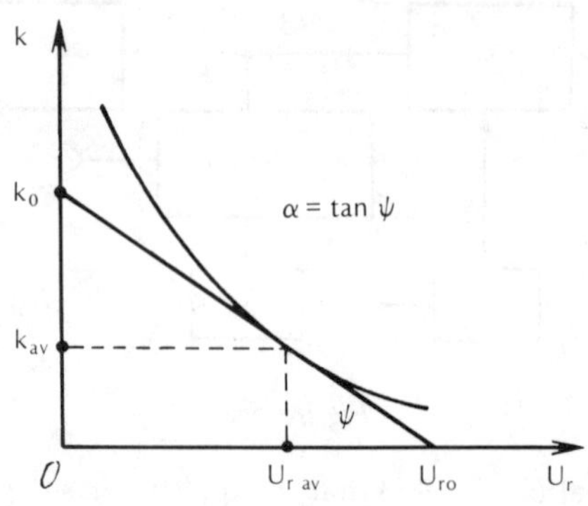

Figure 5.8

We represent the envelope of the input signal as the sum of the mathematical expectation U_{i_0} and increment ΔU_i

$$U_i = U_{i0} + \Delta U_i \qquad (5.1.17)$$

and we assume that the modulus of the maximum increment $|\Delta U_{i\,max}|$ is substantially smaller than the mathematical expectation

$$|\Delta U_{i\,max}| \ll U_{i_0} \qquad (5.1.18)$$

The envelope of the output voltage is also represented as the sum of mathematical expectation U_o and increment ΔU_o. Then, for the envelope of the output voltage, we write

$$U_o = U_{o_0} + \Delta U_o = (k_{av} + \Delta k)(U_{i_0} + \Delta U_i) \qquad (5.1.19)$$

where Δk is the increment of coefficient k_{av}, which depends on the component ΔU_i of the envelope of the input voltage. Ignoring the second-order term $\Delta k \Delta U_{i_0}$, we find

$$U_o = k_{av} U_{i_0} + k_{av} \Delta U_i + \Delta k U_{i_0} . \qquad (5.1.20)$$

Taking the average, we obtain

$$U_o = k_{av} U_i = (k_0 \alpha \overline{U}_r) U_{i_0}. \qquad (5.1.21)$$

On the other hand, when $U_{o_0} > E_d$, in accordance with Equation (5.1.14), we also obtain

$$\overline{U}_r = (U_{o_0} - E_d) k_{av} F(0). \qquad (5.1.22)$$

Considering that $F(0) = 1$, we find from the last two equations for U_{o_0}

$$U_{o_0} = \frac{k_o + E_d \alpha k_a}{1 + \alpha U_{i_0} k_a} U_{i_0} \tag{5.1.23}$$

Comparing this with Equation (5.1.21), we arrive at the conclusion that

$$k_{av} = \frac{k_o + E_d \alpha k_a}{1 + \mu} \tag{5.1.24}$$

where

$$\mu = \alpha U_{i_0} k_a \tag{5.1.25}$$

Consequently, a receiver with an AGC system, in relation to the average input voltage, acts as an amplifier with gain k_{av}, which depends on U_{i_0}. Returning to Equation (5.1.20), we obtain

$$\Delta U_o = k_{av} \Delta U_i + \Delta k U_{i_0} = k_{av} \Delta U_i + \left(\frac{\Delta k}{\Delta U_r} \Delta U_r \right) U_{i_0}. \tag{5.1.26}$$

$$\text{Since } \Delta U_o = \frac{k_{av}}{1 + \alpha k_a U_{i_0} F(D)} \Delta U_i \tag{5.1.27}$$

Hence, it follows that for the increments of the input envelope, the receiver is equivalent to a filter with transfer function

$$H_e(D) = \frac{k_{av}}{1 + \mu F(D)} \tag{5.1.28}$$

These equations correspond perfectly to the zero and first approximations of the solution of the integral equation in [8]. We divide both sides of Equation (5.1.27) by U_{o_0} and assuming that $U_{o_0} = k_{av} U_{i_0}$. Then

$$m_o = \frac{1}{1 + \mu F(D)}$$

$$m_i - \frac{1}{k_{av}} \frac{k_{av} m_i}{1 + \mu F(D)} = \frac{H_e(D)}{k_{av}} m_i \tag{5.1.29}$$

where $m_i = \Delta U_i / U_{i_0}$, $m_o = \Delta U_o / U_o$ are the instantaneous input and output modulation coefficients of the receiver.

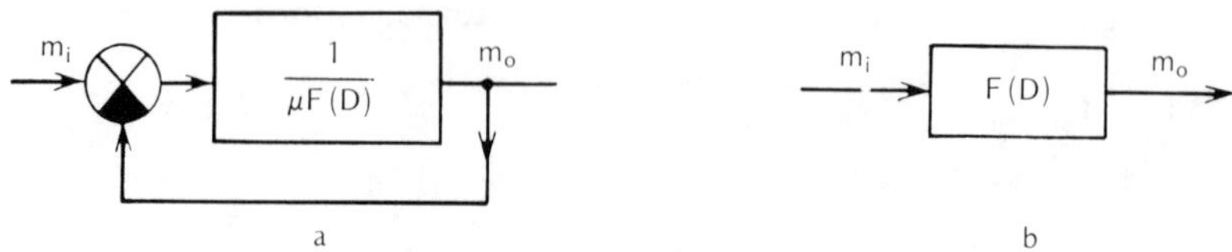

Figure 5.9

Equation (5.1.29) enables us to construct a dynamic structural diagram of an AGC system for the modulation coefficients (Figure 5.9), which is valid for any kind of filter (only if $F(0) = 1$) under the above-stated constraints. Clearly, the structural diagram is not valid for large modulation coefficients.

We note that the speed of an AGC system is not constant, but depends on the average receiver input voltage, and changes in the input voltage are processed increasingly faster as its average value increases. We assume that the receiver input is a signal with constant amplitude U_{i0} and wideband noise with spectral density G_{i0}, which is constant within the amplifier passband. Using the equations derived above, we find all the important relationships between the receiver output signal and noise, since for the spectral density and variance of the noise, and also for the average receiver output voltage, we may write the following expressions:

$$G_o(\omega) = G_{i0} \, | \, H_e(j\omega) \, |^2 \tag{5.1.30}$$

$$\sigma_o^2 = G_i \frac{1}{2\pi} \int_{-\pi B/2}^{\pi B/2} | \, H_e(/\omega) \, |^2 \, d\omega; \tag{5.1.31}$$

$$U_{o0} = k_{av} U_i. \tag{5.1.32}$$

Integration in Equation (5.1.31) is performed within the band B of the IF amplifier, which is considerably wider than double the band of the filter of the AGC system.

In the case of the modulation of the carrier by a sinusoidal signal with frequency ω_s and with modulation coefficient m_i, the amplitude of the detector output voltage of the receiver will be $m_o \times U_o k_d$, where k_d is the gain of the receiver detector.

The time constant of the AGC system of an AM receiver is selected in consideration of the tolerable demodulation of signals in the entire range of modulating frequencies and of the dynamic range of input signals. Since the speed of an AGC system increases with the amplitude of the input signal, the time constant of the filter should be great enough so that the relation

$$A = \frac{|m_o(j\omega_s)|}{|m_i(j\omega_s)|}$$

differs little from unity for the highest modulation frequency ω_s and highest amplitude of the input carrier signal.

If the AGC system incorporates a single-section RC filter with transfer function $F(D) = (TD + 1)^{-1}$, then, in accordance with Equation (5.1.29), we obtain

$$m_o = \frac{1}{\mu + 1} \, \frac{TD + 1}{\tau D + 1} \, m_i \tag{5.1.33}$$

where $\tau = T/(\mu + 1)$ is the time constant of the AGC system.

Consequently

$$A = \frac{1}{\mu + 1} \left| \frac{j\omega_s T + 1}{j\omega_s \tau + 1} \right| = \sqrt{\frac{1 + (\omega_s T)^2}{(1 + \mu)^2 + (\omega_s T)^2}} \tag{5.1.34}$$

It follows from Equation (5.1.34) that when $A = 0.9$ and $\mu = 100$, the product $\omega_s T$ should be of the order of 210, whereas, for $\mu = 50$ this value decreases to values of the order of 100.

3. Response of AGC Receiver to Pulsed Jamming

Powerful intermittent jamming, consisting of long periodic pulses, is a very common kind of jamming intended for a receiver equipped with an AGC system. During the rise time of a pulse the control voltage cannot reach the level sufficient for normal (linear) amplification in the receiver, and it is overloaded for a time. The receiver, during the pauses between the pulses, may not be able to regain sufficient sensitivity to receive a signal which is weaker than the jamming.

If such a receiver is used for automatic conical scanning target tracking, then the modulation caused by the deviation of the target from the axis will disappear during the time of overloading and during pauses between pulses, and the jamming with particular parameters will be effective.

When a receiver is overloaded, the amplitude of its output voltage remains constant and equal to some threshold value u_{th}, regardless of how the input signal changes. The receiver leaves the saturated mode as soon as the amplitude of the output voltage decreases to a value below E_d as a result of an increase of the control voltage and reduction of gain.

It usually may be assumed that an overloaded stage leaves the saturated mode without an appreciable delay. This, of course, is an idealization of actual processes, since discharging the capacitors of overloaded stages takes a certain amount of time, t_d. However, t_d is an order of magnitude less than the time it takes processes associated with recharging of filter capacitors of the AGC system to run their course.

The AGC system is open in the overloaded state and the control voltage is changed under the influence of constant voltage U_{th} E_d.

When the trailing edge of the jamming pulse passes the receivers, the output voltage (from the voltage accumulated during the response to the pulse) may turn out to be so low that it will be less than the delay voltage, and the AGC circuit will again be open: the gain will be insufficient for the reception of a weak signal for some time. Such a state may conditionally be called a temporary loss of sensitivity. Sensitivity is restored as the result of natural discharging of filter capacitors of the AGC system. In this process, gain increases, and at some point reaches the level at which the output voltage begins to exceed E_d. If no pulse arrives before this time the AGC circuit closes and the control process begins.

The process described above is illustrated in Figure 5.10. Illustrated here is the general case when the receiver, during time T_1 of the jamming u_j with amplitude U_{i_1}, is overloaded for time t_{ov} and is able to leave the overloaded state before the end of the jamming pulse. During time T_2 of the signal with amplitude U_{i_2} the AGC system is initially open ("dead" time t_d), and then is closed until the arrival of the next interference pulse.

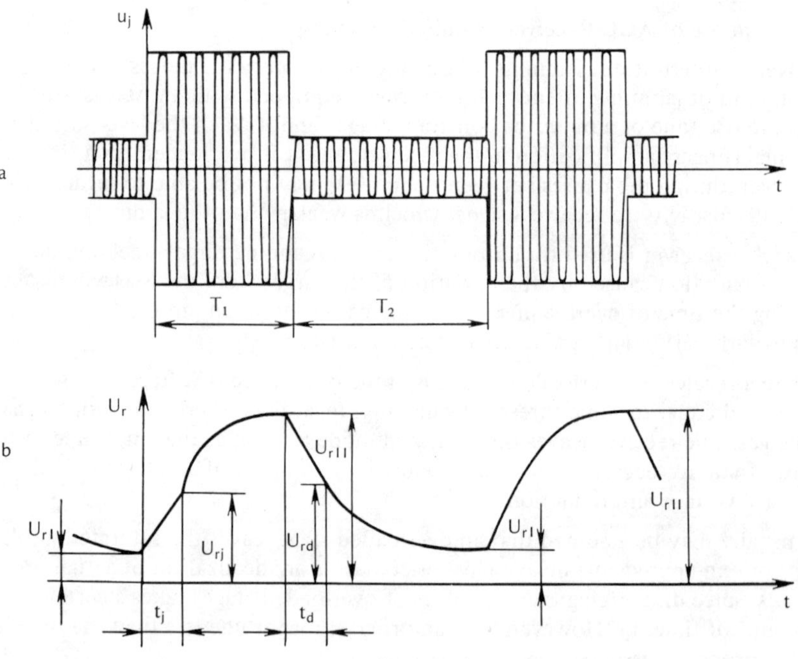

Figure 5.10

The dependence of voltage u_r on time is illustrated in Figure 5.10b. Since we shall be discussing the steady state mode below, the following would be identical in each jamming period:

- voltages $u_r = U_{rI}$ at the times of arrival of the next interference pulse,
- voltages $u_r = U_{rII}$ at the end of the interference pulse,
- voltages $u_r = U_{rj}$ at the points of departure from the overload mode, and
- voltages $u_r = U_{rd}$ at the end of the "dead" time, when sensitivity is recovered.

Let us assume that the AGC system incorporates a signal-section low-frequency RC filter with time constant T. Then voltage u_r in segment 0-t may be written as follows:

$$u_r(t) = (U_{th} - E_d)\, k_a\, (1 - e^{-t/T}) + U_{rI}e^{-t/T}, \tag{5.1.35}$$

In this case, the AGC system is open and the receiver output voltage remains equal to U_{th}.

At time t_j when the AGC system closes, voltage u_r reaches the level

$$U_{rII} = (U_{th} - E_d)k_a(1 - e^{-t_{ov}/T}) + U_{rI}^{-t_j/T}. \tag{5.1.36}$$

After that, the process is described by the equation

$$U_r(t) = A_1(1 - e^{-t/\tau_1}) + U_{rII}\, e^{-t/\tau_1}) + U_{rII}e^{-t/\tau_1} \tag{5.1.37}$$

The above equation was found as the solution of Equation (5.1.35) for $u_r(t)$ with the initial conditions $u_r(0) = U_{rj}$. Here

$$A_1 = \frac{k_0 k_a U_{i_1} - E_d k_a}{1 + \mu_1} \tag{5.1.38}$$

$$\mu_1 = \alpha k_a U_{i_1}$$

$$\tau_1 = T/(1 + \mu_1) \tag{5.1.39}$$

is the equivalent time constant of the AGC system. This equation is valid until the arrival of the trailing edge of pulse u_j, when voltage u_r reaches

$$U_{rII} = A_1(1 - e^{-(T_I - t_j)/\tau_1}) + U_{rII}e^{-(T_I - t_j)/\tau_1} \tag{5.1.40}$$

The AGC system is opened again during time t_d after the trailing edge of the pulse and u_r changes in accordance with the expression

$$u_r = U_{rII}e^{-t/T}. \tag{5.1.41}$$

It is assumed here that the capacitor discharge time constant is equal to the time constant of the filter.

At moment of time $T_1 + t_d$ the filter output voltage u_r reaches $U_{rd} = U_{rII}\exp(-t_d/T)$ (while a signal with amplitude U_{i2} acts upon the input), the system closes again and $u_r(t)$ changes by a law analogous to Equation (5.1.37):

$$U_r(t) = A_2(e^{-t/\tau_1}) + U_{rd}e^{-t/T_1}, \tag{5.1.42}$$

Here

$$A_2 = \frac{k_0 U_{i2} - E_d}{1 + \mu_2}\, k_a, \quad \mu_2 = \alpha k_a\, U_{i2}\, ; \tag{5.1.43}$$

$$\tau_2 = T/(1 + \mu_2) \tag{5.1.44}$$

At the time of arrival of the next interference pulse, voltage u_r reaches the value U_{rI}:

$$U_{rI} = A_2(1 - e^{-(T_2 - t_M)/\tau_1}) + U_{rd}e^{-(T_1 - t_d)/\tau_1}. \tag{5.1.45}$$

We now find the expressions for U_{rj} and U_{rd}. To find U_{rj} we assume that voltage u_r corresponds to the value at which U_o reaches the level U_{th}, when the receiver input is acted upon by interference with amplitude U_{i_1}. Consequently,

$$U_o = U_{th} = (k_0 - \alpha U_{rj})\, U_{i_1}. \tag{5.1.46}$$

We introduce the definition

$$\beta_1 = 1 - U_{th}/k_p U_{1_1}. \tag{5.1.47}$$

This value expresses the overload coefficient of the receiver, where $\beta_1 = 0$ if the receiver is not overloaded and $\beta_1 = 1$ in the case of infinite overloading. Then we obtain

$$U_{rd} = \beta_1 U_{r_0} \tag{5.1.48}$$

where $U_{r_0} = k_0/\alpha$.

Likewise, when voltage $u_r = U_{rd}$, which decreases during time T_2 reaches the value E_d, voltage U_o becomes equal to E_d. Consequently,

$$U_o = E_d = (k_0 - \alpha U_{rd})U_{i_2}. \tag{5.1.49}$$

Introducing the definition

$$\beta_2 = 1 - E_d/k_0 U_{i_2}, \tag{5.1.50}$$

we obtain

$$U_{rd} = \beta_2 U_{r_0}. \tag{5.1.51}$$

The value β_2 expresses the sensitivity loss coefficient. Excluding the values U_{rI} and U_{rII} from the equations derived above, and taking Equations (5.1.48) and (5.1.51) into consideration, we obtain a system of two equations for determining the values t_d and t_j in which we are interested:

$$U_{r_0}\beta_1 = k_a(U_j - E_d)(1 - e^{-t_j/T})$$
$$+ A_2 e^{-t_j/T}(1 - c^{-(T_2 - t_d)/\tau_2}$$
$$+ U_{r_0}\beta_2\, e^{-t_d/T}\, e^{-(T_2 - t_d)/\tau_2}. \tag{5.1.52}$$

$$U_{r_0}\beta_2 = A_1\, e^{-t_d/T}(1 - e^{-(T_1 - t_j)/\tau_1}$$
$$+ U_{r_0}\beta_1\, e^{-t_d/T}\, e^{-(T_1 - t_j)/\tau_1}. \tag{5.1.53}$$

In the absence of a signal, when the only receiver input is jamming, it may be assumed in Equations (5.1.52) and (5.1.53) that $A_2 = 0$, $t_d = 0$, $\tau_2 = T$. Then, after substituting into Equation (5.1.40) the value $U_{r_0}\beta_1$, obtained from Equation (5.1.52), we have

$$U_{r_0}\beta_1(1 - e^{-(T_1 - t_j)/\tau_1} e^{-(t_j + T_1)/T}) = (U_{th} - E_0)k_a\,(1 - e^{-t_j/T})$$
$$+ A_1\, e^{-(t_j + T_1)/T}(1 - e^{-(T_1 - t_j)/\tau_1}). \tag{5.1.54}$$

Hence, it is possible to determine the overload time t_j. We assume, further, that the spaces between the jamming pulses are large enough so that voltage u_r may be assumed equal to zero by the time of arrival of the next pulse. This is equivalent to the assumption that $(T_2/T_1) \to \infty$. Then Equation (5.1.54) is substantially simplified:

$$U_{r_0}\beta_1 = k_a(U_{th} - E_d)(1 - e^{-t_j/T}). \tag{5.1.55}$$

Consequently, we determine for the relative overload time

$$\frac{t_j}{T} = \ln \frac{1}{1 - U_{r_0}\beta_1/(U_{th} - E_d)k_a} = \ln \frac{1}{1 - \beta_1 R}. \tag{5.1.56}$$

Here

$$R = U_{r_0}\beta_1/(U_{th} - E_d)k_a. \tag{5.1.57}$$

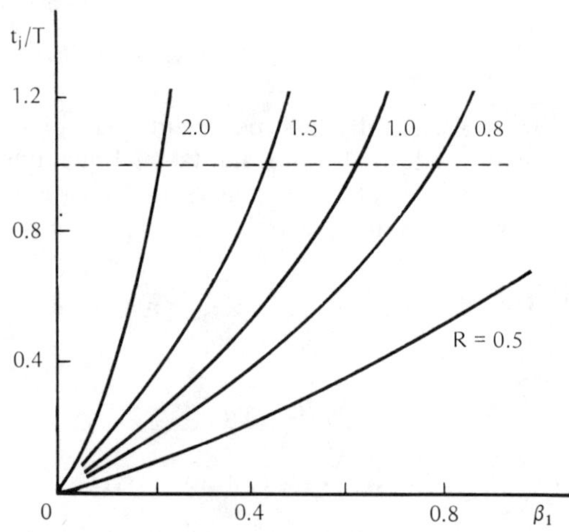

Figure 5.11

Graphs pf the dependence of t_j/T on β_1 are shown in Figure 5.11 for various values of R. The higher the speed of the system (i.e., the smaller T and the larger k_a), the shorter the time period t_j during which the receiver is overloaded.

We now determine the conditions under which there are no constraints on the output voltage during time T_1, assuming that the signal is sufficiently weak during pauses $T_2 (U_{i_2} \ll U_{i_1})$, and the duty factor of the jamming is 0.5. We introduce the definition $T_1 = T_2 = t_p$. Assuming $t_j = 0$ in Equation (5.1.54), we obtain

$$U_{r_0}\beta_1(1 - e^{-t_p/\tau_1} e^{-t_p/T}) = A_1 e^{-t_p/T}(1 - e^{-t_p/\tau_1}). \tag{5.1.58}$$

Using this equation we can determine the conditions under which there are no constraints. For simplicity we assume $\mu \gg 1$ (from which also follows that $(\tau_1/T) \ll 1$). Then we find from the above equation the simple relation $U_{r_0}\beta_1 \approx$

$\approx A_1 e^{-t_p/T}$. Hence, it is obvious that the receiver is not overloaded during reception of jamming, on the assumption that

$$\frac{k_0 U_{r_1}}{U_{th}} < \frac{1 - (E_d/E_j)e^{-t_j/T}}{1 - e^{-t_j/T}} . \tag{5.1.59}$$

By reducing the ratio t_p/T it is possible to eliminate overloading with a strong input signal. Thus, when the jamming period is very short (in comparison with the time constant of the filter) overloading for practical purposes does not occur. Conversely, overloading always occurs when the frequency of the jamming is low, since the filter capacitor is able to discharge during half-period T_2 and each new pulse catches the receiver at high sensitivity.

Let us examine finally the conditions under which the receiver output signal cannot increase to E_d in time T_2. This means that no reception takes place during time T_2 (the output voltage reaches the level E_d when the receiver input signal is at the minimum level). To simplify the analysis we shall assume that overloading does not occur ($t_{ov} = 0$), and the duty factor is 0.5. Then we must assume in Equation (5.1.52) $t_d = t_p$ and $T_1 = T_2 = t_p$. Hence, we find $U_{r_0}\beta_1 = U_{r_0}\beta_2$.

We note that Equation (5.1.59) gives the conditions under which overloading does not occur during half-periods T_1. Thus, the following relation exists between U_{i_1} and U_{i_2}:

$$U_{i_1}/U_{i_2} = U_{th}/E_d. \tag{5.1.60}$$

If U_{i_2} rises above the level $U_{i_1}(E_d/U_{th})$, then the receiver output signal is able to increase above E_d during half-periods T_2.

This technique can be used to find the receiver output voltage at any moment of time, regardless of the relations between $U_{i_1}, U_{i_2}, T_1, T_2, U_{th}, E_d, T$. Without giving the results of the calculations, we shall simply explain how they are obtained. The dependences of voltages $u_r(t)$ are written for four time intervals, as was done during the derivation of Equations (5.1.52) and (5.1.53). Then the values U_{rI} and U_{rII} are excluded. After that the expressions for gains $k = k_0 - \alpha u_r$ are written and the expressions for U_0 are found in each interval in accordance with the equality $U_0 = kU_i$. We note that the output voltage in intervals t_j is U_{th}.

The above analysis shows that it is always possible with an AGC system with fixed parameters to select jamming parameters such that information losses will occur as a result of a change in the amplitude of the signal due to temporary overloading for some period of time after the next jamming pulse, and due to a temporary loss of sensitivity after the end of the jamming pulse.

4. Some Ways of Protecting Receiver from Temporary Overloads and Temporary Sensitivity Loss

There are AGC systems in which measures are taken to minimize overload time t_j and "dead" time t_d, i.e., they have systems that protect them from temporary overloads and sensitivity loss.

FAGC System. Pulse receivers with a fast automatic gain control (FAGC) system have a built-in capability of preventing overloading by jamming pulses longer in duration than signal pulses. A FAGC system is no different in terms of operating principle from the AGC system, described above, except that it has much higher speed by virtue of the use of wide band filters and amplifiers in the feedback circuit. Fast systems are so designed that the gain does not change significantly during the signal pulse, since otherwise the pulse would be severely distorted.

It was demonstrated above that the equivalent time constant of an AGC system (see p. 161) is

$$\tau = T/(1 + \mu), \tag{5.1.61}$$

where $\mu = \alpha k_a U_p$.

However, this equation is valid only on the assumption that the filter time constant, T, is substantially greater then several other short time constants in the system, in particular the time constant T' of the load circuit of the AGC detector, equivalent time constant of the amplifier, parasitic capacitances, etc. An attempt to increase speed by reducing T and increasing k_a magnifies the effect of these time constants. The order of the equations that describe an AGC system increases sharply, and when the AGC circuit has high gain the system may lose sensitivity.

To alleviate the effect of short time constants and to ensure stable operation, an AGC system is designed so that each individual stage Y_1, Y_2, ... of the amplifier is covered by feedback.

The signal from each amplifier is detected and is fed through its AGC amplifier and filter to the control element of that stage in order to change its gain. The result in this case is a multiple loop AGC system. The amplitude response of the amplifier, each stage of which is covered by AGC loops, is close to logarithmic. In other words, an amplifier with a multiple loop AGC system may be viewed as a design modification of a logarithmic amplifier.

Feed-Forward AGC System

A feed-forward AGC system with special control circuits can be used to control video amplifier overloading caused by jamming whose duration substantially exceeds signal pulse duration. Control voltage is supplied in this case to a video detector or video amplifier.

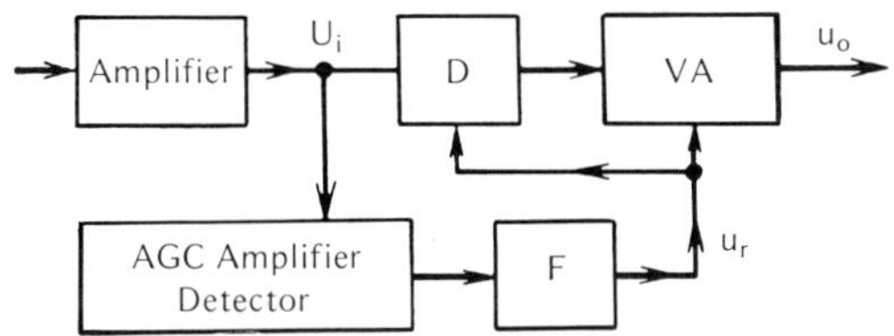

Figure 5.12

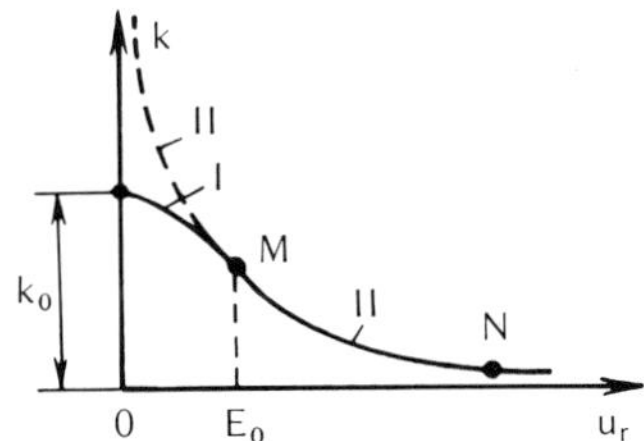

Figure 5.13

A functional diagram of a feed-forward AGC system designed to prevent over-loading is shown in Figure 5.12. The output signal is fed to the AGC detector and amplifier. The output voltage of that system passes through a filter (F) and is applied in the form of control voltage u_r to the video amplifier (VA) and video detector (D). This system is classified as an automatic system without feedback.

The required control characteristics, i.e., the dependence of the gain k of the control stages on voltage u_r, is hyperbolic (curve II, Figure 5.13):

$$k = V_0/u_r \tag{5.1.62}$$

The value V_0 entering in Equation (5.1.62) is defined as follows. This function is valid only in some segment MN, and the greatest difference from the actual function $k(u_r)$, characterized by curve I in Figure 5.13, occurs when $u_r < E_0$. Assuming that Equation (5.1.62) is valid for $E_r \geqslant E_0$, the initial segment $0\text{-}E_0$ (curve II in Figure 5.13) is approximated by the expression

$$k = k_0 e^{-(u_r/B)^2}, \quad 0 \leqslant u_r \leqslant E_0. \tag{5.1.63}$$

When determining the constants k_0 and B in Equation (5.1.62) and (5.1.63), it is necessary to bear in mind that at point M, where $u_r = E_0$, the coefficients k and their derivatives in terms of u_r must coincide for curves I and II. Under these conditions we have, for determining k_0 and B, the following two equations:

$$\frac{V_0}{E_0} = k_0 e^{-(E_0/B)^2},$$

$$\frac{V_0}{E_0^2} = \frac{2k_0 E_0}{B^2} e^{-(E_0/B)^2}$$

$$(5.1.64)$$

from which we obtain

$$B = \sqrt{2} E_0; \quad V_0 = E_0 k_0 / \sqrt{e}.$$

$$(5.1.64)$$

Consequently, the control characteristic is written as

$$k = k_0 e^{-(\mu_r/\sqrt{2}\,B)^2}, \quad 0 \leqslant \mu_r \leqslant E_0.$$

$$k = \frac{E_0 k_0}{\mu_r} e^{-\frac{1}{2}} \text{ for } \mu_r > E_0$$

$$(5.1.65)$$

The control voltage is

$$\mu_r = k_a F(D) U_i$$

$$(5.1.66)$$

where k_a is the gain of the AGC detector and amplifier, U_i is the amplitude of the amplifier input voltage, and $F(D)$ is the transfer function of the filter.

The amplitude characteristic of a video amplifier that uses this control method, i.e., the dependence of the amplitude of output voltage U_o on the amplitude of input voltage U_i in the steady state mode (i.e., when $F(D) = 1$) is determined by the relations

$$U_o = U_i k_0 \exp\left[-\left(\frac{k_a U_i}{1.41 E_0}\right)^2\right];$$

$$U_i \leqslant E_0/k_a;$$

$$U_o = \frac{k_0 E_0}{k_a \sqrt{e}}; \quad U_i \geqslant E_0/k_a$$

and is illustrated in Figure 5.14, where the relative output and input voltages

$$a = \frac{U_o \sqrt{e}\, k_a}{k_0 E_0};$$

$$b = \frac{U_r}{\sqrt{2}E_0} = \frac{k_a U_i}{\sqrt{2}E_0}$$

are plotted on the ordinate and abscissa axes, respectively. Point M corresponds to the conjunction of curves I and II in Figure 5.13.

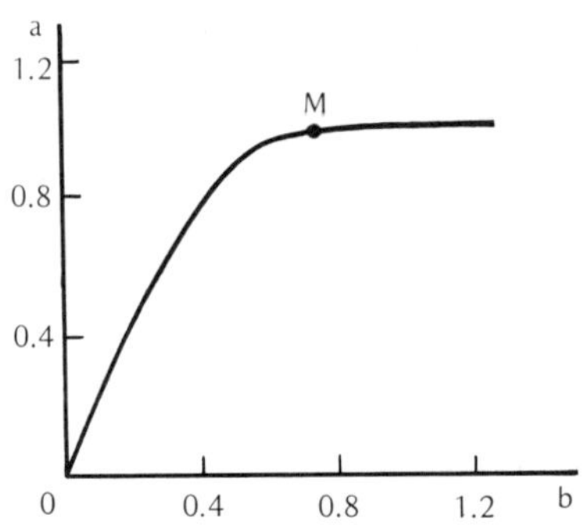

Figure 5.14

The actual characteristic may differ from the one thus determined, but for an efficient AGC system U_0 should change only little with increasing U_i in some operating range. If the AGC system has a simple single-section RC filter, then to prevent signal distortion it is necessary to set the filter time constant $T \gg t_p$, since when this condition is satisfied voltage u_r is virtually constant during the pulses. When a jamming signal of sufficient amplitude and duration (a long pulse, for example) arrives, the output voltage will be $k_0 E_0/\sqrt{e}\,k$ after the initial transient. A short signal pulse, arriving simultaneously with the jamming, obviously will not be limited, since voltage u_r remains constant during the signal pulse width. Jamming pulses of about the same duration as signal pulses will pass undistorted to the output.

AGC Systems with Nonlinear Elements in Filter Circuits

AGC systems have been designed in which nonlinear elements or elements with variable parameters are used in filter circuits to control temporary overloading and sensitivity loss. The operation of these systems is based on the principle that the charging time constant of the capacitor of an AGC filter is sharply reduced because of the fact that the AGC system contains a nonlinear element, and this results in a rapid reduction of amplifier gain and overload time t_j. When the input signal abruptly decreases, because of the nonlinear element, the amplifier regains its sensitivity just as rapidly. Such a system may be designed cited work that an extra limiter be included in the discharge tube control circuit to make the capacitor discharge time independent of signal amplitude.

5.2 Jamming Cancellation with Extra Receiver

1. General Information

Jamming that enters a system from antenna side lobes is most easily cancelled by using an extra receiver. The methods used to cancel such jamming are incoherent (amplitude) and coherent. In the former method, jamming is cancelled in the high frequency or IF channels.

2. Amplitude Jamming Cancellation

The principle of amplitude jamming cancellation is examined below in the analysis of the system illustrated in Figure 5.15 [90]. Active pulse jamming in the receiver of a pulse radar system is assumed to be subjected to cancellation.

The main receiver contains antenna A_m, mixer M_m, IF amplifier IF_m and amplitude detector D_m. The receiver includes the analogous components, designated in Figure 5.15 by the corresponding symbols with the subscript "c". It also has a local oscillator (LO) and a subtracting system (SS).

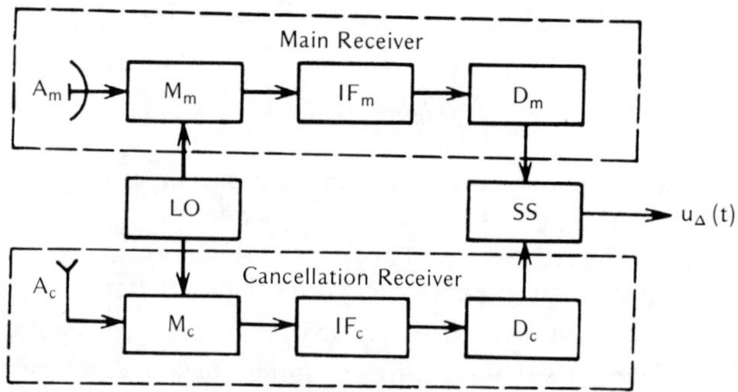

Figure 5.15

Jamming can be cancelled in the subtracting system if the jamming signals, generated by detectors D_m and D_c, begin to act at the same time and have the identical durations and envelopes. For these conditions to be satisfied the like elements of the main and cancellation receivers must be completely identical, and antennas A_m and A_c should have patterns $F_m(\theta)$ and $F_c(\theta)$ which satisfy the equations

$$F_c(\theta) = 0 \qquad \text{for } -0.5\theta_0 \leqslant \theta \leqslant 0.5\theta_0$$

$$F_c(\theta) = F_m(\theta) \quad \text{for } \theta > 0.5\theta_0 < -0.5\theta_0 \qquad (5.2.1)$$

Here θ is the angle between the beam axis of the receiving antenna, A_m, and θ_0 is the width of the mainlobe of the same antenna.

Possible patterns $F_c(\theta)$ and $F_m(\theta)$ are shown in Figure 5.16, where the difference of $F_c(\theta)$ from zero for $-0.5\theta_0 < \theta_0 < 0.5\theta_0$ and from the function $F_m(\theta)$ for $\theta > 0.5\theta_0$ and $\theta < -0.5_0$ is introduced only for clarity.

If the amplitude-frequency responses $W_m(\omega)$ and $W_c(\omega)$ of the linear parts of the main and cancelling receivers satisfy the relation $W_m(\omega) = kW_c(\omega)$, where k is the proportionality coefficient, and detectors D_m and D_c are identical, then the following relation between $F_c(\theta)$ and $F_m(\theta)$ is valid:

$$F_c(\theta) = 0 \qquad \text{for } -0.5\theta_0 \leqslant \theta \leqslant 0.5\theta_0,$$

$$F_c(\theta) = kF_m(\theta) \text{ for } \theta > 0.5\theta_0 < -0.5\theta_0. \tag{5.2.2}$$

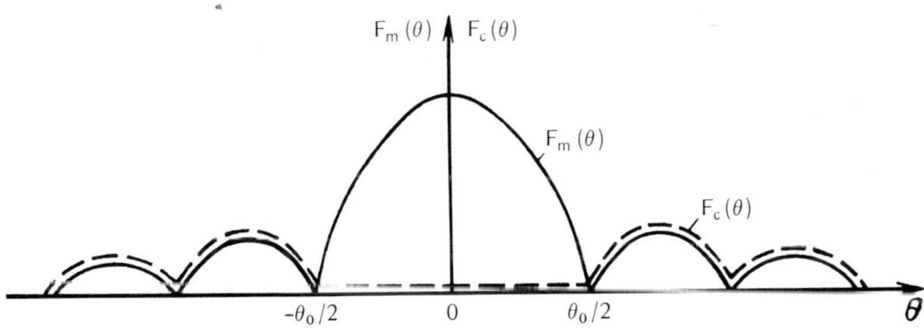

Figure 5.16

When Equations (5.2.1) and (5.2.2) are satisfied not only perfect cancellation of jamming in the side lobes of antenna, A_m, takes place, but there is also no attentuation of useful signal, $u_s(t)$, whose source is located in the vicinity of the main lobe of antenna A_m. However, in the case of the simultaneous appearance of jamming pulses, received through the side lobes of antenna A_m, and of useful signal $u_s(t)$, complete cancellation of jamming does not occur. We shall demonstrate this, assuming that voltages $u_m(t)$ and $u_c(t)$, generated at the outputs of IF_m and IF_c are

$$u_m(t) = [U_{sm}(t) + U_{jm}(t)]\cos\omega_{if}t, \tag{5.2.3}$$

$$u_c(t) = U_{jc}(t)\cos\omega_{if}t. \tag{5.2.4}$$

Here $U_{sm}(t)$ and $U_{jm}(t)$ are the envelopes of the useful signal and of jamming in the cancellation receiver and ω_{if} is the intermediate frequency.

If detectors D_m and D_c are quadratic, then their output voltages u_{dm} and u_{dc} are approximately

$$u_{dm} = 0.5\,k_{dm}[U_{sm}^2(t) + U_{jm}^2(t) + 2U_{sm}U_{jm}], \tag{5.2.5}$$

$$u_{dc} = 0.5 \, k_{dc} U_{dc}^2 (t), \tag{5.2.6}$$

where k_{dm} and k_{dc} are the gains of detectors D_m and D_c, respectively.

It follows from Equations (5.2.5) and (5.2.6) that when $U_{jm}(t) = U_{jc}(t)$ and $k_{dm} = k_{dc}$ voltage $u_\Delta(t)$, generated at the output of the subtracting system (Figure 5.2.15) is

$$u_\Delta(t) = 0.5 \, k_{dc} [U_{sm}^2(t) + 2U_{sm}(t)U_{jm}(t)]$$

and the useful signal is distorted by the jamming. The level of the jamming depends on the level of the side lobes of antenna A_m and on the power of the jamming transmitter. In practice, it is usually not possible to achieve patterns $F_c(\theta)$, described by Equations (5.2.1) and (5.2.2).

Nondirectional cancellation, patterns of which are shown in Figure 5.17 by the dotted line, or antennas with $F_c(\theta)$, illustrated in Figure 5.17 by the dot-dash line, are simpler. Antenna A_c, with the latter kind of pattern, prevents attenuation by the cancellation system of a useful signal, the source of which is located on the axis of A_m ($\theta = 0$). But as θ increases from 0 to $0.5\theta_0$ and decreases from 0 to $-0.5\theta_0$ the attenuation of the useful signal increases.

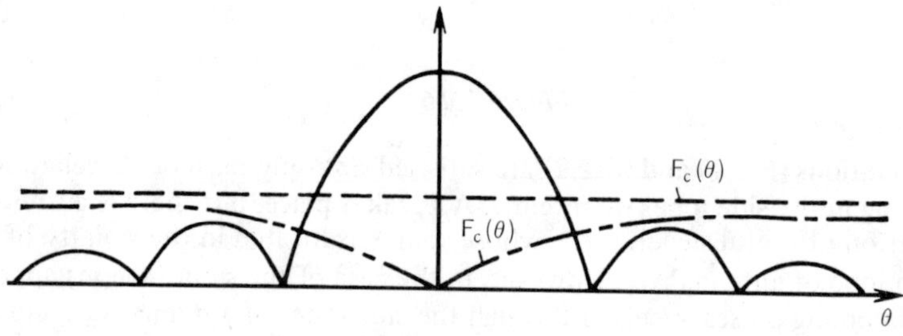

Figure 5.17

When a nondirectional antenna, A_c, is used, systems in which jamming acting through the main lobes of the main antenna is cancelled can be designed in accordance with the diagrams in Figure 5.15 or 5.18 [90].

The system whose diagram is shown in Figure 5.18 contains the main radar receiver and cancellation receiver. But in contrast to the one shown in Figure 5.15, the output signals of antenna A_m go not only to M_m, but also to the input of the cancellation receiver. These signals are fed through a directional coupler (DC), attenuator (At) and phase inverter (PI), to adder (Σ), where they are mixed with the output voltage of A_c.

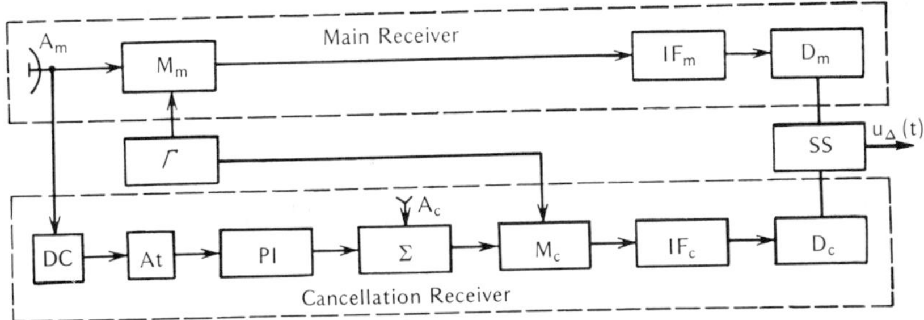

Figure 5.18

The parameters At and PI are so selected that the powers of the signals received by antennas A_m and A_c on the axis of antenna A_m will be identical, and the phases will differ by π. To satisfy these consitions in the entire range of scanning angles of antenna A_m, the antenna should be mounted coaxially with antenna A_c.

By virtue of the DC, At, PI and Σ, with nondirectional antenna A_c the Σ output is changed in accordance with the function $F_c(\theta)$. In this case, however, the radar system becomes extremely complicated, so that the system depicted in Figure 5.15 becomes preferable. The ensuing discussion is presented with specifically this system in mind.

When nondirectional cancellation antennas or antennas with patterns such as the one shown in Figure 5.17 by the dot-dash line, are used, perfect cancellation of jamming is not feasible. Therefore cancellation receivers are designed so that whenever antenna A_m picks up jamming signals from directions corresponding to its maximum side lobe, the inequality $u_{dc} > u_{dm}$ will be satisfied. Under these conditions the effect of overcancellation of jamming occurs and the useful signal is weakened at the output of the subtracting system.

The antenuation of the useful signal is strongest when a nondirectional cancellation antenna is used and the radar receiver is designed in accordance with the diagram in Figure 5.15, and it is related to the simultaneous appearance of the signal at the outputs of detectors D_m and D_c. If, along with useful signal $u_s(t)$, jamming is received, the source of which is located within the range of the side lobes of antenna, A_m, then the overcancellation effect results in additional attenuation, and when the jamming is sufficiently strong the useful signal may be completely suppressed.

Useful signals and jamming comparatively rarely coincide in time in cases when the jamming comes from reflections from local objects, emissions from nearby radar transmitters or random pulse jamming transmitters with a low average pulse duty factor. If the source of jamming is the ground or the surface of a body of water, the efficiency of an amplitude cancellation radar system decreases appreciably. The reason for this is that the interference signals from the ground or from the water surface are essentially random pulses with random amplitudes and durations. The number of such pulses per unit time may be so large that they often will coincide with useful signal pulses, and their power may be sufficient to suppress the useful signal. The efficiency of the amplitude jamming method may also be low in the case of sufficiently strong manmade noise and random pulse jamming; here jamming pulses of the latter kind should have an average repetition frequency many times higher than that of the transmission radar.

Amplitude cancellation also has another deficiency, in addition to the one described above. It consists of additional loss of sensitivity of a radar system due to noise from the cancellation receiver. In fact, if fluctuations of the voltage generated by the local oscillator are ignored, the noises from the main and cancellation receivers are independent. Therefore, the variance, $\sigma^2_{n\Delta}$, of the subtraction system output noise in the absence of useful and external noise signals is

$$\sigma^2_{n\Delta} = \sigma^2_{nm} + \sigma^2_{nc}.$$

Here σ^2_{nm} and σ^2_{nc} are the variances of the noise at the outputs of detectors D_m and D_c, and the gain of the subtracting system is assumed to be unity.

If the amplitude-frequency responses of the main and cancellation receivers are identical and the receiver noise sources have the same power, then the sensitivity of a radar system with a cancellation system is only half of what it would be without it. This leads to the conclusion that it is desirable to turn off the cancellation receiver when the radar system is not acted upon by external radio interference.

Improvement of the sensitivity of a radar system with a cancellation receiver is promoted by using the lowest possible gain, and to achieve the required jamming cancellation the gain of the cancellation antenna must be increased accordingly. Consequently, antenna A_c should be directional and should convert the received signals in such a way as to cancel jamming, the sources of which are located in an *a priori* known sector of space in relation to the radar.

Amplitude jamming cancellation is technically comparatively easy to accomplish and, in spite of its inherent deficiencies, often makes pulse radars highly effective under conditions when radio signals are reflected by local objects. It is also a sufficiently general-purpose method of combatting jamming through the side lobes of receiving antennas and can be used not only in radar, but also in other fields of radio electronics.

3. Coherent Jamming Cancellation

Coherent jamming cancellation, like the amplitude method, uses two receivers, main and cancellation (also called auxilliary). The cancellation receiver receives only jamming and the main one receives a mixture of the useful signal and jamming. In the coherent method jamming acting through the side lobes of the receiving antenna of the main receiver is cancelled.

In coherent jamming cancellation, often called amplitude-phase cancellation signals that are of identical intensity and opposite in phase are obtained by one means or another at the outputs of the high-frequency or intermediate-frequency amplifiers of the main and cancellation receivers. The noise voltage from these amplifiers and the useful signal from the main receivers are supplied to an adder.

Since the high- and intermediate-frequency amplifiers of the main and cancellation receivers are linear converters, the adder output jamming is eliminated, but the signal remains unchanged and is used for further processing. Complete without weakening the useful signal is possible only when cancellation antennas with patterns such as those shown in Figure 5.16 are used. If the cancellation antenna has a pattern different from the one shown in Figure 5.16, then, as in the amplitude method, jamming cancellation is accompanied by the attenuation of the useful signal. However, there will be less attenuation, since the coherent method does not use nonlinear processing of the useful signal and jamming. This question will be examined in greater detail below.

Coherent cancellation is technically more feasible in terms of the processing of IF signals. Therefore, the ensuing discussion refers to signals of this type.

There are various coherent cancellation techniques. The simplest technique consists in the development of a system in which jamming signals $u_{jm}(t)$ and $u_{jc}(t)$ at the IF outputs of the main and jamming receivers are opposite in phase.

Under actual conditions it is virtually impossible to generate antiphase jamming signals at the IF outputs of the main and cancellation receivers, because of differences in the phase-frequency responses of receivers and the frequency instability of jamming transmitters. This means that when the coherent cancellation method is used it is necessary to consider the difference of the envelopes and phases of voltages $u_{jm}(t)$ and $u_{jc}(t)$. Under these conditions, a coherent jamming cancellation system should automatically change the envelope and phase of voltage $u_{jc}(t)$ in such a way that output signal $u_{c\Sigma}(t)$ of this system will be $-u_{jm}(t)$. Interference is cancelled as the result of adding $u_{c\Sigma}(t)$ and $u_{jm}(t)$.

There are at least two ways to obtain the required voltage $u_{c\Sigma}(t)$ from $u_{jc}(t)$. The first is based on the use of quadrature converters [167, 167a], which are well known in optimum reception theory, and the second method calls for the use of an AGC system in the cancellation receiver, which, in the absence of useful signals, operates under the influence of difference voltage $u_{jm}(t) - u_{jc}(t)$, and of a system that automatically controls the phase of voltage $u_{jc}(t)$ [97].

Jamming Cancellation System with Quadrature Converters

The principle of jamming cancellation with quadrature converters is conveniently illustrated by way of the example when $u_{jm}(t)$ and $u_{jc}(t)$ are sinusoidal signals with the identical frequency, but with different amplitudes and initial phases.

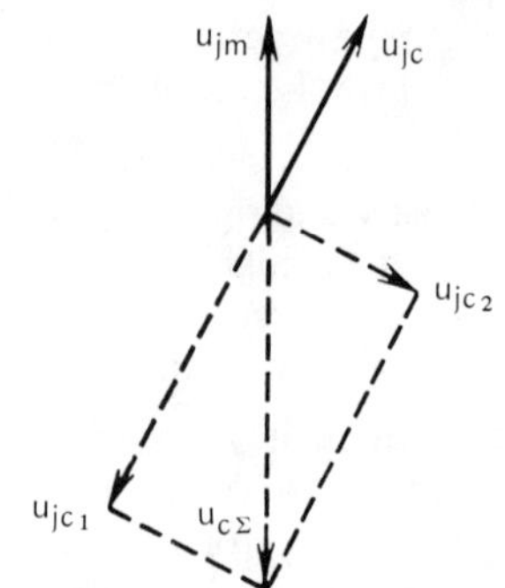

Figure 5.19

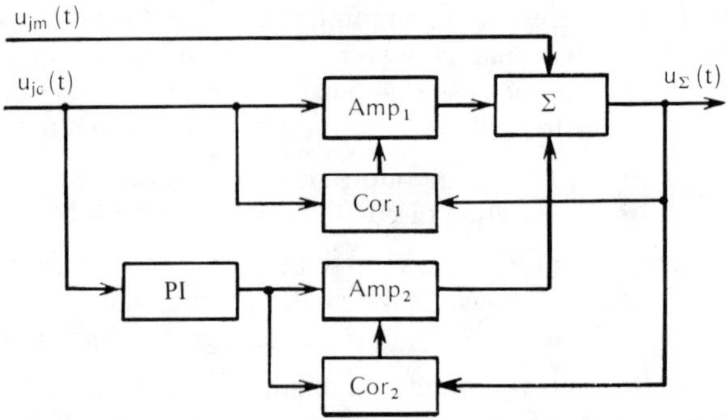

Figure 5.20

These signals may be represented in vector form as shown in Figure 5.19. As
can be seen in Figure 5.19, in order to find the vector $u_{c\Sigma} = -u_{jm}$ it is necessary
to have two voltages which are described by the mutually perpendicular vectors
u_{jc1} and u_{jc_2} which must be equal to

$$u_{jc_1} = -k_1 u_{jc} \quad \text{and} \quad u_{jc_2} = k_2 u_{jcq}$$

where $|u_{jcq}| = |u_{jc}|$, and the angle between vectors u_{jc} and u_{jcq} is $90°$. Here
the proportionality coefficients k_1 and k_2 are selected such that the equation

$$-k_1 u_{jc} + k_2 u_{jcq} = u_{c\Sigma} = -u_{jm}$$

is satisfied.

A diagram of a system that performs coherent jamming cancellation using quadrature converters is illustrated in Figure 5.20. It contains correlators Cor_1 and
Cor_2, amplifiers Amp_1 and Amp_2 with controlled gains, phase inverter PI,
which gives the quadrature component of voltage $u_{jc}(t)$, and adder Σ.

If the IF amplifiers of the main and cancellation generate jamming voltages $u_{jm}(t)$ and $u_{jc}(t)$, which are assumed to be narrow band stationary random processes, then the output voltage, u_Σ, of adder Σ, with a unit gain will be

$$u_\Sigma(t) = u_{jm}(t) + k_{ai}u_{jc}(t) + k_{a_2}u_{jcq}(t). \tag{5.2.7}$$

Here k_{ai} and k_{a_2} are the gains of amplifiers, Amp_1 and Amp_2, which perform broadband conversion of $u_{jc}(t)$ and $u_{jcq}(t)$, respectively. They are changed in proportion to the voltages generated by correlators Cor_1 and Cor_2 and are

$$k_{ai} = k_1 \left[u_{jc}(t)u_\Sigma(t)\right]_{av} \tag{5.2.8}$$

$$k_{a_2} = k_2 \left[u_{jcq}(t)u_\Sigma(t)\right]_{av} \tag{5.2.9}$$

where the symbol $[\bullet]_{av}$ here and in the ensuing analysis signifies computation of the average component; k_1 and k_2 are proportionality coefficients.

After substituting the function $u_\Sigma(t)$, determined by Equation (5.2.7), into Equations (5.2.8) and (5.2.9), assuming that the cross-correlation between the two quadrature voltages $u_{jc}(t)$ and $u_{jcq}(t)$ vanishes at a given moment of time, we obtain

$$k_{a_1} = \frac{k_1 \left[u_{jm}(t)\, u_{jc}(t)\right]_{av}}{1-k_1 \sigma_{jc}^2} \tag{5.2.10}$$

$$k_{a_2} = \frac{k_2 \left[u_{jm}(t)u_{jcq}(t)\right]_{av}}{1-k_2 \sigma_{jc}^2} \tag{5.2.11}$$

Here σ_{jc}^2 is the variance of voltage $u_{jc}(t)$, equal to the variance of voltage $u_{jc}(t)$, and the mathematical expectations of voltages $u_{jc}(t)$ and $u_{jcq}(t)$ are assumed equal to zero.

On the basis of equations (5.2.7), (5.2.10) and (5.2.11) we find the following formula, which determines the instantaneous value of the Σ output interference signal (Figure 5.20):

$$u_\Sigma(t) = u_{jm}(t) + \frac{k_1 \left[u_{jm}(t)u_{jc}(t)\right]_{av}}{1-k_1 \sigma_{jc}^2}\, u_{jc}(t)$$

$$+ \frac{k_2 \left[u_{jm}(t)u_{jcq}(t)\right]_{av}}{1-k_2 \sigma_{jc}^2}\, u_{jcq}(t). \tag{5.2.12}$$

If jamming signals $u_{jm}(t)$ and $u_{jc}(t)$ are narrowband random processes and differ from each other in terms of envelopes at each mome nt of time, and in terms of phase shift, then

$$u_{jm}(t) = U_{jm}(t) \cos \left[\omega_{if}t + \phi_m(t)\right], \qquad (5.2.13)$$

$$u_{jc}(t) = U_{jc}(t) \cos \left[\omega_{if}t + \phi_c(t)\right]$$

$$= U_{jc}(t) \cos \left[\omega_{if}t + \phi_m(t) + \Delta\phi(t)\right]. \qquad (5.2.14)$$

$$u_{jcq}(t) = U_{jc}(t) \sin \left[\omega_{if}t + \phi_m(t) + \Delta\phi(t)\right]. \qquad (5.2.15)$$

Here $U_{jm}(t)$ and $U_{jc}(t)$ are random envelopes and $\phi_m(t)$ and $\phi_c(t)$ are the randomly changing phases of voltages $u_{jm}(t)$ and $u_{jc}(t)$, respectively; $\Delta\phi = \phi_c(t) - \phi_m(t)$, ω_{if} is the angular intermediate frequency.

Introducing the definitions

$$\left[u_{jm}(t)u_{jc}(t)\right]_{av} = \sigma_{jm}\sigma_{jc}\rho_{jm}$$

$$\left[u_{jm}(t)u_{jcq}(t)\right]_{av} = \sigma_{jm}\sigma_{jc}\rho_{jc}$$

where ρ_{jm} and ρ_{jc} are the crosscorrelation coefficients of voltages $u_{jm}(t)$, $u_{jc}(t)$ and $u_{jm}(t)$, $u_{jcq}(t)$, respectively, and σ_{jm}^2 is the variance of voltage $u_{jm}(t)$, we transform Equation (5.2.12) using Equations (5.2.13) through (5.2.15) to the form

$$u_{\Sigma}(t) = U_{jm}(t) \cos \left[\omega_{if}t + \phi_m(t)\right]$$

$$+ B \sin \left[\omega_{if}t + \phi_m(t) + \Delta\phi(t) + \phi_1\right] \qquad (5.2.16)$$

for $k_1\sigma_{jc}^2 < 1$ and $k_2\sigma_{jc}^2 < 1$,

$$u_{\Sigma}(t) = U_{jm}(t) \cos \left[\omega_{it}t + \phi_m(t)\right]$$

$$- B \sin \left[\omega_{it}t + \phi_m(t) + \Delta\phi(t) + \phi_1\right] \qquad (5.2.17)$$

for $k_1\sigma_{jc}^2 > 1$ and $k_2\sigma_{jc}^2 > 1$.

Here

$$B = \sigma_{jm}\sigma_{jc}U_{jc}(t) \left[\frac{k_1^2\rho_{jm}^2}{(1-k_1\sigma_{jm}^2)^2} + \frac{k_2^2\rho_{jc}^2}{(1-k_2\sigma_{jc}^2)^2}\right]^{0.5}; \qquad (5.2.18)$$

$$tg\phi_1 = \frac{k_1\rho_{jm}(1-k_2\sigma_{jc}^2)}{k_2\rho_{jc}(1-k_1\sigma_{jc}^2)} \qquad (5.2.19)$$

Analysis of Equations (5.2.16) through (5.2.19) shows that jamming cancellation is impossible so long as $k_1 \sigma_{jc}^2 > 1$ and $k_2 \sigma_{jc}^2 < 1$. When $u_{jc}(t) = au_{jm}(t)$, however, which leads to the equations

$$\sigma_{jc} = a\sigma_{jm},$$

$$[u_{jm}(t)u_{jcq}(t)]_{av} = 0,$$

$$[u_{jm}(t)u_{jc}(t)]_{av} = a\sigma_{jm}^2$$

and also when $\Delta\phi(t) = 0$, $k_1 \sigma_{jc}^2 \gg 1$ and $k_2 \sigma^2 \gg 1$, voltage $u_\Sigma(t)$ vanishes at any moment of time. Consequently, by changing gains k_{a_1} and k_{a_2} in accordance with

$$k_{a_1} = -\frac{[u_{jm}(t)u_{jc}(t)]_{av}}{\sigma_{jc}^2} \tag{5.2.20}$$

$$k_{a_2} = -\frac{[u_{jm}(t)u_{jcq}(t)]_{av}}{\sigma_{jc}^2} \tag{5.2.21}$$

and by guaranteeing identical phase-frequency responses and amplitude-frequency responses, different only in terms of scale, for the main and cancellation receivers, it is possible to cancel jamming completely in the absence of a useful signal.

In reality, as was mentioned above, these conditions are not valid and only minimzation of the variance of the output noise voltage of adder Σ (Figure 5.20) is feasible.

If, on the basis of Equation (5.2.17),and on the assumption that there is no cross correlation between quadrature random signals $u_{jc}(t)$ and $u_{jcq}(t)$ at coinciding moments of time, and that the mathematical expectations for voltages $u_{jm}(t)$, $u_{jc}(t)$ and $u_{jcq}(t)$ are zero, it is possible to find the variance, σ_Σ^2, of voltage $u_\Sigma(t)$, then it can be proved that σ_Σ^2 is minimized when $k_1 \sigma_{jc}^2 = k_2 \sigma_{jc}^2 = \infty$, i.e., when k_{a_1} and k_{a_2} change in accordance with Equations (5.2.20) and (5.2.21). In this case, σ_Σ^2 will be

$$\sigma_\Sigma^2 = \sigma_{jm}^2 (1 - \rho^2), \tag{5.2.22}$$

where

$$\rho^2 = \rho_{jm}^2 + \rho_{jc}^2. \tag{5.2.23}$$

The analogous formula can be derived with some approximation when the inequalities $k_1 \sigma_{jc}^2 \gg 1$ and $k_2 \sigma_{jc}^2 \gg 1$ are satisfied.

It follows from Equation (5.2.22) and from the above analysis that in the coherent jamming cancellation method with quadrature converters it is essential to achieve the best possible correlation between the jamming signals acting on the IF outputs of the main and cancellation receivers.

In cases when the jamming is accompanied by a useful signal coming from the main lobe of the antenna of the main receiver, and picked up simultaneously by the antenna of the cancellation receiver, then, instead of $u_{jm}(t)$ and $u_{jc}(t)$, it is necessary to analyze the voltages

$$u_1(t) = u_{sm}(t) + u_{jm}(t), \tag{5.2.24}$$

$$u_2(t) = u_{sc}(t) + u_{jc}(t) \tag{5.2.25}$$

respectively. Here $u_{sm}(t)$ and $u_{sc}(t)$ are the useful signal voltages at the IF outputs of the main and cancellation receivers.

To obtain more understandable results, characterizing the performance of a noise canceller with quadrature converters in the presence of a useful signal, we will assume that the latter is deterministic and varies at the IF output of the main receiver according to $u_{sm}(t) = U_{sm} \cos(\omega_{if}t + \phi_m)$. Here U_{sm} and $\phi_m(t)$ occurs at the time of appearance of a jamming signal and

$$u_{sc}(t) = U_{sc} \cos(\omega_{if}t + \phi_0 + \phi_c),$$

where $U_{sc} = bU_{sm}$, $b < 1$, ϕ_c is additional phase shift of voltage $u_{sc}(t)$ in the cancellation receiver

When useful and jamming signals arrive simultaneously, voltage $u_\Sigma(t)$, as follows from Figure 5.20, is

$$u_\Sigma(t) = u_{sm}(t) + u_{jm}(t) + k_{a_1}(t) + k_{a_1} u_{jc}(t)$$

$$+ k_{a_2} u_{scq}(t) + k_{a_2} u_{jcq}(t). \tag{5.2.26}$$

Here $u_{scq}(t) = U_{sc} \sin(\omega_{if}t + \phi_m + \phi_c)$ is voltage $u_{sc}(t)$, phase-shifted by $-90°$.

Gains k_{a_1} and k_{a_2} change in proportion to the voltages generated by correlators Cor_1 and Cor_2 and are

$$k_{a_1} = k_1 \left\{ [u_{sc}(t) + u_{jc}(t)] \, u_\Sigma(t) \right\}_{av} \tag{5.2.27}$$

$$k_{a_2} = k_2 \left\{ [u_{scq}(t) + u_{jcq}(t)] \, u_\Sigma(t) \right\}_{av} \tag{5.2.28}$$

Using Equations (5.2.26) through (5.2.28), we obtain

$$k_{a_1} = k_1 \frac{\left\{ u_{sm}(t)u_{sc}(t) \right\}_{av} + \left\{ u_{jm}(t)u_{jc} \right\}_{av}}{1 - k_1 \left\{ u_{sc}^2(t) \right\}_{av} - k_1 \sigma_{jc}^2}, \tag{5.2.29}$$

$$k_{a2} = k_2 \frac{\left\{u_{sm}(t)u_{scq}(t)\right\}_{av} + \left\{u_{jm}(t)u_{jcq}(t)\right\}_{av}}{1 - k_2 \left\{u_{scq}^2(t)\right\}_{av} - k_2 \sigma_{jc}^2}. \tag{5.2.30}$$

As before, we can show that the canceller operates in the best mode when

$$k_1 \left\{u_{sc}^2(t)\right\}_{av} + k_1 \sigma_{jc}^2 \gg 1, \tag{5.2.31}$$

$$k_2 \left[u_{scq}^2(t)\right]_{av} + k_2 \sigma_{jc}^2 \gg 1. \tag{5.2.32}$$

Under these conditions we find, on the basis of Equations (5.2.26), (5.2.29) and (5.2.30),

$$u_\Sigma(t) = u_{sm}(t) + u_{jm}(t) - \frac{\left\{u_{sm}(t)\,u_{sc}(t)\right\}_{av} + \left\{u_{jm}(t)u_{jc}(t)\right\}_{av}}{\left\{u_{jcq}^2(t)\right\}_{av} + \sigma_{jc}^2} \left[u_{sc}(t) + u_{jc}(t)\right]$$

$$- \frac{\left\{u_{sm}(t)u_{scq}(t)\right\}_{av} + \left\{u_{jm}(t)u_{jcq}(t)\right\}_{av}}{\left\{u_{jcq}^2(t)\right\}_{av} + \sigma_{av}^2} \left[u_{scq}(t) + u_{jcq}(t)\right]. \tag{5.2.33}$$

In accordance with the equations derived above for $u_{sm}(t)$, $u_{sc}(t)$, $u_{scq}(t)$, $u_{jm}(t)$, $u_{jc}(t)$ and $u_{jcq}(t)$, Equation (5.2.33), after simple conversions, acquires the following form:

$$u_\Sigma(t) = U_{sm} \frac{\sigma_{jc}^2 - b\sigma_{jm}\sigma_{jc}\,\rho \sin(\phi_c + \Delta\psi)}{b^2 \sigma_{sm}^2 + \sigma_{jc}^2} \cos(\omega_{if}t + \phi_m)$$

$$- \frac{b\sigma_{jm}\sigma_{jc}\rho U_{sm} \cos(\phi_c + \Delta\psi)}{b^2 \sigma_{sm}^2 + \sigma_{jc}^2} \sin(\omega_{if}t + \phi_m)$$

$$+ \left\{ u_{jm}(t) - \frac{b\sigma_{sm}^2}{b^2 \sigma_{sm}^2 + \sigma_{jc}^2} U_{jc}(t) \cos[\Delta\phi(t) - \phi_c] \right.$$

$$\left. - \frac{\sigma_{jm}\sigma_{jc}\rho}{b^2 \sigma_{sm}^2 + \sigma_{jc}[} U_{jc}(t) \sin[\Delta\phi(t) + \Delta\psi] \right\} \cos\left[\omega_{if}t + \phi_m(t)\right]$$

$$+ \left\{ \frac{b\sigma_{sm}^2}{b\sigma_{sm}^2 + \sigma_{jc}} U_{jc}(t) \sin|\Delta\phi(t) - \phi_c| \right.$$

$$\left. + \frac{\sigma_{jm}\sigma_{jc}\rho}{b^2 \sigma_{sm}^2 |\sigma_{jc}} U_{jc}(t) \cos|\Delta\phi(t) + \Delta\phi| \right\} \sin\left[\omega_{if}t + \phi_m(t)\right]. \tag{5.2.34}$$

Here

$$\sigma^2_{sm} = 0.5\, U^2_{sm} \quad \text{and} \quad \Delta\psi = \text{arctg}\, \frac{\rho_{jm}}{\rho_{jc}}. \tag{5.2.35}$$

It follows from Equation (5.2.34) that voltage $u_{\Sigma}(t)$ contains the component

$$u_s(t) = U_{sm}\, \frac{\sigma^2_{jc} - b\sigma_{jm}\sigma_{jc}\,\rho\, \sin\,(\phi_c + \Delta\psi)}{b^2\sigma^2_{sm} + \sigma^2_{jc}} \cos\,(\omega_{if}t + \phi_m), \tag{5.2.36}$$

which changes in time the same way as useful signal $u_{sm}(t)$ at the IF of the main receiver, and the component

$$u_{j\Sigma}(t) = u_{\Sigma}(t) - u_s(t), \tag{5.2.37}$$

which is the jamming component.

Analysis of Equations (5.2.34) through (5.2.36) leads to the following conclusion. If antenna A_c of the cancellation receiver does not see the useful signal arriving from the main lobe of the antenna of the main receiver, i.e., when $b = 0$, then, when $U_{jc}(t) = aU_{jm}(t)$, which leads to the equality $\sigma_{ic} = a\sigma_{jm}$, voltage $u_s(t)$ is equal to useful signal $u_{sm}(t) = U_{sm} \cos\,(\omega_{if}t + \phi_m)$, which appears at the IF output of the main receiver. Under the same conditions

$$u_{j\Sigma}(t) = U_{jm}(t) \cos\,[\omega_{if}t + \phi_m(t)] - \rho U_{jm}(t) \sin\,[\omega_{if}t + \phi_m(t) + \Delta\phi(t) + \Delta\psi].$$

Hence, it follows that useful signal $u_{sm}(t)$ for $b = 0$ and $u_{jc}(t) = aU_{jm}(t)$ is distorted by interference if $\Delta\phi(t) = 0$ and the correlation coefficient $\rho \neq 1$. But if $\rho = 1$, which leads to the equality $\Delta\psi = 0.5\pi$, and equation $\Delta\phi(t) = 0$ is satisfied at the same time, which is equivalent to the identity of the phase-frequency responses of the main and cancellation receivers, $u_{j\Sigma}(t) = 0$ at any moment of time, and consequently perfect jamming cancellation is assured. If the interference in the main and cancellation receivers is not correlated ($\rho = 0$), the cancellation system will have practically no effect on the jamming voltage.

When antenna A_c receives a useful signal ($b \neq 0$), then the strength of the useful signal will decrease and incomplete jamming cancellation will take place, even when the phase-frequency responses of the main and cancellation receivers are perfectly identical ($\Delta\phi(t) = 0$, $\phi_c = 0$ and $\rho = 1$).

Actually, if we assume $b \neq 0$, $\phi_c = 0$, $\Delta\phi = 0$, $\rho = 1$ and $\Delta\psi = 0.5\pi$, we find on the basis of Equations (5.2.34) through (5.2.37)

$$u_s(t) = U_{sm} \frac{a\sigma_{jm}^2(a-b)}{b^2\sigma_{sm}^2 + a^2\sigma_{jm}^2} \cos(\omega_{if}t + \phi_m), \qquad (5.2.38)$$

$$u_{j\Sigma}(t) = \frac{b(b-a)\sigma_{sm}^2}{b^2\sigma_{sm}^2 + a^2\sigma_{im}^2} U_{jm}(t)\cos[\omega_{if}t + \phi_m(t)] \qquad (5.2.39)$$

Equation (5.2.38) shows a reduction of the amplitude of voltage $u_s(t)$ due to a difference of b from zero, which occurs when antenna A_c receives signals, the sources of which are located within the range of the main lobe of antenna A_m, which is a part of the main receiver. This reduction increases with b. The level of jamming $u_{j\Sigma}(t)$ is also a strong function of coefficient b.

To evaluate the degree of suppression of jamming by a coherent canceller with quadrature converters, we compare the ratios of the effective powers of the useful signal and jamming at the IF output of systems with and without a canceller. Here the performance of a system with a canceller will be determined on the basis of Equations (5.2.38) and (5.2.39).

If a system has no canceller, then voltages $u_{sm}(t)$ and $u_{jm}(t)$ of the useful signal and jamming at the IF output of the receiver are

$$u_{sm}(t) = U_{sm}\cos(\omega_{if} + \phi_m),$$

$$u_{jm}(t) = U_{jm}(t)\cos[\omega_{if}t + \phi_m(t)],$$

and their effective powers are proportional to the variances of their voltages $\sigma_{sm}^2 = 0.5U_{sm}^2$ and σ_{jm}^2, respectively. The ratio of σ_{sm}^2 to σ_{jm}^2 is denoted through the symbol q_{bc}.

When a system has a canceller, the effective power of the useful signal is proportional to the mean value of σ_s^2 minus the square of voltage $u_s(t)$, which, as follows from Equation (5.2.38), is

$$\sigma_s^2 = \sigma_{sm}^2 \frac{a^2(a-b)^2\sigma_{jm}^2}{(b^2\sigma_{sm}^2 + a^2\sigma_{jm}^2)^2},$$

The effective jamming power in a system with a canceller is proportional to the variance $\sigma_{j\Sigma}^2$ of voltage $u_{j\Sigma}(t)$. On the basis of Equation (5.2.39) we find

$$\sigma_{j\Sigma}^2 = \sigma_{jm}^2 \frac{b^2(a-b)^2\sigma_{sm}^4}{(b^2\sigma_{sm}^2 + a^2\sigma_{jm}^2)^2}$$

Therefore,

$$q_{sc} = \frac{\sigma_s^2}{\sigma_{j\Sigma}^2} = \frac{a^2\sigma_{jm}^2}{b^2\sigma_{sm}^2} = \frac{a^2}{b^2 q_{bc}}, \qquad q = \frac{q_{sc}}{q_{bc}} = \frac{a^2}{b^2 q_{bc}}$$

$$(5.2.40)$$

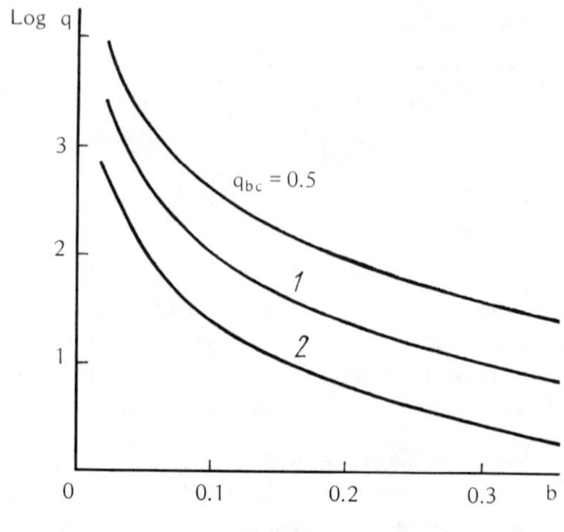

Figure 5.21

Calculation of the dependences of q on b and q_{bc} for a = 1, carried out in accordance with Equation (5.2.40), yields the graphs shown in Figure 5.21. It follows from Equation (5.2.40) and Figure 5.21 that as $1/q_{bc}$ increases and b decreases, the ratio q increases and may reach several tens of thousands.

The feasibility of substantially reducing b depends on the development of antennas A_m, the side lobes of which have a substantially lower level than the main lobe.

In many cases, signal duration, τ_p, may be vastly shorter than the duration of jamming, and the time constants of the averaging filters of correlators Cor_1 and Cor_2 (Figure 5.20) are several times greater than τ_p.

Under this condition $b^2 \sigma_{sm}^2 \ll \sigma_{jc}^2$ and, on the basis of Equations (5.2.34), (5.2.36) and (5.2.37), we obtain for $u_s(t)$ and $u_{j\Sigma}(t)$

$$u_s(t) = U_{sm} \left[1 - b \ \frac{\sigma_{jm}}{\sigma_{jc}} \ \rho \sin(\phi_c + \Delta\psi) \right] \cos(\omega_{if}t + \phi_m),$$

$$(5.2.41)$$

$$u_{j\Sigma}(t) = \left\{ U_{jm}(t) - \frac{\sigma_{jm}}{\sigma_{jc}} \ \rho U_{jc}(t) \sin[\Delta\phi(t) + \Delta\psi] \right\} \cos[\omega_{if}t + \phi_m(t)]$$

$$+ \ \frac{\sigma_{jm}}{\sigma_{jc}} \ \rho U_{jc}(t)\cos[\Delta\phi(t) + \Delta\psi] \ \sin[\omega_{if}t + \phi_m(t)]$$

$$- \ b \ \frac{\sigma_{jm}}{\sigma_{jc}} \ \rho U_{sm}\cos(\phi_c + \Delta\psi) \sin(\omega_{if}t + \phi_m).$$

$$(5.2.42)$$

Equation (5.2.42) shows that for antenna A_m of the main receiver, with low-level side lobes, when $b = 0.01$-0.001 and the jamming levels are approximately identical at the IF outputs of the main and cancellation receivers, the reduction of the amplitude of the useful signal by the jamming is negligible. It is $b(\sigma_{jm}/\sigma_{jc}) \sin(\phi_c + \Delta\psi)$ and does not exceed $b(\sigma_{jm}/\sigma_{jc}) \ll 1$.

The interference level $u_{j\Sigma}(t)$ is also negligible, particularly when the phase-frequency responses of the main and cancellation receivers are identical, i.e., when $\Delta\phi(t) = \phi_c = 0$. Under these conditions, when $U_{jm}(t) = U_{jc}(t)$ and $b \ll 1$,

$$u_{j\Sigma}(t) = (1 - \rho \sin \Delta\psi)U_{jc}(t) \cos[\omega_{if}t + \phi_m(t)]$$
$$+ \rho U_{jc}(t) \cos \Delta\psi \sin[\omega_{if}t + \phi_m(t)] . \tag{5.2.43}$$

When $\rho = 1$ and $\Delta\psi = \pi$ complete jamming cancellation is achieved.

Thus, high-quality jamming cancellation and negligible attenuation of a known useful signal are guaranteed in coherent cancellation systems with quadrature converters under the following conditions: the useful signal entering the cancellation receiver has low power; the amplitude-frequency and phase-frequency responses of the IF of the main and cancellation receivers are identical; the pattern of antenna A_c of the receiver is satisfactorily matched with the side lobes of antenna A_m of the main receiver; the cross correlation coefficient of jamming signals at the IF outputs of the main and cancellation receivers is equal to unity. Under these conditions, the degree of suppression of jamming increases as b^2 decreases and the reproduction quality of the useful signal is improved.

If the useful signal changes randomly in time and is not correlated with the jamming, then relations analogous to Equations (5.2.34) through (5.2.42) can be derived, i.e., the same conclusions can be drawn in relation to the jamming cancellation and reproduction of the useful signal as in the case of the combined action of jamming and a deterministic useful signal.

We mention in conclusion that the crosscorrelation coefficient, ρ, is influenced not only by the parameters of the main and cancellation receivers, but also by the difference of the paths of the electromagnetic wave from the jamming transmitter to the antennas of the main and cancellation receivers.

Interference Cancellation System with AGC and PLL (Phase-Lock Loop) Systems

A functional diagram of such a system is shown in Figure 5.22 [97]. It contains a main receiver and a cancellation system. The main receiver includes antenna A_m, mixer M_m, local oscillator LO_m and intermediate-frequency amplifier IF_m with an AGC_m system, which maintains a

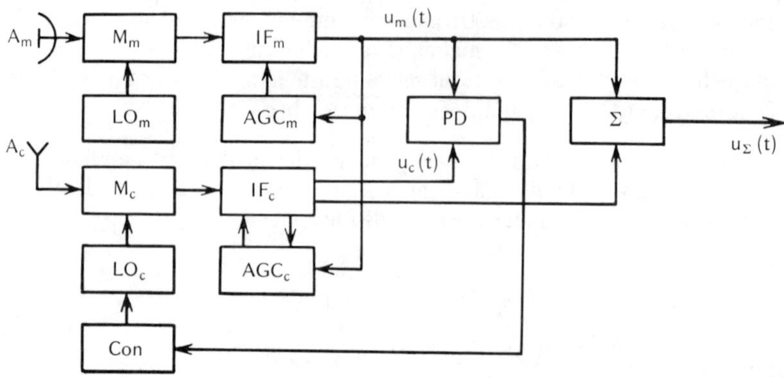

Figure 5.22

constant output voltage level. The main parts of the cancellation system
are antenna A_c, intermediate-frequency amplifier IF_c, AGC_c system, mixer
M_c, local oscillator LO_c, control system Con, and phase detector PD.
Adder Σ adds up the voltages generated in IF_m and IF_c.

The AGC_c system is actuated by a signal, equal to the difference of the
IF_m and IF_c output voltages, thereby equalizing the instantaneous values
of the input jamming signals of adder Σ.

The PLL system is comprised of phase detector PD, control system Con,
LO_c, mixer M_c and IF amplifier IF_c. Voltage u_{pd}, generated by PD and
dependent only on the difference of phases, ϕ, of signals $u_s(t)$ and $u_c(t)$,
is supplied to the control system. The latter changes the frequency and
phase, which is uniquely related to it, of the oscillations of LO_c and
hence the phase of the voltage generated by mixer M_c and IF amplifier
IF_c.

When the phases of PD input voltages become identical, u_{pd} ceases to
change in time and is proportional to $\phi = \int_0^t \Delta\omega dt$, where $\Delta\omega = \omega_{if\,m}$
$- \omega_{if\,c}$, and $\omega_{if\,m}$ and $\omega_{if\,c}$ are the frequencies of voltages $u_m(t)$ and
$u_c(t)$. The LO frequency will also be constant, along with u_{pd}. Con-
sequently, it is possible to take from IF_c a voltage which is opposite
in phase to signal $u_m(t)$. This is easily accomplished, for example, with
an inverter, connected between IF_c and the phase detector.

The combined operation of the AGC_c and PLL systems provides the required
phase and amplitude relations between the jamming signals entering adder
Σ from IF_m and IF_c. When these systems perform ideally and antennas A_m

and A_c receive only jamming signals, the latter do not pass to the adder output. Under actual conditions complete jamming cancellation is not achieved because of the imperfect performance of the AGC and PLL systems. These systems can be calculated in accordance with given requirements with the aid of presently available textbooks and manuals [31, 86, 165].

The AGC_c and PLL systems that are used in jamming cancellation systems should be extremely fast. The speed of the AGC system should be such that its output voltage will reach the required level in considerably less time than the correlation time for the envelope of the signal, and the PLL system should be able to track all phase changes of the jamming acting upon IF_m. Furthermore, it is essential the AGC_c system not alter the gain of the cancellation receiver when it is acted upon by a useful signal. This requires that the time constant, T_{ac} of the AGC_c system satisfy the inequality $T_{ac} \gg \tau_p$, where τ_p is the pulse duration of the useful signal.

Concluding this examination of coherent jamming cancellation, for which quadrature converters or AGC and PLL systems are used, it should be pointed out that this method, like the amplitude cancellation method, reduces the sensitivity of the receiver. This happens because the variance of the output fluctuations of adder Σ (Figures 5.20 and 5.22) increases due to cancellation receiver noise. This phenomenon is counteracted with the same measures that are used in amplitude jamming cancellation.

If it is necessary, when using quadrature cancellers, to supress jamming from the side lobes of the main receiver from not one, but from n_d directions $(n_d > 1)$, then n_d cancellers of the type shown in Figure 5.20, should be used. The method described above may be used to design an adaptive antenna of the phased array type [178].

5.3. INTERPERIOD CLUTTER CANCELLATION

1. Interperiod Fixed Clutter

Interperiod cancellation systems are a modification of moving target indication systems and find application in pulse radars with a low pulse duty factor for supressing in the receiver signals produced by reflections from chaff, from the underlying surface and from various stationary installations.

By analyzing signals that reach a radar receiver from a moving target and fixed clutter, it can be proved that when pulse voltage

$$u(t) = U \cos \omega_0 t \quad \text{for } kT_p \leqslant t \leqslant \tau_p + kT_p \tag{5.3.1}$$

is generated in a radar transmitter, where U and ω_0 are amplitude and angular frequency, τ_p and T_p are the pulse duration and repetition period, and $k =$

= 0, 1, 2, ..., signal $u_{st}(t)$, appearing at the receiver input as a result of reflection from a fixed object, will change according to

$$u_c(t) = U_c \cos (\omega_0 t - \phi_c). \tag{5.3.2}$$

Here U_c is the amplitude of voltage $u_c(t)$ and ϕ_c is the phase, determined by the properties of the clutter and by the propagation time of radio waves from the radar to the clutter and back.

At the same time, the receiver input voltage $u_t(t)$, produced by the moving target, without consideration of usual amplitude and phase fluctuations, and on the assumption that the radial velocity of the target v_t relative to the radar is constant, is

$$u_t (t) = U_t \cos [(\omega_0 \pm \omega_d)t - \phi_t], \tag{5.3.3}$$

where U_t is amplitude, ω_d is the Doppler frequency, and ϕ_t is phase resulting from properties of the target and propagation from the radar to the moving target and back.

Since the carrier frequency of pulses $u_c(t)$ remains equal to ω_0 and the carrier frequency of voltage $u_t(t)$ differs from ω_0 by ω_d or $-\omega_d$, depending on the direction of target motion, it becomes possible to cancel the signals arriving from the stationary clutter, which for the radar is interference. Such cancellation, called interperiod cancellation, takes place in the IF stages of the receiver or at video frequency.

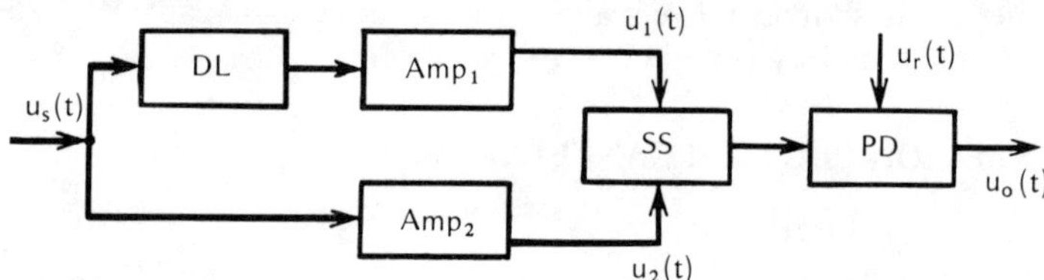

Figure 5.23

A functional diagram of a system that performs interperiod jamming or clutter cancellation at IF is shown in Figure 5.23 [147]. Here voltage $u_s(t)$, generated by the IF amplifier (not shown in Figure 5.23), is supplied to ultrasonic delay line DL and amplifier Amp_2. The delay line delays the output pulses by time T_p, equal to the repetition period of the radar pulses. The DL output is connected to amplifier Amp_1. Voltages from Amp_1 and Amp_2 are sent to a subtracting system (SS), and from there to a phase detector (PD), which also receives reference voltage $u_r(t)$. The latter must be coherent with the intermediate-frequency signal $u_s(t)$.

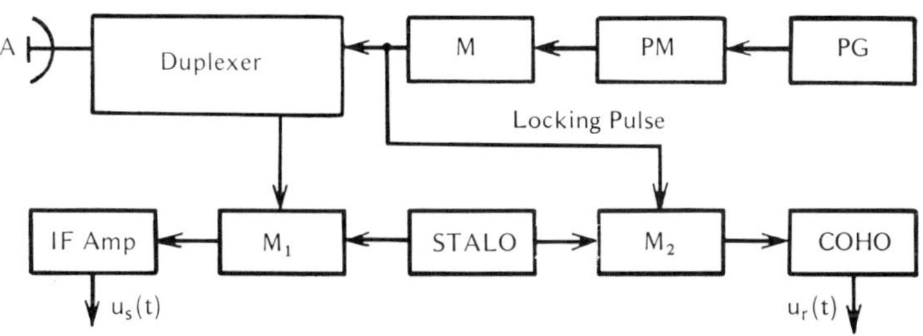

Figure 5.24

Voltages $u_s(t)$ and $u_r(t)$ can be made coherent with oscillators internal to the radar system or by using signals from stationary objects. Accordingly, there are radars with internal and with external coherence [147]. A typical diagram of a radar transceiver with internal coherence, in which high-frequency signals are generated by a magnetron, is shown in Figure 5.24.

The part of the signal generated by the magnetron oscillator (M), the operation of which is controlled by a pulse modulator (PM) and pulse generator (PG) is sent to mixer M_2, which also receives voltage from a stable local oscillator (STALO). Intermediate-frequency signal, ω_{if}, which synchronizes the phase of the coherent oscillator (COHO) appears on the output of mixer M_2. At the same time the STALO is connected to mixer M_1, the second input of which is connected to a duplexer which passes the signals picked up by antenna A to M_1. The output voltage of M_1 has frequency $\omega_{if} \pm \omega_d$. This voltage is supplied to the IF amplifier.

If the radar illuminates a moving target, then high-frequency signal $u_t(t)$ at the receiver input is determined by Equation (5.3.3). This signal, after passing through the mixer and IF, is converted to intermediate-frequency voltage:

$$u_s(t) = U_s \cos\left[(\omega_{if} \pm \omega_d)t - \phi_1\right] \tag{5.3.4}$$

Here U_s and ϕ_1 are the amplitude and initial phase of voltage $u_s(t)$ and ϕ_1 is a function of ϕ_t and of the phase-frequency characteristic of the RF and IF receiver channels

At the outputs of Amp_1 and Amp_2, on the assumption that the DL of Amp_1, on the one hand, and the Amp_2, on the other hand, have identical gains and do not alter the phases of signal $u_s(t)$, will appear voltages $u_1(t)$ and $u_2(t)$, equal to

$$u_1(t) = U_1 \cos\left[(\omega_{if} \pm \omega_d)(t - T_p - \phi_1\right], \tag{5.3.5}$$

$$u_2(t) = U_1 \cos\left[(\omega_{if} \pm \omega_d)t \quad \phi_1\right]. \tag{5.3.6}$$

Here $u_2(t)$ is a pulse, which acts on the IF output at any given moment of time, t, and $u_1(t)$ describes a pulse, which coexists with $u_2(t)$, but which is produced by the preceding radar transmitting signal.

The subtraction system produces the voltage

$$u_{ss}(t) = u_1(t) - u_2(t) = 2U_1 \; \sin \; [(\omega_{if}\pm\omega_d \; \frac{T_p}{2} \; \omega_d)\,(t - 0.5T_p) - \phi_1]\,.$$

$$(5.3.7)$$

Assuming that reference voltage $u_r(t)$ is equal to $u_r(t) = U_r \sin\omega_{if}t$, and the phase detector performs multiplication of the input signals and extraction of the voltage of the phase difference we obtain

$$u_{pd}(t) = k_{pd}\,U_1 \; \sin\,(\omega_{if}\pm\omega_d)\,\frac{T_p}{2}\,\cos\left[\pm\,\omega_d t - (\omega_{pd} \pm \omega_d)\,\frac{T_p}{2} - \phi_1\right].$$

Here k_{pd} is the gain of the phase detector. $$(5.3.8)$$

The above expression is sinusoidal within pulse duration τ_p and has frequency ω_d and amplitude equal to

$$U_{pd} = k_{pd}\,U_1 \; \cdot \; \sin\,(\omega_{if} \pm \omega_d)\,\frac{T_p}{2}\,.$$

$$(5.3.9)$$

But if the radar receiver input receives a signal from a stationary target with the same intensity as from a moving target, then $\omega_d = 0$ and

$$u_{pd}(t) = k_{pd}\,U_1 \sin\,\frac{\omega_{if}T_p}{2}\,\cos\left(\frac{\omega_{if}T_p}{2} - \phi_1\right).$$

$$(5.3.10)$$

Voltage $u_{pd}(t)$, determined by Equation (5.3.10), does not depend on time within the pulse width and vanishes when $0.5\omega_{if}T_p = \pi f_{if}T_p = k\pi$ ($k = 0$, $1, 2, ...$), where $f_{if} = \omega_{if}/2\pi$. The last condition means that complete cancellation of signals produced by stationary objects is possible only in cases when the intermediate frequency ω_{if} is a multiple of the transmitted pulse repetition period.

It was assumed in the derivation of Equations (5.3.8) and (5.3.10) that the DL delays the input signals by precisely their repetition period, T_p. This assumption usually is not valid in practice. High-quality clutter suppression is achieved when the input delay error $\Delta\tau_d$ does not exceed a few percent of period $T_{if} = 1/f_{if}$. If, for example, error $\Delta\tau_d = 0.06T_{if}$, $f_{if} = 30$ MHz and $T_p = 1,000$ μs, then $\Delta\tau_d/T_p$ should be $2 \cdot 10^{-6}$. In order to satisfy such a requirement and the equation $\pi f_{if}T_p = k\pi$ ($k = 0, 1, 2, ...$) it is necessary in practice to develop an extremely sophisticated cancellation system, equipped with precision systems for automatically controlling its parameters

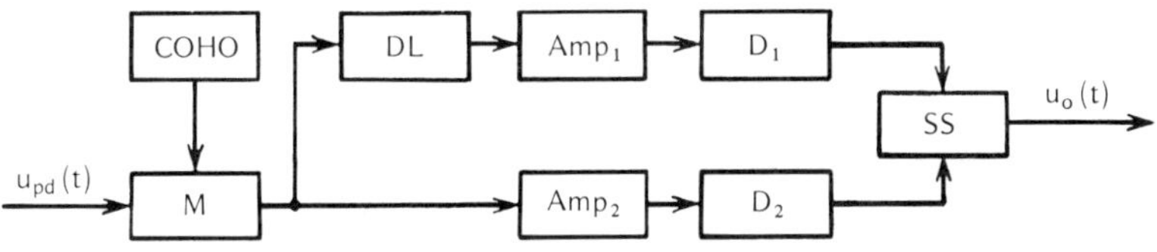

Figure 5.25

and the characteristics of the radar system. This limits the application of interperiod cancellation systems in the IF channel.

Interperiod video frequency interference cancellation systems are developed in accordance with Figure 5.25. Here voltage $u_{pd}(t)$, which is supplied to modulator (M), is generated by the phase detector. This detector receives reference voltage $u_r(t) = U_r \cos \omega_{if}t$, and signal $u_c(t)$ is produced at the IF output by radio waves reflected from a stationary object, equal to

$$u_c(t) = U_c \cos (\omega_{if} - \phi_c)$$

where U_c and ϕ_c are the amplitude and initial phase of voltage $u_c(t)$, then the phase detector will produce DC pulses

$$u_{pd}(t) = 0.5U_c k_{pd} \cos \phi_c \qquad (5.3.11)$$

with a constant $0.5U_c k_{pd} \cos \phi_c$, duration τ_p and repetition period T_p.

When signals are received from a moving target the IF generates a voltage, given by Equation (5.3.4). Therefore, the pulse

$$u_{pd}(t) = 0.5U_c k_{pd} \cos (\pm\omega_d t - \phi_1) \qquad (5.3.12)$$

with duration τ_p, is formed at the phase detector output.

If pulse duration τ_p is sufficiently small in comparison with $1/F_d = 2\pi/\omega_d$, then voltage $u_{pd}(t)$, produced by a moving target, consists of video pulses with different frequencies, which depend on ω_d and ϕ_1. The envelope of such a pulse is determined by the function $\cos (\pm\omega_d t - \phi_1)$. The approximate equation $\cos (\pm\omega_d t - \phi_1) \approx$ const is valid within τ_p, and the height of the pulse with the number n (n = 1, 2,) may be assumed equal to $0.5k_{pd}U_s \cos [\pm\omega_d(n - 1)T_p - \phi_1]$.

An external coherence radar has an amplitude detector instead of a phase detector. When it simultaneously receives signals from a moving target and from clutter, it generates amplitude-modulated pulses, and when the radar picks up pulses only from stationary clutter, it generates signals with constant amplitude. Radars with an amplitude detector are often called noncoherent.

Analysis of Equations (5.3.11) and (5.3.12) leads to the conclusion that by delaying pulses from the phase detector by time T_p, and by creating a difference between the delayed and undelayed pulses of the phase detector, it is possible to cancel signals which are generated in a radar system by reflections from stationary objects. Time T_p is a few hundreds and more microseconds.

Therefore, in order to delay the pulses it is necessary to use ultrasonic delay lines operating with carrier frequencies of 5-60 MHz [147]. Such signals are generated by modulator (M) in the diagram shown in Figure 5.25. Sinusoidal voltage $u_1(t)$ with a frequency of 5-60 MHz, produced by the COHO, is mixed with the input signal.

Voltage $u_1(t)$, amplitude-modulated in the general case by bipolar pulses, and consisting of bursts of sine waves with constant duration and repetition period T_p, is supplied to the ultrasonic DL and amplifier Amp_2. The DL delays the output signals by time T_p, and amplifiers Amp_1 and Amp_2 perform the same function as in Figure 5.23.

Detectors D_1 and D_2 convert the output voltages of amplifiers Amp_1 and Amp_2 to video pulses, which pass to the subtraction system SS. The latter, when all components perform perfectly, generates video pulses, which characterize moving targets, and produces zero output voltage when the radar receives signals from stationary objects. The SS output pulses are bipolar and their amplitude changes in time. If unipolar pulses are required for radar operation, the appropriate rectifier is installed after the SS.

Less rigid requirements are imposed on video frequency noise cancellation systems than on interperiod cancellation systems in the IF channels of receivers. The major requirements are the following [33, 147]:

• error $\Delta\tau_d$ in the time lag of a signal should be a small fraction of pulse duration (of the order of 1%), but not of period T_{if}, as in an IF cancellation system;

• a slight difference is permitted in the gains of systems that convert undelayed pulses and perform signal delay.

The second requirement is satisfied with an AGC system having as its input the SS output voltage and controlling the gain of amplifier, Amp_1; the first requirement is satisfied by using a controllable synchronizing pulse generator [147]. The latter generates periodic pulses whose generation times are controlled automatically by the SS voltage. These pulses trigger modulator (M) in Figure 5.25 and synchronize the operation of the entire radar.

According to published data [147], the uncancelled residue of fixed clutter may only be 1%.

The basic properties of the video frequency interperiod cancellation technique can be seen in greater detail by examining a cancellation system with a delay line as a linear filter. It is helpful to assume here, for simplicity, that pulses $u_{pd}(t)$ from the phase detector go to the input of delay line DL (Figure 5.25). Amplifier Amp_2 performs distortion-free transmission of undelayed pulses $u_{pd}(t)$, and DL, along with Amp_1, only delays $u_{pd}(t)$ by time T_p. Voltage $u_{pd}(t)$, in the case when signals are received from a target traveling at a constant radial velocity relative to the radar, is determined by Equation (5.3.12). Pulses $u_{pd}(t)$, delayed by DL and Amp_1, may be written as

$$u_{a1}(t) = u_{pd}(t - T_p) = 0.5k_{pd}\, U_s \cos\left[\pm\omega_d(t - T) - \phi_1\right]. \tag{5.3.13}$$

The subtraction system (SS) generates pulses

$$u_{ss}(t) = u_{a1}(t) - u_{pd}(t) = k_{pd}\, U_s \sin\left(\pm\frac{\omega_d T_p}{2}\right) \cos\left[\omega_d(t - 0.5T_p) - \phi_1\right]$$

$$\tag{5.3.14}$$

Comparison of Equations (5.3.12) and (5.3.14) shows that the amplitude ratio for voltages $u_{ss}(t)$ and u_{pd} is

$$q(\omega_d) = 2\left|\sin\left(\pm\frac{\omega_d T_p}{2}\right)\right|.$$

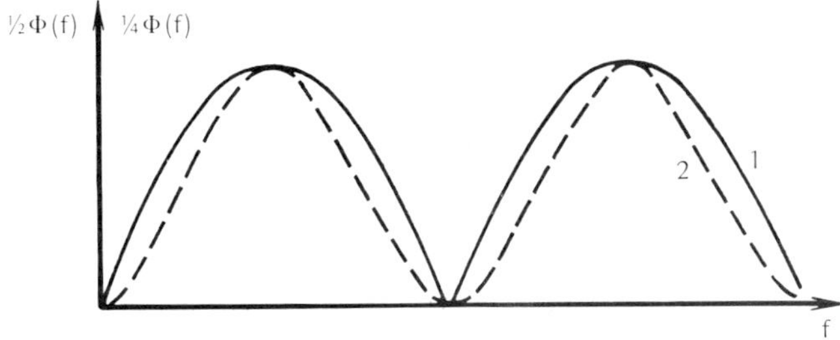

Figure 5.26

Substituting ω_d in this equation for the current frequency f, we obtain the amplitude-frequency characteristic

$$\phi_1(t) - 2\,|\sin \pi f T_p| \tag{5.3.15}$$

of a filter, which describes the selective properties of an interperiod clutter cancellation system. The function $\phi(f)/2 = |\sin \pi f T_p|$ is depicted graphically in Figure 5.26 (curve 1), from which it follows that the constant voltage and the harmonics of the pulse repetition frequency, equal to $1/T_p$, are not passed by this filter. However, nonfluctuating pulse signals of infinite duration, produced by radio waves reflected from stationary clutter, have a linear amplitude spectrum with frequency components at n/T_p, $n = 0, 1, 2, \ldots$. Pulses produced by moving targets are amplitude modulated and their spectral components are not suppressed. However, non-suppression of useful signals by an interperiod canceller does not always occur. This happens when a target travels relative to the radar at the so-called blind speed v_b. The latter is determined on the basis of the condition $f_d = (2v_b/\lambda) = (n/T_p)$ and is

$$v_b = n\lambda/2T_p . \tag{5.3.16}$$

Blind speeds limit the performance of radars with a moving target indication system, but there are ways to counteract this effect [134, 147].

When a radar operates in the scanning mode in a given sector of space, the receiver picks up bursts of pulses from small stationary objects. In view of the amplitude difference of adjacent pulses in a burst it is not possible to cancel completely the clutter from pulses on the edges of a burst. The reason for this, from the spectral viewpoint, is that the spectrum of a burst is not linear, and lower and upper side components appear near each harmonic of the pulse repetition frequency, with the result that continuous amplitude-frequency spectra are formed with central frequencies n/T_p, and with a width that increases as the number of pulses in the burst decreases. Therefore, a filter with the amplitude-frequency characteristic shown in Figure 5.26 (curve 1) cannot completely suppress signals from stationary objects.

The suppression of clutter whose spectrum is not purely linear can be improved by using a double interperiod canceller. Such a system consists of two series-connected systems, a diagram of which is shown in Figure 5.25.

When double interperiod cancellation is used the amplitude-frequency characteristic $\Phi(f)$ of the compensating filter acquires the form [167]

$$\Phi(f) = 4 \sin^2 \pi f T_p . \tag{5.3.17}$$

The graphical function $(1/4)\,\Phi(f) = \sin^2 \pi/T_p$, depicted in Figure 5.26 (curve 2), shows that this filter has a wider frequency range in which $\Phi(f) \approx 0$.

The double interperiod cancellation system, in spite of its obvious advantages, has two serious deficiencies: it is extremely complex in design, and the number of blind speeds increases when it is used, i.e., target detection conditions deteriorate in comparison with ordinary (single) interperiod cancellation.

The efficiency of single and double interperiod cancellation systems can be further improved by using negative feedback circuits [147, 167] and by developing systems with correlation feedback. Systems of the latter type are examined in reference [167] and are analogous to the interference cancellation systems with quadrature converters, analyzed in the preceding section.

Spectral analysis of the interperiod cancellation system shows that it is equivalent in terms of properties to a filter with a comb-shaped amplitude-frequency characteristic. In a "comb" with zero values near frequencies n/T_p, the closer to rectangular each tooth is, the more effective the MTI system. This equivalence provides opportunities for developing so-called filter moving target indication systems on the basis of resistors, capacitors and inductors, without using subtraction systems.

The quality of interperiod clutter cancellation systems for internal coherence radars is also influenced by the following basic factors:

● the frequency instability of the magnetron transmitter and of the STALO and COHO.

● the instabilities of the repetition frequency and duration of the transmitted pulses,

● fluctuations of the received radio signals, caused by the properties of the underlying surface (wave action, oscillations of the vegetation cover caused by wind, etc), and

● fluctuations of the received signals due to the scanning of the radar antenna.

A detailed analysis of the influence of these factors on the performance of radars with MTI systems is given in all modern radar textbooks and manuals and is not presented in this book in view of its limited scope; only the basic results of analysis, borrowed from references [33, 147], are given.

The efficiency of interperiod clutter cancellation systems is usually expressed in terms of the subclutter visibility factor [167]. For a virtually linear system this factor is usually defined as the number

$$k_{sv} = k_s^2 k_j \ , \tag{5.3.18}$$

where $k_s = U_{ss}/U_{pd}$ is the transfer coefficient of the canceller system, U_{ss} and U_{pd} are the amplitudes of the voltages produced by the subtraction system and phase detector when they receive signals from moving and stationary targets, respectively, $k_j = C_{in}/C_{out}$ is the clutter attenuation ratio of the canceller system, and C_{in} and C_{out} are the input and output clutter powers of the canceller system, respectively.

Sometimes the interperiod cancellation ratio determined by the relation

$$k_{ic} = u_{av}/U_{pd} \ , \tag{5.3.19}$$

where u_{av} is the average amplitude of the output pulse of the subtraction system, is also included in the analysis along with k_{sv}.

Analysis of the influence of frequency instabilities in a magnetron oscillator, stable local oscillator and coherent oscillator of an internal coherence radar with a coherent oscillator (Figure 5.24), using the interperiod cancellation ratio k_{ic}, yields the following results for video frequency clutter cancellation systems, in which unipolar pulses are generated after the subtraction system [147].

The STALO exhibits frequency instability Δf_s and the magnetron oscillator and COHO operate perfectly stably for time T_p, then $k_{ic} = 6Bf_sT_p$. Consequently, the toierable Δf_{st} of instability Δf_s should not exceed $k_{ic}/6T_p$, and, for example, when $T_p = 10^{-3}$ s and $k_{ic} = 0.01$ the inequality $\Delta f_{st} \leqslant 1.66$ Hz should be satisfied. These are exactly the requirements that are imposed on the frequency stability of the COHO. If only the magnetron oscillator performs unstably, then $k_{ic} = 4\Delta f_m T_p$, where Δf_m is the frequency drift of the magnetron oscillator in the pulse repetition period T_p. But when all oscillators are perfectly stable and the COHO is out of tune by Δf_{if}, then $k_{ic} = 4\Delta f_{if}T_p$.

Different results will be obtained when the design of the radar transceiver is altered. However, highly stable oscillators are nearly always required in internal coherence radar with MTI systems.

The instability of the repetition frequency $1/T_p$ of the transmitted pulses, by a fixed time lag, T_{lag}, leads to the appearance of uncancelled clutter residue during time $2\Delta\tau$, where $\Delta\tau = |T_p - T_{lag}|$. Therefore, when the phase detector generates rectangular pulses $k_{ic} = 2\Delta\tau/T_p$ and, for example, when $k_{ic} = 0.01$ and $\tau_p = 10^{-6}$ s, $\Delta\tau = 0.005$ μs is acceptable.

The instability of the transmitted pulse duration is determined in like manner. The phase and amplitude of signals from such stationary objects as a sea surface, vegetation cover, etc., fluctuate from pulse to pulse. This causes the spectrum to expand in the Doppler frequency domain and produces uncancelled clutter at the output of the cancelling system. If the performance of the latter in the case of single interperiod cancelling is estimated through the clutter attenuation ratio, then we obtain [147]

$$k_c = \frac{0.5}{1 - \exp(-\pi^2 f_0^2 / a_1 f_r^2)} , \tag{5.3.20}$$

where f_0 is the radar carrier frequency and a_1 is a parameter whose magnitude is determined by the kind of clutter. Thus, $a_1 = 3.9 \times 10^{19}$ for a sparsely wooded hill on a calm day, and $a_1 = 1.41 \times 10^{16}$ for a sea surface on a windy day.

Analysis of Equation (5.3.20) for different values of a_1 shows that k_c may reach several tens of decibels. Here the smallest values of k_c are obtained on rain clouds, and the largest are obtained on sparsely wooded hills on a calm day.

As the antenna beam scans through space, the total number of elementary clutter reflectors illuminated by the antenna remains practically constant from pulse to pulse, and their position changes relative to the antenna. Consequently, the incoming radar signals will fluctuate in amplitude and phase, which will cause the spectrum of the clutter signals to expand, leading to uncancelled clutter residue. As the number of pulses, n_p per beam width increases, the coefficient k_c increases. If single interperiod clutter cancellation is used, then as n_p changes from 5 to 100, the value of k_c for a Gaussian radiation pattern changes in the range of 10 to 55 dB [147]. A double interperiod canceller under otherwise identical conditions virtually doubles k_c (in decibels).

To eliminate the influence of scanning fluctuations of the signal, it is helpful to use the following scanning mode: the antenna beam remains stationary at each angle for a time long enough to receive the number of pulses required for target detection, and then it jumps to the next angle position. Another way of reducing uncancelled clutter residue with a scanning antenna, different from the method described above, is described in [147].

For external coherence radars with amplitude detectors, designed for converting IF signals to video pulses and for controlling the interperiod cancellation system, it is necessary, in addition to having amplifiers Amp_1 and Amp_2 with the required DL transfer functions (Figure 5.23), to satisfy the following requirement. During simultaneous reception of clutter and target signals, the detector output pulses will have an envelope that changes in time. This means that an ordinary amplitude detector in this case functions as a phase-sensitive element.

If the radar input picks up signals reflected only by a stationary object or only by a moving target, then, assuming the received signals have no amplitude fluctuations and the radar has an AGC system, the amplitude detector will generate pulses of constant amplitude, which are suppressed by the interperiod canceller. Consequently, a moving target flying through a cloudless sky (clouds produce clutter) may be lost by the radar. To prevent this the radar should be equipped with a clutter gate. The existence of clutter is established when the amplitudes of the pulses generated by the amplitude detector exceed a prescribed level for a certain period of time. When there is no clutter the canceller is turned off, and when clutter appears, it is turned on.

All of the above analysis was based on the fact that a radar with an MTI is stationary. If the radar is installed aboard a ship or an airplane, then the radar approaches a fixed clutter at a radial closing velocity. Consequently, the radio signal reflected by a fixed clutter acquires a Doppler frequency shift, and the stationary object may appear mobile. However, the Doppler frequency of a signal from a moving target is attributed both to its velocity and to the velocity of the radar. At the same time the Doppler frequency

shift of signals reflected by fixed clutter is determined only by the travel
speed of the radar and by its position relative to the clutter. This difference
is used to identify signals reflected by a clutter and to cancel them. The
greater the difference between the Doppler frequency of signals from moving
and fixed objects, the easier it is to solve the clutter cancellation problem.

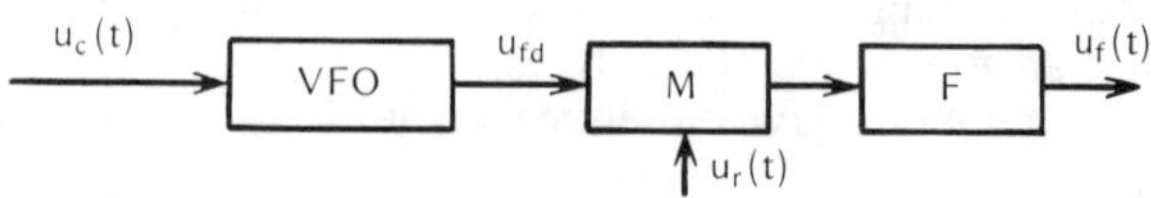

Figure 5.27

Before a mobile internal coherence radar can perform interperiod cancellation
of fixed clutter it must first compensate for the Doppler frequency shift of
the clutter signals. The solution of this problem in a radar with a COHO
amounts to changing its frequency in accordance with the radial closing
speed of the radar and clutter. To do this, the output voltage $u_r(t)$ of the
COHO is switched not to the phase detector, but to the system whose
block diagram is shown in Figure 5.27.

As can be seen in Figure 5.27, voltage $u_r(t)$ and the signal from the compensating Doppler frequency oscillator VFO reach the mixer (M) simultaneously. The compensating oscillator generates waves whose frequency is
changed under control voltage $u_c(t)$, which changes in time in such a way
that by sinusoidal voltage u_{fd} with Doppler frequency F_d, caused by the
relative motion of the radar and clutter appears on the VFO output.

The mixer, which multiplies voltages u_{fd} and $u_r(t)$, generates a signal, which
contains two components. One is sinusoidal with frequency $f_{if} + F_d$, and
the other with frequency $f_{if} - F_d$, where f_{if} is the intermediate frequency
of the radar receiver.

Filter F extracts the component of voltage $u_f(t)$ with the frequency $f_{if} + F_d$.
Consequently, the radar phase detector, which receives the IF output
voltages of the receiver and $u_f(t)$, generates pulses with amplitudes that remain
constant on signals from the clutter. The velocity of the radar itself relative
to the clutter, with the receiving antenna in a fixed position, is thereby
cancelled.

When a radar operates in the scanning mode the angle of its antenna changes,
which makes F_d a function of time, even when the radar and clutter have a
constant radial closing speed. This happens when the antenna has a comparatively wide radiation pattern and makes it difficult to cancel Doppler frequencies, caused by reflections from clutter when a mobile radar is used. It

follows from the above that it is necessary to generate $u_c(t)$ in consideration of the translatory velocity of the radar and of the angle of the antenna during the reception of clutter which is subjected to interperiod or filter cancellation.

If the external coherence principle is used in a mobile radar with an MTI system, then the conditions for cancellation of F_d are satisfied automatically without having to resort to Doppler frequency compensation. External coherence radar is preferable from this standpoint to internal coherence radar. However, external coherence radar requires that the signals reaching the receiver input from fixed clutter be sufficiently strong.

The radar oscillator instabilities and signal fluctuations described earlier also influence the performance of mobile radars, especially airborne systems. The performance of airborne radars with an MTI system is also influenced by oscillations of the aircraft around its center of mass, which are always of a fluctuating nature. These oscillations cause the spectrum of the received signals to expand, so that uncancelled clutter residue increases.

The strength of the output signals of MTI radar depends on the radar-target closing speed. When the envelope of the phase detector output pulses changes with frequency $f = 1/2T_p$, the output voltage of the canceller system is maximum, and when $f = 1/T_p$, it vanishes. This means that an MTI radar has worse target detection characteristics than a radar without canceller systems. In many cases, however, this deficiency may not be of deciding importance.

2. Interperiod Cancellation of Chaff

Radars may be acted upon, in addition to clutter from fixed objects, by chaff intentionally employed by an enemy, against which canceller or filter MTI systems are used. Because of the specific properties of chaff the resulting jamming immunity characteristics often differ from those determined for fixed clutter.

The main sources of passive jamming are chaff clouds, a feature of which is that they cannot be considered stationary. Individual reflectors in a cloud move randomly relative to each other due to atmosphereic turbulence and the cloud itself is propelled by the wind as a single body.

Random oscillations of individual reflectors produce amplitude and phase fluctuations, and the translatory motion of the cloud gives rise to a Doppler frequency shift in the reflected signals.

The pulse envelopes turn out to be different because of the fluctuations. Consequently, clutter residue occurs, the strength of which increases as the spectrum of the clutter signals broadens.

In internal coherence radar, the Doppler frequency produced by the translatory motion of a chaff cloud must be compensated by regulating the frequency of the COHO in consideration of the speed of the radar and wind velocity. Assuming that such a compensation system performs ideally, we shall examine the influence of amplitude and phase fluctuations of a chaff cloud on an internal coherence radar with an MTI system [24, 167].

If the transmitted radar signal is monochromatic, then, after being reflected by a chaff cloud, it generates the voltage

$$u_c(t) = U_c(t) \cos \left[\omega_0 t + \phi_c(t) \right] \tag{5.3.21}$$

in the receiver output circuits. Here $U_c(t)$ and $\phi_c(t)$ are the random envelope and phase, which change slowly in comparison with $\omega_0 t$, and ω_0 is the angular carrier frequency.

The efficiency of the MTI system is estimated in terms of the sub clutter visibility factor k_{sv}, determined by Equation (5.3.18).

The signal transmission coefficient k_s is characterized, obviously, by Equation (5.3.15), in which $\Phi_1(f)$ is replaced by k_s and f by $F_{d\,r}$, where $F_{d\,r} = = F_d - F_{d\,c}$, $F_d = \omega_d/2\pi$, $F_{d\,c}$ is the Doppler frequency produced by the motion of the radar and wind velocity. The mathematical expectation of voltage $u_0(t)$, produced by the phase detector of the radar when it is acted upon by passive jamming, is equal to zero. Therefore, assuming that $u_0(t)$ is an approximately stationary random signal, we obtain the chaff cancellation ratio:

$$k_c = \frac{\sigma_0^2}{2[\sigma_0^2 - R_0(T_p)]} ,$$

where $R_0(T_p) = \sigma_0^2 \rho(T_p)$ and σ_0^2 are the correlation function and the variance of voltages, $u_0(t)$ and $u_0(t - T_p)$, respectively, and $\rho(T_p)$ is the pulse-to-pulse correlation coefficient for chaff at the phase detector output. Therefore,

$$k_c = \frac{1}{2[1 - \rho(T_p)]} \tag{5.3.22}$$

and consequently,

$$k_{sv} = \frac{2\sin^2 \pi F_{dr} T_p}{1 - \rho(T_p)} . \tag{5.3.23}$$

If we determine the coefficient k_{sv} for a radar with double canceller MTI in the same manner, we obtain

$$k'_{sv} = \frac{8 \sin^4 \pi F_{dr} T_p}{3 - 4\rho(T_p) + \rho(2T_p)} . \tag{5.3.24}$$

In calculations the function $G_0(f)$, which describes the spectral density of voltage $u_0(t)$ for $f > 0$, is often determined as

$$G_c(F) = G_{co}\, \exp(-f^2/2\sigma_f^2)$$

where σ_f is the mean square scatter of Doppler frequencies (one-half of the band of the mean square energy spectrum of passive jamming $u_c(t)$ at the 0.61 level). Then

$$\rho(T_p) = \frac{\displaystyle\int_0^\infty G_c(f)\cos 2\pi f T_p\, df}{\displaystyle\int_0^\infty G_c(f)\, df} = \exp\left(-2\pi^2\,\sigma_f^2\, T_p^2\right).$$

Under this condition

$$k_{sv} = \frac{2\sin^2 \pi F_{dr} T_p}{1 - \exp(-2\pi^2\,\sigma_f^2\, T_p^2)}, \tag{5.3.25}$$

$$k'_{sv} = \frac{8\sin^4 \pi F_{dr} T_p}{3 - 4\exp(-2\pi^2\,\sigma_f^2 T_p^2) + \exp(-4\pi^2\,\sigma_f^2\, T_p^2)}. \tag{5.3.26}$$

It follows from Equations (5.3.25) and (5.3.26) that the clutter visibility factor depends on frequency $F_{d\,r}$, which is related to the velocities of the target, radar system and wind, and also on $\sigma_f = 2\sigma_v/\lambda$, where σ_v is the mean square scatter of the radial velocities of dipole reflectors. Here, the larger σ_f is, the smaller k_{sv} and k'_{sv}. Comparison of Equations (5.3.25) and (5.3.26) shows that double canceller MTI suppresses clutter better than the single version. It is also important to consider blind velocities, for which k_{sv} and k_{sv} are equal to zero. Controlling blind speeds, which can be done by a variety of methods, under passive jamming conditions is an extremely urgent problem.

It is possible to increase k_{sv} and k'_{sv} for given values of $F_{d\,r}$ and σ_f by shortening period T_p. Under actual conditions, therefore, it may be advantageous to use radar systems with high pulse repetition frequencies, in spite of the fact that such a measure violates the conditions of unambiguous determination of target range.

Passive jamming suppression can be made even more effective by using cancellation systems with correlation feedback [167]. These systems are similar to quadrature jamming which suppress jamming from the side lobes of the receiving antenna. The basic properties of such cancellers can be determined on the basis of analysis of a system described in reference [167]. This analysis can be performed in accordance with the procedure examined in the previous section.

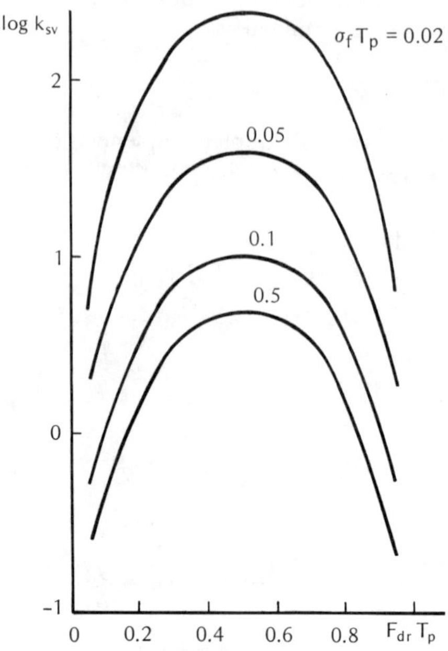

Figure 5.28

The dependence of $\log k_{sv}$ on $F_{d\,r}T_p$ is illustrated graphically in Figure 5.28. Analysis of Figure 5.28 shows that for small $\sigma_f T_p (\sigma_f T_p \ll 1)$ and $F_{d\,r}T_p \approx 0.5$ the clutter visibility factor may reach a few hundreds. But if $\sigma_f T_p > 0.1$ (for example when $T_p = 1,000\ \mu s$, which corresponds to $\sigma_f > 100$ Hz), then k_{sv} does not exceed a few tens.

We should like to mention in conclusion that the formulas derived above for k_{sv} and k'_{sv} are approximate for at least two reasons. First, the combined action of clutter and useful signals on radar systems was ignored, and second, the voltage produced by chaff is assumed to be stationary.

In reality, chaff produces virtually stationary random voltages in a radar receiver only after all chaff packages have been opened for a considerable period of time. During the opening, or blooming process, the clouds are small and the slip stream of the jamming vehicle (an airplane for example) has a considerable influence on the statistical characteristics of the signals.

As the altitude of the passive jamming vehicle increases their spectrum broadens, as a rule, because of the increasing spread of wind velocities due to stronger atmospheric turbulence.

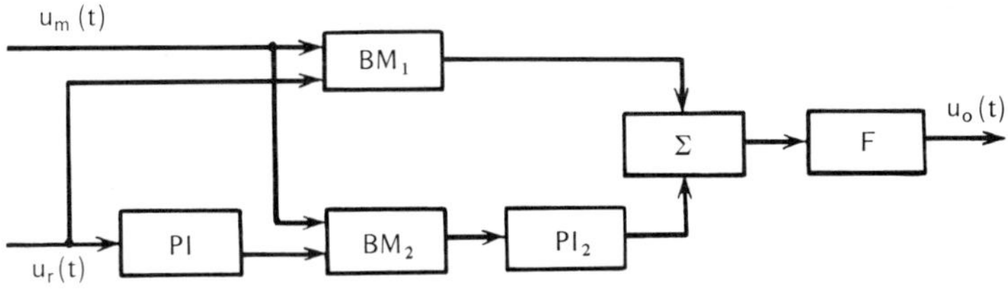

Figure 5.29

5.4 JAMMING CANCELLATION WITH DUAL BALANCED MIXERS

1. Design of Dual Balanced Mixer and Principle of Jamming Cancellation

A dual balanced mixer, a functional diagram of which is illustrated in Figure 5.29, is essentially a single sideband detection system, which our compatriot Ye. G. Momot proposed in 1937 [115].

This mixer controls wideband noise and is used extensively in radar systems with continuous monochromatic and frequency-modulated transmitting signals [30, 116, 147]. It contains balanced mixers BM$_1$, BM$_2$, phase inverters PI$_1$, PI$_2$, adder Σ (or subtractor SS) and filter F, which extracts the Doppler frequency signal.

Balanced mixer BM$_1$ receives mixture $u_m(t) = u_s(t) + u_j(t)$ of useful signal $u_s(t)$ and jamming $u_j(t)$, produced at the IF ω_{if}, which describes the signal radiated by a radar transmitter. The input voltage of BM$_2$ is mixture $u_m(t)$ and signal $u_{r1}(t)$, generated as a result of the passage of $u_r(t)$ through PI$_1$. Phase inverters PI$_1$ and PI$_2$ change the phases of the incoming signals by -0.5π.

A single balanced mixer differs from a dual balanced mixer in that it has no PI$_1$, BM$_2$, PI$_2$ and Σ (or SS).

We shall examine the jamming cancellation principle of a dual balanced mixer in an example of a radar system that radiates a monochromatic signal, reflected by a target which approaches the radar at a constant radial velocity. Under these conditions

$$u_m(t) = U_s \cos(\omega_{if} t + \omega_d + \psi) + u_j(t), \tag{5.4.1}$$

$$u_r(t) = U_r \cos \omega_{if} t, \tag{5.4.2}$$

where U_s and U_r are the amplitudes of the useful and reference voltages; ϕ is the initial phase.

To clarify the presentation we shall assume spectral density $G_j(\omega)$ of jamming $u_j(t)$, acting on the IF output, to be constant within the IF passband, and we write the instantaneous voltage $u_j(t)$ in the following form:

$$u_j(t) = U_j\left\{\sum_{i=1}^{0.5n} \cos\left[(\omega_{if}t + i\delta)t + \psi_1\right] + \sum_{i=1}^{0.5n} \cos\left[(\omega_{if} - i\delta)t + \phi_i\right]\right\}$$

$$(5.4.3)$$

Here $n = \Delta\omega/\delta$, $\Delta\omega$ is the IF passband, δ is the frequency interval between adjacent harmonic components of voltage $u_j(t)$, U_j is the amplitude of the harmonic component of voltage $u_j(t)$, and ψ_i and ϕ_i are the initial phases.

Since the voltage variance of the jamming is $\sigma_j^2 = G_j(\omega)\Delta\omega$, and, as follows from Equation (5.4.3), $\sigma_j^2 = (n/2)U_j$, U_j is

$$U_j = \sqrt{2G_j(\omega)\delta} \qquad\qquad (5.4.4)$$

Hence U_j essentially characterizes the intensity of the jamming in frequency band δ.

Balanced mixers BM_1 and BM_2 perform multiplication of the input signals and generate a voltage, the phase of which is equal to the phase difference of the inputs. Consequently, we will have at the output of BM_1, the gain of which is k_{b1},

$$u_1(t) = k_{b1}\left[U_s\cos(\omega_d t + \phi) + U_j\sum_{i=1}^{0.5n}\cos(i\delta t + \psi_i) \right.$$
$$\left. + U_j\sum_{i=1}^{0.5n}\cos(i\delta t - \phi_i)\right] \quad . \qquad (5.4.5)$$

Phase inverter PI_1, which, like PI_2, is considered to be ideal, generates the voltage

$$u_{r1}(t) = U_r\sin\omega_{if}t.$$

Therefore, for voltage $u_2(t)$, generated by balanced mixer BM_2 with gain $k_{b2} = k_{b1}$, we obtain

$$u_2(t) = -k_{b1}\left[U_s\sin(\omega_d t + \phi) + \sum_{i=1}^{0.5n}U_j\sin(i\delta t + \psi_i) \right.$$
$$\left. - \sum_{i=1}^{0.5n}U_j\sin(i\delta t - \phi_i)\right]. \qquad (5.4.6)$$

Consequently, the output voltage $u_d(t)$ of phase inverter PI_2 will be

$$u_d(t) = k_{b1}\left[U_s\cos(\omega_d t + \phi) + U_j\sum_{i=1}^{0.5n}\cos(i\delta t + \psi_i) \right.$$
$$\left. - U_j\sum_{i=1}^{0.5n}\cos(i\delta t - \phi_i)\right].$$

The output signal of adder Σ is

$$u_\Sigma(t) = u_1(t) + u_d(t) = 2kb_1 \left[U_s\cos(\omega_d t + \phi) \right.$$
$$\left. + U_j \sum_{i=1}^{0.5n} \cos(i\delta t + \psi_i) \right] . \tag{5.4.8}$$

Analysis of this expression shows that jamming components with frequencies $\omega_{if} - i\delta$ are cancelled by the dual balanced mixer. But if a subtractor were used instead of an adder, then noise components with frequencies $\omega_{if} + i\delta$ and the useful signal of the Doppler frequency along with them would be cancelled.

To compare the noise immunity of dual balanced and single balanced mixers, we shall find the effective output signal-to-noise ratio of each of them.

The effective useful signal power $u_\Sigma(t)$, after the signal passes through the filter, is proportional to

$$\sigma_{sd}^2 = 2k_f^2 K_{b1}^2 U_s^2 , \tag{5.4.9}$$

where k_f is the gain of the filter, for which the amplitude-frequency characteristic $\Phi_f(\omega)$ for $0 \leqslant \omega_f \leqslant \Delta\omega_f$ is determined by $\Phi_f(\omega) = k_f$, and $\Delta\omega_f$ is the passband of the filter.

The output noise power of filter F is a linear function of the variance σ_{jd}^2 of the noise voltage $u_{jf}(t)$, generated by that filter. In consideration of passbands $\Delta\omega$ and $\Delta\omega_f$, in order to determine $u_{jf}(t)$ in accordance with Equation (5.4.8) it is necessary to consider changes of i not to 0.5n, but to $n_1 = \Delta\omega_f/\delta$; it is assumed here that δ is the value by which $\Delta\omega$ and $\Delta\omega_f$ are divisible without a remainder. Therefore,

$$\sigma_{jd}^2 = 2k_f^2 k_{b1}^2 n_1 U_j^2$$

Substituting into this expression the values of U_j^2 and n_1 we obtain

$$\sigma_{jd}^2 = 4k_f^2 k_{b1}^2 G_j(\omega) \Delta\omega_f \tag{5.4.10}$$

Consequently,

$$q_d = \frac{\sigma_{sd}^2}{\sigma_{jd}^2} = \frac{U_s^2}{2G_j(\omega)\Delta\omega_f} . \tag{5.4.11}$$

The effective useful signal power in a single balanced mixer is determined by Equation (5.4.5) and is proportional to

$$\sigma_s^2 = (1/2)k_f^2 k_{b1}^2 U_s^2 \tag{5.4.12}$$

The filter output noise voltage variance σ_j^2 of the same single balanced mixer, also calculated by Equation (5.4.5), is

$$\sigma_j^2 = k_f^2 k_{b1}^2 \, n_1 \, U_j^2 = 2k_f^2 k_{k1}^2 \, G_j(\omega) \, \Delta\omega_f \; . \tag{5.4.13}$$

Therefore,

$$q_0 = \frac{\sigma_s^2}{\sigma_j^2} = \frac{U_s^2}{4G_j(\omega)\Delta\omega_f} \; . \tag{5.4.14}$$

Comparison of Equations (5.4.11) and (5.4.14) shows that the noise immunity of the dual balanced mixer in relation to the signal-to-noise ratio is twice that of the single balanced mixer due to the cancellation of jamming with a uniform spectral density.

Under actual conditions, the spectral density of the IF output jamming voltage of a radar receiver cannot usually be assumed to be uniform, and a radar transmitter often radiates an FM signal. Under these conditions approximate analysis of the influence of jamming is insufficient, and more exact methods of analyzing the noise immunity of single balanced and dual balanced mixers must be employed. In the ensuring analysis, therefore, we shall compare the noise immunity of these two kinds of mixers on the assumption that the radar transmitter generates voltage $u_t(t)$ with double frequency modulation, i.e.,

$$u_t(t) = U \cos(\omega_0 t + \beta_1 \sin \Omega_1 t + \beta_2 \sin \Omega_2 t), \tag{5.4.15}$$

where U and ω_0 are the amplitude and carrier frequency of voltage $u_t(t)$, β_1 and β_2 are the frequency modulation indices, and Ω_1 and Ω_2 ($\Omega_2 \gg \Omega_1$) are the frequencies of the two sinusoidal modulating voltages.

By analyzing the influence of jamming on single balanced and dual balanced mixers, and in consideration of the signal in Equation (5.4.15), it is possible to obtain the necessary results for the same mixers for radar systems in which the carrier is frequency-modulated by one sinusoidal signal ($\beta_2 = 0$ or $\beta_1 = 0$) and in radar systems without frequency modulation ($\beta_1 = \beta_2 = 0$).

2. Influence of Noise on Single Balanced and Dual Balanced Mixers in Radar with Continuous Double Frequency-Modulated Transmission

General Premises

To solve this problem we shall proceed from the results published in [109] and from the following assumptions:

- signals $u_s(t)$ and $u_r(t)$ do not fluctuate;

- interference $u_j(t)$ is produced by white noise at the radar receiver input and consists of a narrow band stationary noise voltage with zero mean and with a known correlation function;

- each balanced mixer multiplies the incoming signals and extracts from the product the component whose phase is equal to the phase difference of the inputs;

- phase inverters PI_1 and PI_2 do not alter the input signal level and only change the phase of each harmonic component -0.5π.

The above assumptions simplify the solution of the problem while taking no consideration the basic characteristics of actual systems.

It may be assumed, as a rule, that intermediate and modulating frequencies ω_{if}, Ω_1 and Ω_2 are harmonically related. Under this assumption voltage $u_r(t)$ is periodic.

Periodic nonstationary random processes take place in balanced mixers due to the multiplication of periodic reference signals by the jamming voltage. Their initial phases usually are unimportant for practical purposes. Therefore, time-averaging operations may be performed to determine the correlation functions for such processes [48].

Variances $\sigma_{n\Sigma}^2$ and $\sigma_{n\Delta}^2$, proportional to the effective fluctuation powers of noises $u_{n\Sigma}(t)$ and $u_{n\Delta}(t)$ which are present in the output voltage, $u_o(t)$, of a linear filter, and which appear when adder Σ and subtractor (SS), respectively, are used in a dual balanced mixer, are comparatively easy to determine if the time-average correlation functions $R_\Sigma(\tau)$ and $R_\Delta(\tau)$ of noises $u_\Sigma(t)$ and $u_\Delta(t)$ at the outputs of Σ and SS are known.

Functions $R_\Sigma(\tau)$ and $R_\Delta(\tau)$, on the assumption that the gains of adder Σ and the subtracting system are equal to unity, are determined by the following relation:

$$R_i(\tau) = R_1(\tau) + R_{f_2}(\tau) + (-1)^{i-1}\left[R_{1f}(\tau) + R_{f_2 1}(\tau)\right]. \qquad (5.4.16)$$

Here $R_i(\tau) = R_\Sigma(\tau)$ for $i = 1$, $R_i(\tau) = R_\Delta(\tau)$ for $i = 2$ and $R_1(\tau)$ and $R_{f_2}(\tau)$ are the time-averagecorrelation functions of random voltages $u_{j_1}(t)$ and $u_{j_2}(t)$ at the outputs of balanced mixer BM_1 and phase inverter PI_2, respectively; $R_{1f_2}(\tau)$ and $R_{f_2 1}(\tau)$ are the time-averaged cross correlation functions of the same random voltages.

The function $R_1(\tau)$ is found by analyzing processes which take place in balanced mixer BM_1, and in order to determine $R_{f_2}(\tau)$, $R_{1f_2}(\tau)$ and $R_{f_2 1}(\tau)$ it is necessary to analyze the response of phase inverter PI_2 to signal $u_{j_2}(t)$, generated by balanced mixer BM_2.

It may be assumed, in consideration of assumptions concerning the performance of phase inverters PI_1 and PI_2 (see above) that the functions $u_{jf_2}(t)$ and $u_{j_2}(t)$ are determined by the Hilbert transform which, as is known [99], describes the performance of a perfect phase inverter.

However, the correlation functions of the initial random process and its Hilbert transform are equal [55]. Therefore

$$R_{f_2}(\tau) = R_2(\tau), \qquad\qquad (5.4.17)$$

where $R_2(\tau)$ is the time-averaged correlation function of voltage $u_{j2}(t)$.

Assuming that $u_{jf_2}(t)$ is the Hilbert transform of voltages $u_{j1}(t)$, we obtain [55]

$$R_{jf_2}(\pm\tau) = \frac{1}{\pi} \int_{-\infty}^{\infty} \frac{R_{1,2}(\pm\beta)}{\tau - \beta}\, d\beta, \qquad\qquad (5.4.18)$$

where $R_{1,2}(\beta)$ is the time-averaged crosscorrelation function of voltages $u_{j1}(t)$ and $u_{j2}(t)$.

It follows from Equation (5.4.18) that $R_{1f_2}(\tau)$, for all τ, is the Hilbert transform of the crosscorrelation function of voltages $u_{j1}(t)$ and $u_{j2}(t)$. In consideration of the basic properties of the crosscorrelation functions, $R_{f_2 1}(\pm\tau) = R_{1 f_2}(\pm\tau)$. Therefore, as follows from Equations (5.4.16) through (5.4.18), in order to calculate $R_\Sigma(\tau)$ and $R_\Delta(\tau)$ it is necessary to know the correlation functions $R_1(\tau)$ and $R_2(\tau)$ and to determine the crosscorrelation function $R_{1 f_2}(\tau)$ for various values of τ.

To find the squares of the effective values, which are proportional to the effective powers, for the useful voltages contained in signal $u_o(t)$ and generated by a dual balanced mixer with an adder or subtraction system, it is necessary to analyze the response to $u_s(t)$ of BM_1, BM_2, PI_2, Σ (or SS) and F. In this case, the useful output voltage of phase inverter PI_2 should be the Hilbert transform for the useful signal from balanced mixer BM_2.

Evaluation of the influence of jamming on a single balanced mixer amounts to excluding phase inverters PI_1, PI_2, balanced mixer BM_2, and the adder from the diagram in Figure 5.29 and finding the variance of voltage $u_{j1}(t)$ at the output of the filter and the square of the effective useful signal.

Assuming that the IF of a radar receiver performs quasistatic conversion of the useful signal and that the radial target-radar closing rate is constant, and ignoring the usually unimportant initial phase of voltage $u_s(t)$, we obtain

$$u_m(t) = U_s \cos\left[\omega_{if} t + \omega_d t + \beta_1 \sin\Omega_1(t - \tau_d) + \beta_2 \sin\Omega_2(t - \tau_d) + \phi_r\right]$$
$$+ U_j(t)\cos\left[\omega_{if}t - \phi_j(t)\right]. \qquad\qquad (5.4.19)$$

Here U_s is the amplitude of voltage $u_s(t)$, τ_d is the range delay time, ϕ_r is the phase, which depends on the range r between the radar and target, and $U_j(t)$ and $\phi_j(t)$ are the envelope and phase of random jamming voltage $u_j(t)$.

At the same time, reference voltage $u_r(t)$ is the analog of signal $u_t(t)$ at intermediate frequency and is

$$u_r(t) = U_r \cos (\omega_{if}\, t + \beta_1 \sin \Omega_1 t + \beta_2 \sin \Omega_2 t), \qquad (5.4.20)$$

where U_r is the amplitude of voltage $u_r(t)$.

Influence of Jamming on Dual Balanced Mixer

By analyzing the interaction of signals $u_m(t)$ and $u_r(t)$ in balanced mixer BM_1 (Figure 5.29), and assuming that the components with frequencies close to $2\omega_{if}$ may be ignored, we find the output voltage of BM_1:

$$u_1(t) = u_{s2}(t) + u_{j2}(t), \qquad (5.4.21)$$

where

$$u_{s1}(t) = U_{s1} \cos \left[\omega_d t - 2\beta_1 \sin \frac{\Omega_1 \tau_d}{2} \times \cos (\Omega_1 t - 0.5\Omega_1 \tau) \right.$$

$$\left. - 2\beta_2 \sin \frac{\Omega_2 \tau_d}{2} \times \cos (\Omega_2 t - 0.5\Omega_2 \tau_d) + \phi_r \right] \qquad (5.4.22)$$

is the useful signal;

$$u_{j1}(t) = k_{b1} U_j(t) \cos [\beta_1 \sin \Omega_1 t + \beta_2 \sin \Omega_2 t + \phi_j(t)] \qquad (5.4.23)$$

is jamming.

$$U_{s1} = k_{b1} U_s$$

Balanced mixer BM_2 receives voltages $u_m(t) = u_s(t) + u_j(t)$ and $u_{r1}(t)$, where $u_{r1}(t)$ is the Hilbert transform of the signal $u_r(t)$.

Because of the fact that ω_{if} is substantially larger than the passband of the system in which $u_r(t)$ is generated, the Hilbert transfrom for $u_r(t)$ is, for practical purposes, equivalent to a change of the argument of the trigonometric function $\cos (\omega_{if}t + \beta_1 \sin \Omega_1 t + \beta_2 \sin \Omega_2 t)$ in Equation (5.4.20) by -0.5π. This can be proved by comparatively simple calculations, by first expanding the function $u_r(t)$ into a Fourier series and then shifting the phase of each harmonic component by 0.5π.

The interaction of signals $u_m(t)$ and $u_{r1}(t)$ in balanced mixer BM_2 leads to the formation at its output of the useful signal jamming mixture

$$u_2(t) = u_{s2}(t) + u_{j2}(t), \qquad (5.4.24)$$

where

$$u_{S2}(t) = - U_{S2} \sin \left[\omega_d t - 2\beta_1 \sin \frac{\Omega_1 \tau_d}{2} \cos (\Omega_1 t - 0.5\Omega_1 \tau_d) \right.$$

$$\left. - 2\beta_2 \sin \frac{\Omega_2 \tau_d}{2} \cos (\Omega_2 t - 0.5\Omega_2 \tau_d) + \phi_r \right]$$

(5.4.25)

is the useful signal;

$$u_{j_2}(t) = k_{b_2} U_j(t) \sin [\beta_1 \sin \Omega_1 t + \beta_2 \sin \Omega_2 t + \phi_j(t)] \qquad (5.4.26)$$

is the interference;

$$U_{S2} = k_{b_2} U_s$$

We now examine the response to the useful and jamming voltages separately, which is permissible in view of the linearity of all the elements illustrated in Figure 5.29.

After simple transformations Equations (5.4.22) and (5.4.25) may be written as follows [109] :

$$u_{S1}(t) = U_{S1} \sum_{\ell=0}^{\ell_1} a_\ell J_\ell \left(2\beta_2 \sin \frac{\Omega_2 \tau_d}{2} \right)$$

$$\times \cos \left[\omega_d t - 2\beta_1 \sin \frac{\Omega_1 \tau_d}{2} \cos (\Omega_1 t - 0.5\Omega_1 \tau_d) \right.$$

$$\left. - \frac{\ell\pi}{2} + \phi_r \right] \cos \ell \; (\Omega_2 t - 0.5\Omega_2 \tau_d), \qquad (5.4.27)$$

$$u_{S2}(t) = U_{S2} \sum_{\ell=0}^{\ell_1} a_\ell J_\ell \left(2\beta_2 \sin \frac{\Omega_2 \tau_d}{2} \right)$$

$$\times \sin \left[\omega_d t - 2\beta_1 \sin \frac{\Omega_1 \tau_d}{2} \cos (\Omega_1 t - 0.5\Omega_1 \tau_d) \right.$$

$$\left. - \frac{\ell\pi}{2} + \pi + \phi_r \right] \cos \ell \; (\Omega_2 t - 0.5\Omega_2 \tau_d), \qquad (5.4.28)$$

where $J_\ell(x)$ is an ℓ- order Bessel function; $a_\ell = 1$ for $\ell = 0$ and $a_\ell = 2$ for $\ell \neq 0$.

Only the maximum $\ell = \ell_1$, corresponding to components with frequencies equal to the IF passband, need be taken into consideration in Equations (5.4.27) and (5.4.28).

Analysis of Equations (5.4.27) and (5.4.28) shows that each voltage $u_{S1}(t)$ and $u_{S2}(t)$ is determined by two groups of terms. The first groups in Equations (5.4.27) and (5.4.28) are found for $\ell = 0$ and describe frequency-modulated (FM) waves with angular carrier frequency ω_d. In this case, a radar system is always designed so as to satisfy the inequality $\omega_d > \Omega_1$. The second groups of terms in Equations (5.4.27) and (5.4.28) are obtained for $\ell \neq 0$ and represent combinations ℓ_1 of amplitude-modulated waves with the suppressed carrier frequencies $\ell\Omega_2$ ($\Omega_2 \gg \Omega_1$, $\Omega_2 > \omega_d$), i.e., balanced amplitude modulated waves. Here FM waves with carrier frequencies ω_d are the modulating waves.

In FM and AM systems the frequency bands occupied by components with most of the power are usually much smaller than the carriers themselves. Under these conditions, as is shown by comparatively simple calculations, the Hilbert transform of the voltage $u_{S2}(t)$, performed by phase inverter PI_2, is equivalent with a high degree of accuracy to the shifting of the phases of components with the carrier frequencies ω_d and $\ell\Omega_2$ ($\ell = 1$, 2, . . .) in Equation (5.4.28) by -0.5π. Therefore, on the basis of Equation (5.4.28), voltage $u_{f2}(t)$, generated at the output of phase inverter PI_2 and characterizing the useful signal, is calculated without any particular difficulty.

By knowing $u_{S1}(t)$ and determining $u_{f2}(t)$, we find voltage $u_{S\Sigma}(t) = u_{S1}(t)$ + $u_{f2}(t)$ and difference $u_{S\Delta}(t) = u_{S1}(t) - u_{f2}(t)$.

Analysis of the results of addition and subtraction of voltages $u_{S1}(t)$ and $u_{f2}(t)$ shows that when an adder is used in a dual balanced mixer, the components that are grouped near the frequencies $\ell\Omega_2 - \omega_d$ ($\ell = 1, 2,$ . . .) are cancelled, and when a subtraction system is used in the same converter components concentrated near frequencies $\ell\Omega_2 + \omega_d$ ($\ell = 0,$ 1, 2, . . .) are cancelled. Consequently, the noise reduction performance of the filter, which is capable of extracting from $u_{S\Sigma}(t)$ the FM voltage $u_{\Sigma\ell}(t)$ with carrier frequency $\ell\Omega_2 + \Omega_d$ ($\ell = 0, 1, 2,$ . . .), and from $u_{S\Delta}(t)$ an FM signal with carrier $\ell\Omega_2 + \omega_d$ ($\ell = 1, 2,$ . . .), are improved.

From the relations characterizing $u_{S\Sigma}(t)$ and $u_{S\Delta}(t)$, we find

$$u_{\Sigma\ell}(t) = k_{\Sigma\ell}(U_{S1} + U_{S_2})J_\ell\left(2\beta_2 \sin\cdot\frac{\Omega_2\tau_d}{2}\right)$$

$$\times \cos\left[(\ell\Omega_2 + \omega_d t) - 2\beta_1 \sin\frac{\Omega_1\tau_d}{2}\right. \tag{5.4.29}$$

$$\left.\times \cos\Omega_1(t-0.5\,\tau_d) - 0.5\ell\pi - 0.5\ell\Omega_2\tau_d) + \phi_r\right] \quad \text{for } \ell = 0, 1, 2, \ldots,$$

$$u_{\triangle\ell}(t) = k_{\triangle\ell}(U_{S_1} + U_{S_2})$$

$$\times J_\ell\left(2\beta_2 \sin\frac{\Omega_2\tau_d}{2}\right)\cos\Bigg[(\ell\Omega_2 - \omega_d)t$$

$$+ 2\beta_1 \sin\frac{\Omega_1\tau_d}{2}\cos\Omega_1(t-0.5\tau_d) + 0.5\ell\pi$$

$$- 0.5\ell\Omega_2\tau_d - \phi_r\Bigg] \quad \text{for } \ell = 1, 2, \ldots \tag{5.4.30}$$

Here $k_{\Sigma\ell}$ and $k_{\triangle\ell}$ are the gains of the filter in the given frequency range for voltages $u_{\Sigma\ell}(t)$ and $u_{\triangle\ell}(t)$, respectively.

It follows from Equation (5.4.29) and (5.4.30) that the squares of the effective values of $\sigma_{\Sigma\ell}^2$ and $\sigma_{\triangle\ell}^2$ of voltages $u_{\Sigma\ell}(t)$ and $u_{\triangle\ell}(t)$ are

$$\sigma_{\Sigma\ell}^2 = \frac{1}{2}k_{\Sigma\ell}^2(U_{S_1} + U_{S_2})^2 J_\ell^2\left(2\beta_2 \sin\frac{\Omega_2\tau_d}{2}\right)$$

$$\text{for } \ell = 0, 1, 2, \ldots, \tag{5.4.31}$$

$$\sigma_{\triangle\ell}^2 = \frac{1}{2}k_{\triangle\ell}^2(U_{S_1} + U_{S_2})^2 J_\ell^2\left(2\beta_2 \sin\frac{\Omega_2\tau_d}{2}\right)$$

$$\text{for } \ell = 1, 2, \ldots. \tag{5.4.32}$$

We now determine $R_1(\tau)$, $R_2(\tau)$, and $R_{12}(\tau)$. On the basis of Equation (5.4.23), in which trigonometric functions of the form cos (y cos x) and sin (y cos x) are replaced by Fourier series, we obtain

$$R_1(\tau) = k_{b_1}^2 \sigma^2 \rho(\tau) J_0\left(2\beta_1 \sin\frac{\Omega_1\tau}{2}\right) J_0\left(2\beta_2 \sin\frac{\Omega_2\tau}{2}\right). \tag{5.4.33}$$

Likewise, by determing correlation function $R_2(\tau)$ on the basis of Equation (5.4.26), in consideration of Equation (5.4.17), we prove that

$$R_{f_2}(\tau) = \frac{k_{b_2}^2}{k_{b_1}^2} R_1(\tau). \tag{5.4.34}$$

Now, using Equations (5.4.23) and (5.4.26), we can prove that the time-averaged crosscorrelation function of voltages $u_{j_1}(t)$ and $u_{j_2}(t)$ for $\Omega_2/\Omega_1 \gg 1$ is virtually equal to zero. But if $\beta_1 = 0$ and $\beta_2 = 0$, then the equality $R_{12}(\tau) = 0$ is exactly satisfied. Consequently, phase inverter PI_2 decorrelates jamming generated at the outputs of balanced mixers BM_1 and BM_2.

Since $R_{12}(\tau) \approx 0$ the crosscorrelation functions $R_{1f_2}(\tau) = R_{f_21}(\tau)$ are also equal to zero and, on the basis of Equation (5.4.16), we may write

$$R_\Sigma(\tau) = R_\Delta(\tau) = \left(1 + \frac{k_{b2}^2}{k_{b1}^2}\right) R_1(\tau). \tag{5.4.35}$$

For given τ and $k_{b1} = k_{b2}$ the moduli of the functions $R_\Sigma(\tau)$ and $R_\Delta(\tau)$ are maximum and equal to $2|R_1(\tau)|$.

If the transfer function of the filter in the frequency band is given, the variances $\sigma_{n\Sigma\ell}^2$ and $\sigma_{n\Delta\ell}^2$ of the filter output jamming for the useful FM signals $u_{\Sigma\ell}(t)$ and $u_{\Delta\ell}(t)$ are determined, respectively, by the formulas

$$\sigma_{n\Sigma\ell}^2 = \frac{2}{\pi} \left(1 + \frac{k_{b2}^2}{k_{b1}^2}\right) \int_0^\infty \int_0^\infty |\phi_{\Sigma\ell}(j\omega)|^2 \, R_1(\tau) \cos \omega\tau d\tau d\omega, \tag{5.4.36}$$

$$\sigma_{n\Delta\ell}^2 = \frac{2}{\pi} \left(1 + \frac{k_{b2}^2}{k_{b1}^2}\right) \int_0^\infty \int_0^\infty |\phi_{\Delta\ell}(j\omega)|^2 \, R_1(\tau) \cos \omega\tau d\tau d\omega. \tag{5.4.37}$$

Here $\phi_{\Sigma\ell}(j\omega)$ are the transfer functions of the filter in the frequency band for useful signals with the carrier frequencies $\ell\Omega_2 + \omega_d$ ($\ell = 0, 1, 2, \ldots$) and $\ell\Omega_2 - \omega_d$ ($\ell = 1, 2, \ldots$), respectively.

If, finally, we use Equations (5.4.31), (5.4.32), (5.4.36) and (5.4.37), then we can find the desired effective output signal-to-noise ratios $q_{\Sigma\ell}$ and $q_{\Delta\ell}$ of a dual balanced mixer with an adder and subtractor. These ratios are

$$q_{\Sigma\ell} = \frac{\sigma_{\Sigma\ell}^2}{\sigma_{n\Sigma\ell}^2} = \frac{\pi k_{\Sigma\ell}^2 k_{b1}^2 (U_{s1} + U_{s2})^2}{4(k_{b1}^2 + k_{b2}^2)}$$

$$\times \frac{J_\ell^2 (2\beta_2 \sin 0.5\Omega_2\tau_d)}{\displaystyle\int_0^\infty \int_0^\infty |\phi_{\Sigma\ell}(j\omega)|^2 \, R_1(\tau) \cos \omega\tau d\tau d\omega}, \tag{5.4.38}$$

$$q_{\Delta\ell} = \frac{\sigma_{\Delta\ell}^2}{\sigma_{n\Delta\ell}^2} = \frac{\pi k_{\Delta\ell}^2 k_{b1}^2 (U_{01} + U_{02})^2}{4(k_{b1}^2 + k_{b2}^2)}$$

$$\times \frac{J_\ell^2 (2\beta_2 \sin 0.5\Omega_2\tau_d)}{\displaystyle\int_0^\infty \int_0^\infty |\phi_{\Delta\ell}(j\omega)|^2 \, R_1(\tau) \cos \omega\tau d\tau d\omega} \tag{5.4.39}$$

When the functions $\rho(\tau)$, $\phi_{\Sigma\ell}(j\omega)$ and $\phi_{\Delta\ell}(j\omega)$ are given in explicit form the specific values of $q_{\Sigma\ell}$ and $q_{\Delta\ell}$ are computed on the basis of Equations (5.4.38) and (5.4.39).

Influence of Jamming on Single Balanced Mixer

The voltages of useful signal $u_{ss}(t)$ and of interference $u_{js}(t)$, generated by a single balanced mixer, are determined by Equations (5.4.27) and (5.4.23), respectively, in which U_{s1} and k_{b1} are replaced by U_{ss} and k_{bs}, where $U_{ss} = k_{bs} U_{s1}$ and k_{bs} is the gain of the balanced mixer. The filter extracts from $u_{ss}(t)$ the FM voltages $u_{h\ell}(t)$ and $u_{low\ell}(t)$ with the carrier frequencies $\ell\Omega_2 + \omega_d$ ($\ell = 0, 1, 2, \ldots$) and $\ell\Omega_2 - \omega_d$ ($\ell = 1, 2, \ldots$).

The squares of the effective values of $\sigma^2_{h\ell}$ and $\sigma^2_{low\ell}$ of these voltages are approximately

$$\sigma^2_{h\ell} = 0.5\, k^2_{h\ell}\, U^2_{so}\, J^2_\ell\, (2\beta_2 \sin 0.5\, \Omega_2\, \tau_d)$$
$$\text{for } \ell = 0, 1, 2, \ldots, \tag{5.4.40}$$

$$\sigma^2_{low\ell} = 0.5\, k^2_{low\ell}\, U^2_{so}\, J^2_\ell\, (2\beta_2 \sin 0.5\, \Omega_2\, \tau_d)$$
$$\text{for } \ell = 1, 2, \tag{5.4.41}$$

where $k_{h\ell}$ and $k_{low\ell}$ are the gains of the filter on frequencies $\ell\Omega_2 + \omega_d$ and $\ell\Omega_2 - \omega_d$ for generation of voltages $u_{h\ell}(t)$ and $u_{low\ell}(t)$.

If the transfer functions $\phi_{h\ell}(j\omega)$ and $\phi_{low\ell}(j\omega)$ in the frequency band of the filter, which generates the signals $u_{h\ell}(t)$ and $u_{low\ell}(t)$, are known, then the variances $\sigma^2_{nh\ell}$ and $\sigma^2_{n\,low\ell}$ of the noises generated from mixture $u_{jo}(t)$ with one of the voltages $u_{h\ell}(t)$ or $u_{low\ell}(t)$, will be determined, obviously, by the following formulas:

$$\sigma^2_{nh\ell} = \frac{2}{\pi} \int_0^\infty \int_0^\infty |\phi_{h\ell}(j\omega)|^2\, R_0\,(\tau)\, \cos\, \omega\tau d\omega d\tau, \tag{5.4.42}$$

$$\sigma^2_{n\,low\ell} = \frac{2}{\pi} \int_0^\infty \int_0^\infty |\phi_{low\ell}(j\omega)|^2\, R_0\,(\tau)\, \cos\, \omega\tau d\omega d\tau. \tag{5.4.43}$$

Here $R_0(\tau)$ is the correlation function of voltage $u_{jo}(t)$:

$$R_0(\tau) = k^2_{bo}\, \sigma^2\, \rho(\tau)\, J_0\left(2\beta_1 \sin \frac{\Omega_1 \tau}{2}\right) J_0\left(2\beta_2 \sin \frac{\Omega_2 \tau}{2}\right). \tag{5.4.44}$$

Therefore,

$$q_{h\ell} = \frac{\sigma^2_{h\ell}}{\sigma^2_{nh\ell}} = \frac{\pi k^2_{h\ell} \, U^2_{so} \, J^2_\ell \, (2\beta_2 \sin 0.5 \, \Omega_2 \tau_d)}{4 \displaystyle\int_0^\infty \int_0^\infty |\phi_{h\ell}(j\omega)|^2 \, R_0(\tau) \cos \omega\tau d\tau d\omega} \qquad (5.4.45)$$

$$q_{low\ell} = \frac{\sigma^2_{low\ell}}{\sigma^2_{nlow\ell}}$$

$$= \frac{\pi k^2_{low\ell} \, U^2_{so} \, J^2_\ell \, (2\beta_2 \sin 0.5 \, \Omega_2 \tau_d)}{4 \displaystyle\int_0^\infty \int_0^\infty |\phi_{low\ell}(j\omega)|^2 \, R_0(\tau) \cos \omega\tau d\tau d\omega} \, . \qquad (5.4.46)$$

Thus, the ratios $q_{h\ell}$ and $q_{low\ell}$, as in a dual balanced mixer, are comparatively easy to calculate when the parameters of the system and the response to the signal and jamming are known.

Comparison of Influence of Jamming on Single Balanced and Dual Balanced Mixers

It is most often pointed out in the literature that single balanced and dual balanced mixers can operate either in the homodyne Doppler frequency mode, where the carrier frequency of the output FM voltage is ω_d, or in the offset Doppler frequency extraction mode with the carrier frequency $\ell\Omega_2 \pm \omega_d$ ($\ell = 1, 2, \ldots$).

Therefore, it is worthwhile to determine the ratios $q_1 = q_{h\ell}/q_{ho}$, $q_2 = q_{low\ell}/q_{ho}$, $q_3 = q_{h\ell}/q_{lows}$, $q_4 = q_{\Sigma\ell}/q_{\Sigma_0}$, $q_5 = q_{\Delta\ell}/q_{\Sigma_0}$, $q_6 = q_{\Sigma\ell}/q_{\Delta\ell}$, $q_7 = q_{\Sigma\ell}/q_{hs}$ and $q_8 = q_{\Delta\ell}/q_{lows}$. Here q_{ho} and q_{Σ_0} are the ratios, $q_{h\ell}$ and $q_{\Sigma\ell}$, for $\ell = 0$. The values of ℓ and s for finding $q_1 - q_6$ and q_8 may be $1, 2, 3, \ldots$. The symbol ℓ in ratio q_7 may be $0, 1, 2, \ldots$ etc. It follows from equation (5.4.45) that

$$q_1 = \frac{k^2_{h\ell} \, J^2_\ell \, (2\beta_2 \sin 0.5\Omega_2\tau_d)}{k^2_{ho} \, J^2_0 \, 2\beta_2 \sin \dfrac{\Omega_2\tau_d}{2}}$$

$$\times \frac{\displaystyle\int_0^\infty \int_0^\infty |\phi_{ho}(i\omega)|^2 \, R_0(\tau) \cos \omega\tau d\tau d\omega}{\displaystyle\int_0^\infty \int_0^\infty |\phi_{h\ell}(i\omega)|^2 \, R_0(\tau) \cos \omega\tau d\tau d\omega} \, . \qquad (5.4.47)$$

Analysis of this equation shows that when the frequency response of the receiver IF is uniform, and the spectral density of jamming voltage $u_j(t)$ is constant within the IF passband, $\Delta\omega$, and the effective passbands and gains of filters with transfer functions $\phi_{h_0}(j\omega)$ and $\phi_{h\ell}(j\omega)$ are identical, the variances $\sigma^2_{nh_0}$ and $\sigma^2_{nh\ell}$, determined by Equation (5.4.42), will be equal. Under these assumptions

$$q_1 = \frac{J_\ell \left(2\beta_2 \sin \dfrac{\Omega_2 \tau_d}{2} \right)}{J_0^2 \left(2\beta_2 \sin \dfrac{\Omega_2 \tau_d}{2} \right)} \qquad (5.4.48)$$

represents the filter output signal power ratio for homodyne and offset Doppler frequency voltages.

Hence, it follows that q_1 may acquire values that are either larger and smaller than unity, depending on the magnitude of ℓ and $x = 2\beta_2 \sin (\Omega_2 \tau_d/2)$.

If the modulation index, β_2 is selected such that a single balanced mixer in a given operating mode gives the highest possible signal strength, then β_2 should be small when a homodyne Doppler frequency signal is extracted, but when an offset Doppler frequency is generated, the function $J_\Sigma^2(2\beta_2 \sin 0.5 \, \Omega_2 \tau_{de_1})$ should be maximized. Under these conditions q_1 is always smaller than unity. This does not mean, however, that a single balanced mixer should not be used in the offset Doppler frequency extraction mode.

As a matter of fact, both external jamming and receiver noise and the transmitter feedthrough signal exert a considerable influence on the receiver of a CW radar system. The magnitude of the filter output signal in a single balanced mixer is also determined, in the first approximation, by zero order and ℓ ($\ell = 1$, 2, . . .) -order Bessel functions if the signals are generated without offset and with offset, respectively. In this case, however, $\tau_{de_1} = 0$, and the feedthrough signal theoretically has no effect on radar performance in the offset Doppler voltage mode.

The ratio q_1 changes as a function of the argument $\Omega_2 \tau_d$ for $\beta_2 = $ const and $\ell = $ const from zero to some maximum value. Shown in Figure 5.30 as an illustration is the graph of the dependence of q_1 on x for $\ell = 1$. As can be seen by the graph, when $x < 1.4$ for an arbitrary $\Omega_2 \tau_d$ the inequality $q_1 < 1$ is satisfied. Consequently, a single balanced mixer has better noise immunity in the homodyne mode than in the offset mode. When $x > 1.4$ we obtain $q_1 > 1$ for certain values of $\Omega_2 \tau_d$.

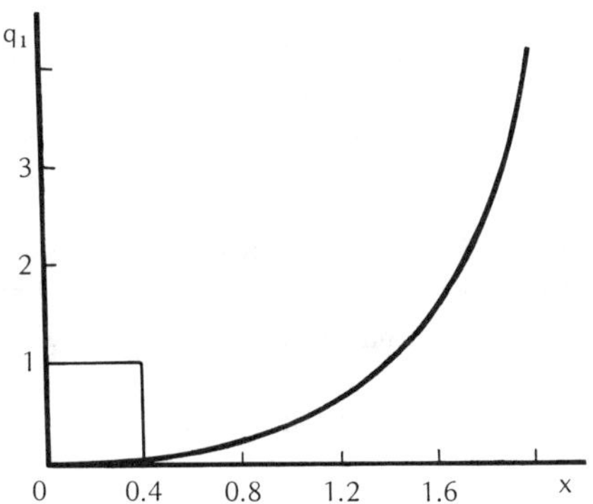

Figure 5.30

If the frequency response of the IF differs from uniform and, for example, is Gaussian, then for identical gains and effective passbands of filters with the transfer functions $\phi_{ho}(j\omega)$ and $\phi_{h\ell}(j\omega)$, the inequality

$$\int_0^\infty \int_0^\infty |\phi_{ho}(j\omega)|^2 \, R_0(\tau) \cos \omega\tau \, d\tau \, d\omega$$

$$> \int_0^\infty \int_0^\infty |\phi_{h\ell}(j\omega)|^2 \, R_0(\tau) \cos \omega\tau \, d\tau \, d\omega \tag{5.4.49}$$

is satisfied. This inequality increases with Ω_2 and ℓ because of a reduction of the spectral density of jamming as ω increases. When Equation (5.4.49) is satisfied $q_1 > 1$ will be obtained for smaller values of x in comparison with the case of an IF with a uniform response.

The ratio $q_2 = q_{low\ell}/q_{ho}$ is transformed in accordance with Equations (5.4.45) and (5.4.46) to Equation (5.4.47), in which $k_{h\ell}$ is substituted for $k_{low\ell}$ and $\phi_{h\ell}(j\omega)$ for $\phi_{low\ell}(j\omega)$. Consequently, q_2 is a function of x, junst as is q_1. The ratio $q_3 = q_{h\ell}/q_{lows}$ is determined by dividing Equation (5.5.45) by Equation (5.4.46). By analyzing the resulting ratio q_3 we can prove that it is possible to obtain values of q_3 either larger and smaller than unity. But if $\ell = s$ the frequency response of the IF is uniform, $k_{h\ell} = k_{lows}$ and the effective passbands of filters with transfer functions $\phi_{lows}(j\omega)$ and $\phi_{h\ell}(j\omega)$ are identical, then $q_3 = 1$.

The ratio $q_4 = q_{\Sigma\ell}/q_{\Sigma n}$, determined on the basis of Equation (5.4.38), is characterized the same as q_1.

By using Equations (5.4.38) and (5.4.39) we can find $q_5 = q_{\triangle\ell}/q_{\Sigma_0}$. It turns out here that everything said above in relation to q_2 applies also to q_5. A dual balanced mixer, however, performs cancellation of several components, which improves useful signal filtering performance.

The ratio $q_6 = q_{\Sigma\ell}/q_{\triangle\ell}$ is determined by dividing Equation (5.4.38) by Equation (5.4.39) and may be both larger and smaller than unity. The ratio $q_7 = q_{\Sigma\ell}/q_{hs}$ is determined by Equations (5.4.39) and (5.4.45), and its analysis leads to the conclusion that when single balanced and dual balanced mixers generate voltages with identical carrier frequencies $(\ell = s)$, and when identical output filters are used, i.e., when

$$|\phi_{hs}(j\omega)|^2 = |\phi_{\Sigma\ell}(j\omega)|^2 ,$$

$$q_7 = \frac{(k_{b_1} + k_{b_2})^2}{k_{b_1}^2 + k_{b_2}^2}$$

Under these conditions, as we can see, q_7 is always greater than unity, is a function of k_{b_1}/k_{b_2}, and is maximum when $k_{b_1} = k_{b_2}$. The maximum value of q_7 is two. Consequently, a dual balanced mixer, operating in identical modes and having the same elements with identical parameters, has twice the noise immunity of a single balanced mixer.

If $\ell \neq s$, then q_7 may be either larger or smaller than unity.

The ratio $q_8 = q_{\triangle\ell}/q_{lows}$, determined by dividing Equation (5.4.39) by Equation (5.4.46), is similar in terms of properties to q_7.

In summary, we may formulate the following basic conclusions:

- a dual balanced mixer, generating signals with identical carrier frequencies, always has better noise immunity than a single balanced mixer, here the maximum gain in noise immunity in terms of the effective signal to noise ratio, is 2;

- the noise immunity of single balanced and dual balanced mixers, operating in the offset Doppler frequency mode, may be either higher or lower than in the homodyne mode;

- single balanced and dual balanced mixers weaken the influence of feed-through, in the offset Doppler frequency mode, but not in the homodyne mode;

- phase inverter PI_2, illustrated in Figure 5.29, should have a wide band, which poses certain difficulties in its design.

Chapter 6
Spatial, Polarization, Frequency and Phase Selection

6.1. Spatial Selection and Jamming Suppression with Shaped Radiation Patterns

At the present time there are two basic ways to improve spatial selection:

change the distribution of the amplitude and phase of the antenna aperture illumination;

use antennas with nonlinear signal processing.

1. Choice of Amplitude and Phase Distributions Over Antenna Aperture

The spatial selectivity of antenna systems is improved by selecting the proper field amplitude distribution over the aperture. In order to eliminate ambiguous measurements of the angular coordinates of sources of radiation it is necessary to reduce the side lobe level of the radiation pattern. This is done for antennas of given sizes by selecting an amplitude distribution that falls off gradually toward the edges of the aperture. Such a distribution, however, causes the width of the main lobe of the pattern to increase.

According to antenna theory it is feasible to build an optimum antenna, the pattern of which, for a given main lobe width, has the minimum side lobe level. Such a pattern is described by a Chebyshev polynomial [2, 187]. To protect radars, for example, from interference created by objects that are located at various distances from a target, when the side lobe level must be reduced considerably (to 40 dB and more), it is helpful to use Chebyshev radiation patterns. The resulting increase of the width $\theta_{0.5}$ of the main beam causes the effective area of the antenna to decrease and is the penalty for improving the noise immunity in relation to interference acting through the side lobes of the pattern. The best way to solve the problem of improving the resolving power while simultaneously reducing the side lobe level for the linear cophased class of antennas is to select the correct antenna size and Chebyshev field distribution over the aperture.

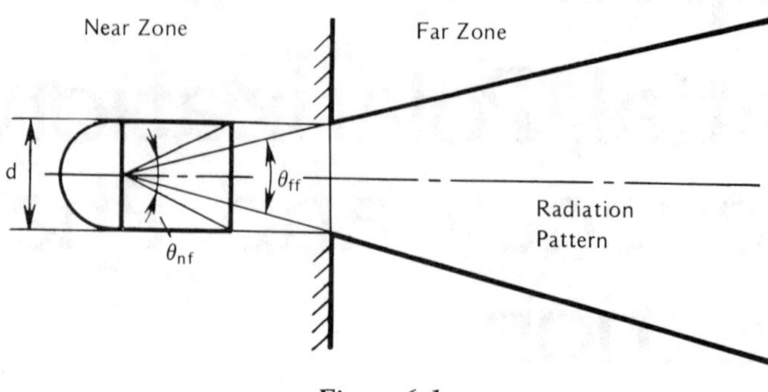

Figure 6.1

For large antenna systems, the dimensions d of which are commensurable with target range r, coordinate measurements are performed in the near field of the diffraction pattern, where the angular dimensions θ_{nf} of the main beam of the pattern increase in comparison with the width of the radiation pattern θ_{ff} in the far field (Figure 6.1). In the near field, therefore the resolving power of cophased antennas decreases. Resolving power in terms of angular coordinates cannot be improved simply by increasing the antenna aperture, since the cross section of the beam in the near field is commensurable with the dimensions of the antenna. Ray width θ_{nf} increases because of quadratic phase distortions of the field in the antenna aperture [136, 187]. By correcting these distortions it is possible, just as in optics, to focus the beam at a given range. This measure substantially improves the resolving power of the radar.

An antenna can be focused by building it as a sphere. The curvature of the antenna determines its focal range. It is especially easy to focus phased array antennas, in which undesired phase shifts are compensated by phase shifters.

Focusing an antenna increases its gain G_f, and the amount of improvement can be estimated using the formula [136, 187]

$$K_f = G_f/G_{co} = d^2/\lambda r,$$

where G_{co} is the gain of a cophased antenna with a uniform aperture field distribution; r is the focal range.

Focusing is used extensively in synthetic aperture radars and in holographic radars.

By selecting the proper amplitude and phase distributions for a given aperture it is possible, in principle, to attain exceedingly narrow radiation patterns ("superdirectional" antennas). S.A. Shchelkunov demonstrated theoretically that this is feasible by establishing an aperture field distribution that oscillates sharply in phase and amplitude [161, 205]. The fields that are created simultaneously by each element of such an antenna are combined in space as a result of constructive interference, and this is responsible for the phenomenon of superdirectivity.

When the aperture amplitude and phase distributions oscillate the amount of reactive energy that accumulates near the antenna increases sharply. Consequently radiated power decreases.

Superdirectivity is achieved in antenna arrays by placing the radiators closer together (at distances less than $\lambda/4$). The resulting enhancement of the directional properties of an antenna leads to a reduction of its efficiency, since the number of feed lines that transmit energy with losses increases. Furthermore, when the distance between the elementary antennas is reduced the passband of the antenna system is narrowed.

The general drawbacks of superdirectional antennas include the following:

reduced efficiency;

narrow operating frequency band;

the spatial position of the beam cannot be controlled electrically;

feed systems are complicated and hard to design.

The superdirectivity phenomenon may be used for comparatively small antenna systems or antenna arrays with a few radiators.

2. Nonlinear Processing Antennas

There is a great variety of nonlinear processing antenna systems. The output signal of such an antenna system is characterized by a nonlinear signal function (product, power, etc.) of elementary antennas [15, 136, 204]. By using nonlinear processing it is possible to synthesize any radiation pattern, which could be generated by selecting the desired amplitude-phase distribution in an ordinary linear array. Nonlinear processing is usually performed for the purposes of improving the directivity of an antenna (for a given size) and reducing the side lobe level. A multiplicative interferometer with two antennas, separated by distance 2d, is a typical representative of this class of antennas.

Suppose an elementary antenna of an interferometer has radiation pattern $F(\theta)$, where $\theta = \sin\,\theta$. Then the multiplier output signal will be

$$u_n = k_n U F^2(\theta) \cos(kd\theta),$$

where $k = 2\pi/\lambda$, and k_n is the proportionality coefficient. The analogous linear antenna has the radiation pattern

$$u_\varrho = UF(\theta) \cos(0.5kd\theta).$$

Comparison of u_n and u_ϱ shows that nonlinear processing narrows the radiation pattern.

Directivity can be improved substantially by using multiple multiplication of the signals from several antennas. Various modifications of nonlinear antenna systems are described in the literature [15, 136, 187].

Nonlinear processing antennas have several drawbacks:

the signal/noise ratio decreases in proportion to the ratio of the number of elements in a nonlinear processing array to the number of elements in an ordinary uniform linear array.

resolving power depends on relative signal strength, which is related to the effect of the suppression of a weak signal by a stronger one in a nonlinear system.

If the increase of the noise factor in nonlinear antenna systems is ignored, then such antennas provide a gain in resolving power. This gain is estimated to be a factor of 1.2 for certain types of antennas [15].

Nonlinear antenna systems may be used to improve resolving power for the possibility of long-term signal storage, when the reduction of the signal to noise ratio, associated with nonlinear processing, is a minor factor.

6.2. Polarization Selection

1. Basic Definitions

Polarization is the space-time characteristic of an electromagnetic wave; it determines the spatial orientation of the electric (or magnetic) field intensity during the period of the carrier. For plane uniform waves, to which the ensuing discussion will refer, the electrical and magnetic field intensity vectors lie in a plane perpendicular to wave propagation.

To analyze wave polarization it is sufficient to know the orientation of just one of the electromagnetic field intensity vectors. Polarization is usually estimated on the basis of the orientation of electrical field intensity vector **E**.

Wave polarization is described by a polarization diagram, which represents the projection of the curve described by the end of vector **E** onto a plane perpendicular to wave propagation. There are three basic kinds of polarization: linear, circular and elliptical. In linear polarization the spatial orientation of the **E** vector remains constant and the polarization diagram is a straight line.

The distinguishing feature of circular polarization is the fact that the **E** vector, which has a constant amplitude, rotates at a constant angular velocity around the propagation axis. In this case the end of the vector describes a circle. The rotation period is equal to the period of the electromagnetic wave. In the case of elliptical polarization the end of the **E** vector, when rotated, describes an ellipse. Here the modulus of the **E** vector and its angular rotation velocity experience periodic changes during the rotation period.

Electromagnetic waves are divided into three groups, depending on whether the parameters of the polarization diagram change in time or remain constant: completely polarized, partially polarized and unpolarized. A wave, whose polarization parameters remain constant in time, is called completely polarized. An electromagnetic wave, the parameters of whose polarization diagram are continuous and relatively slowly changing (the time necessary for an appreciable change of a parameter vastly exceeds the period of high-frequency waves) is called partially polarized. An unpolarized or randomly polarized wave is characterized by fast fluctuations of the **E** vector in terms of both modulus and direction of rotation. In this case the polarization diagram acquires all possible shapes and orientations, so that it does not exhibit a preferential position.

The projections e_x and e_y of the **E** vector onto orthogonal coordinate axes $0x$ and $0y$ (Figure 6.2) may be written as

$$e_x = E_x \cos(\omega_0 t + \phi_x), \tag{6.2.1a}$$

$$e_y = E_y \cos(\omega_0 t + \phi_y), \tag{6.2.1b}$$

Any kind of polarization can be obtained by changing the ratio of the amplitude components of the waves E_x and E_y) and the phase shift between them ($\Delta\phi = \phi_x - \phi_y$). For example, if $\Delta\phi$ is 0 or π, then linear polarization occurs. And if $E_x = 0$, $E_y \neq 0$ or $E_x \neq 0$, $E_y = 0$, then vertical and horizontal polarization, respectively, are obtained. But when $E_x \neq 0$, $E_y \neq 0$, and $\Delta\phi = 0$, the orientation of α_E will acquire linear polarization and angle α_E between the $0x$ axis and the **E** vector is determined by the equality $\tan \alpha_E = E_y/E_x$. If the phase difference of the waves is equal to odd number $\pi/2$, i.e., $\Delta\phi = (2n + 1)\pi/2$, where n is an integer, and $E_x = E_y$, then circular polarization occurs. In cases when $0 < \Delta\phi < 0.5\pi$ or $E_x \neq E_y$, and also when the two indicated conditions are satisfied simultaneously, the result is elliptical polarization. Random fluctuations of $\Delta\phi$, E_x and E_y result in random polarization.

When determining the polarization of any signals or of a signal and jamming it is possible, using polarization selectors (filters), to separate one signal from another or to substantially weaken the jamming, and thus to achieve more reliable extraction of the useful information.

2. Parameters of Elliptical Polarization

Elliptical polarization of an electromagnetic wave is the most common kind of polarization, from which linear and circular polarization can be derived as special cases.

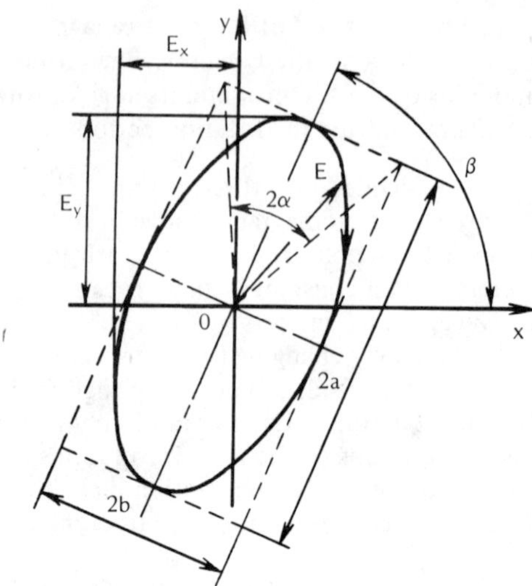

Figure 6.2

A polarization ellipse is described by its shape, by the orientation of its axes
relative to the selected coordinate system x0y (Figure 6.2), and by the
direction of rotation of the **E** vector. The shape of an ellipse is usually characterized by the ellipticity factor k_{el}, the absolute value of which is equal to
the ratio of the minor b and major a semi-axes of the ellipse

$$k_{el} = b/a. \tag{6.2.2}$$

The values of the modulus of k_{el} are confined to obvious limits: $0 \leqslant |k_{el}| \leqslant 1$.
The ellipticity factor has different signs, depending on the direction of rotation
of the **E** vector. If, when viewed from the source in the direction of propagation
of the wave, the **E** vector rotates clockwise, then the wave is called right-polarized
and k_{el} is assumed to be positive. When the vector rotates counter-clockwise
the wave is called left-polarized and k_{el} is negative.

Also used to describe the shape of an ellipse and the direction of rotation of the
E vector, in addition to the coefficient k_{el}, is the ellipticity angle

$$\alpha = \arctan k_{el} \tag{6.2.3}$$

where $-0.25\pi \leqslant \alpha \leqslant 0.25\pi$.

Angle α is equal to one-half the angle between the diagonals of a rectangle, the
sides of which are tangent to the ellipse and parallel to its axes (Figure 6.2).
When the above condition is satisfied angle α uniquely describes the shape of
the ellipse and its sign uniquely determines the direction of its rotation.

The orientation of the ellipse is determined by angle β, which is formed by the 0x axis of the selected coordinate system and the major axis of the ellipse; when the position of the ellipse is uniquely determined the values of β fall within the limits $0 \leqslant \beta \leqslant \pi$.

Conversion from field components in Cartesian coordinates to ellipse parameters and back is accomplished with the formulas presented below, which can be derived with the aid of simple trigonometric conversions. If the parameters a, k_{el} and β of an ellipse are given, then

$$E_x = a(\cos^2 \beta + k_{el} \sin^2 \beta)^{0.5},$$

$$E_y = a(\sin^2 \beta + k_{el} \cos^2 \beta)^{0.5},$$

$$\Delta\phi = \phi_y - \phi_x = \text{arctg} \frac{2k_{el}}{(1 - k_{el}^2) \sin 2\beta}, \tag{6.2.4}$$

$$\sin \Delta\phi = 2 k_{el}/(1 + k_{el}^2),$$

$$\cos \Delta\phi = (1 - k_{el}^2)/(1 + k_{el}^2).$$

For given E_x, E_y and $\Delta\phi$ we have

$$k_{el} = \left(\frac{E_x^2 \sin^2 \beta - E_x E_y \sin 2\beta \cos \Delta\phi + E_y^2 \cos^2 \beta}{E_x^2 \cos^2 \beta + E_x E_y \sin 2\beta \cos \Delta\phi + E_y^2 \sin^2 \beta} \right)^{0.5} \tag{6.2.5a}$$

$$\beta = \frac{1}{2} \text{arctg} \frac{2E_x E_y \cos \Delta\phi}{E_x^2 - E_y^2}. \tag{6.2.5b}$$

The above conversions are useful for the solution of numerous problems of practical importance (for example, evaluation of the parameters of the resulting wave at the input of a receiving antenna when two independent, differently polarized waves exist simultaneously).

3. Polarization Selectors

Any receiving antenna feed channel is a polarization selector. The output signal power of the selector depends (under otherwise identical conditions) on the polarization of the arriving wave. We shall examine two typical variations of an antenna feed channel as examples. The first version has a parabolic reflector, which is illuminated by a circular waveguide. The component parts of the waveguide feed (Figure 6.3) are: circular waveguide segment (a) with a dielectric lamellar insert, the length of which is 0.25λ; transition section (b) between the circular and rectangular waveguides; rectangular waveguide segment (c), to which a radar transmitter and receiver may be connected by means of a duplexer. Let us assume that the rectangular waveguide is connected to the

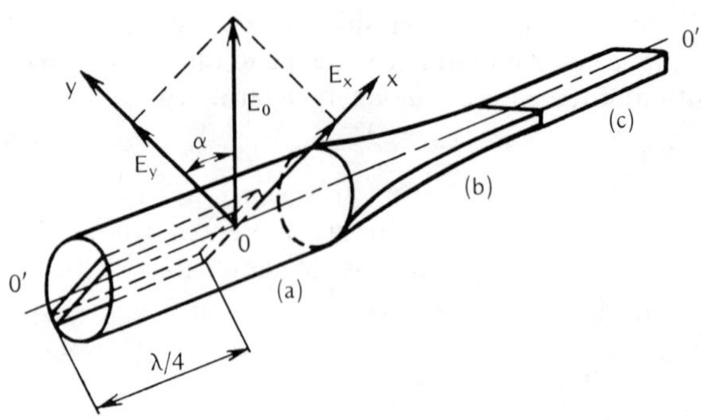

Figure 6.3

transmitter and it generates a vertically polarized wave of the type H_{01} (the $\mathbf{E_0}$ vector is perpendicular to the wide wall of the waveguide). The circular wave-guide generates a wave of the H_{11} type. In order to determine the influence of the section with the dielectric insert the $\mathbf{E_0}$ vector must be expanded into two orthogonal components. One of them ($\mathbf{E_x}$) lies in the plane of the insert, and the other ($\mathbf{E_y}$) is perpendicular to $\mathbf{E_x}$. The $\mathbf{E_y}$ component is virtually constant (if a slight reduction of amplitude due to energy losses in the dielectric is ignored) as it passes through the section with the insert. However, the $\mathbf{E_x}$ component acquires a phase lag of 0.5π. Thus, if the electrical field intensity in the input of the antenna feed channel is characterized by a vector with linear polarization and with the instantaneous value $E_0 \cos \omega_0 t$, then, after the phase-shifting section, the orthogonal field components will be, respectively,

$$E_x = E_0 \sin \alpha \sin \omega_0 t, \tag{6.2.6a}$$

$$E_y = E_0 \cos \alpha \cos \omega_0 t, \tag{6.2.6b}$$

where α is the angle of the E vector relative to the dielectric plate.

The field determined by formulas (6.2.6) is, in the general case, elliptically polar-ized. As angle α changes gradually from 0 to 0.5π (the section with the insert rotates around the $00'$ axis) the ellipticity factor of the wave changes from 0 to 1. When $\alpha = 0$ and $\alpha = 0.5\pi$ linear vertical polarization occurs, and when $\alpha = 0.25\pi$ polarization is circular. It may be assumed that the polarization of the signal radiated by the antenna is not altered by the reflector during the radiation pattern shaping process (if only within the main lobe).

It may be assumed on the basis of what we have said above that an antenna is characterized by the polarization diagram that corresponds to the polarization of the radiated wave. The polarization reciprocity principle is valid for the antenna examined here; i.e., the polarization diagram remains constant during reception and transmission, and waves with the same polarization as the radi-ated waves are received best.

This antenna feed system performs as a polarization selector when receiving waves with different polarizations. For example, if $\alpha = 0.25\pi$, then, according to formulas (6.2.6), a right-hand circularly polarized wave is radiated. This system does not admit a left-hand circularly polarized wave. Actually, let us assume that a symmetrical body, for example a metal sphere, the radius of which is substantially longer than wavelength, stands in the path of the radiated wave. On reflection from the sphere the phase shift between field components E_X and E_Y does not change and the wave remains circularly polarized, but because of a 180° rotation of the wave propagation axis the wave becomes left-hand polarized and orthogonal with respect to the wave that was originally radiated. The $\mathbf{E_y}$ component of the $\mathbf{E_0}$ vector passes unaltered through the section with the dielectric insert; it may be written as

$$E_y' = kE_0 \cos \frac{\pi}{4} \cos \omega_0 t.$$

Here k is a coefficient that considers the change of the amplitude of $\mathbf{E_y}$ during propagation. The $\mathbf{E_x}$ component acquires an additional 0.5π phase lag and will be

$$E_x' = - kE_0 \sin \frac{\pi}{4} \cos \omega_0 t.$$

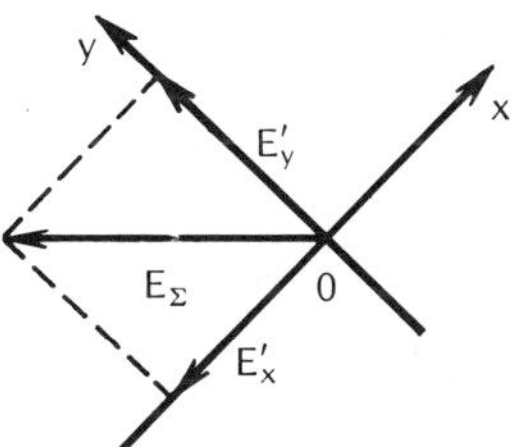

Figure 6.4

The resulting field $\mathbf{E_\Sigma}$ (Figure 6.4) at the output of the circular waveguide will have linear polarization and will be oriented horizontally. Therefore the wave cannot pass through the rectangular waveguide, which ends with the selector.

Conversely, if the received wave were a right-rotating wave, then it would pass to the selector output without losses. For the case, for instance, when the radiated wave is characterized for formulas (6.2.6), it would be necessary, for the right-rotating reflected wave, to change the phase, for example, of the $\mathbf{E_x}$ component, by π; i.e.,

$$E_y' = kE_0 \cos \frac{\pi}{4} \cos \omega_0 t,$$

$$E_x' = kE_0 \sin \frac{\pi}{4} \sin (\omega_0 t - \pi).$$

After the phase-shifting section of the selector

$$E'_y = kE_0 \cos \frac{\pi}{4} \cos \omega_0 t,$$

$$E'_x = kE_0 \sin \frac{\pi}{4} \cos \omega_0 t.$$

The resulting field is vertically polarized and the signal passes to the receiver
without losses.

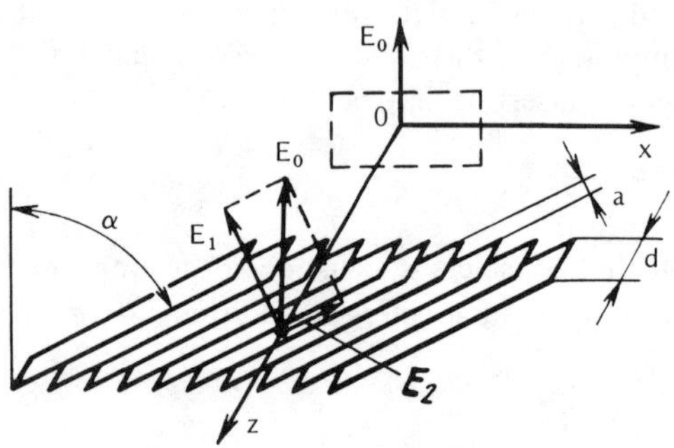

Figure 6.5

The second of the designs to be discussed represents a combination of any
antenna with linear polarization and a polarization grating, also called a
polarization filter. A polarization grating is a set of thin parallel metal plates
with width d, separated from each other by distance a (Figure 6.5). The propagation axis coincides with the 0z axis.

Let the radiated wave have vertical polarization (vector $\mathbf{E_0}$). The polarization
of the wave changes as it passes through the polarization grating. In the grating
the $\mathbf{E_0}$ vector may be broken down into the $\mathbf{E_1}$ component, which is perpendicular to the plane of the plates, and the $\mathbf{E_2}$ component which is parallel to
the plates. The $\mathbf{E_1}$ component passes through the grating virtually unaltered,
but the $\mathbf{E_2}$ component travels with a higher phase velocity (just as it does in
the waveguide) and at the output of the grating acquires phase shift $\Delta\phi$ relative
to the first component. The equality $\Delta\phi = \pi/2$ can be satisfied by selecting the
correct dimension d. Here any polarization can be acquired after the polarization grating, depending on the size of angle α between the plane of the plates
and the electrical field intensity vector of the incident wave.

The polarization grating may be inserted between the feed and reflector of the
antenna (in the primary radiation field) or in the aperture of the antenna itself
(in the secondary radiation field). Again, it is obvious that an antenna of this
type is a polarization selector; it reacts differently to differently polarized
received waves.

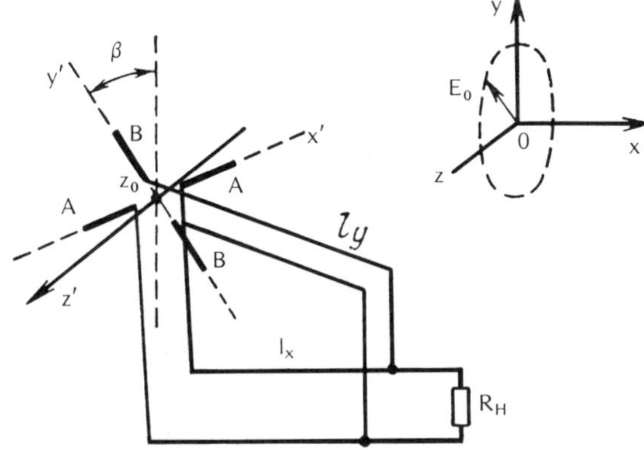

Figure 6.6

4. Reception Polarization Coefficient

The power losses of a wave of arbitrary polarization, striking a receiving antenna, are customarily determined by polarization coefficient γ_p, which is the ratio of the power S_r of the signal that actually reaches the receiver input to the maximum possible input power $S_{r\,max}$, which would exist if the polarization characteristics of the incident wave and receiving antenna were perfectly matched. The values of γ_p fall within the limits of 0 to 1.

Let us find the relationship between γ_p and the polarization parameters of a wave and antenna. Suppose we have an elliptically polarized plane wave, propagating in the direction of the 0z axis (Figure 6.6). In coordinate system 0xyz the electrical field intensity vector of the wave may be written as

$$\mathbf{E}_0 = \mathbf{i}_x E_0 \exp\left[j \left(\omega_0 t - \frac{2\pi z}{\lambda}\right)\right]$$

$$+ \mathbf{i}_y E_0 \exp\left[j \left(\omega_0 t - \frac{2\pi z}{\lambda} + \Delta\phi\right)\right], \tag{6.2.7}$$

where $\mathbf{i}_x$ and $\mathbf{i}_y$ are unit vectors on the 0x and 0y axes; $\Delta\phi$ is phase shift which determines the ellipticity of the wave.

At point $z = z_0$ are located two orthogonal receiving dipoles A-A and B-B, the axes of which $z_0 x\,'$ and $z_0 y'$ are rotated relative to the 0x and 0y axes by angle β. Both the dipoles are connected by long line segments to a common load, consisting of active resistance R_p. The length of the line segments ℓ_x and ℓ_y are different ($\ell_x - \ell_y = \Delta\ell$). Difference $\Delta\ell$ enables us to change the polarization diagram of the receiving antenna:

when $\Delta\ell > 0$ we will have right elliptical polarization;

when $\Delta\ell < 0$ the antenna will have left polarization.

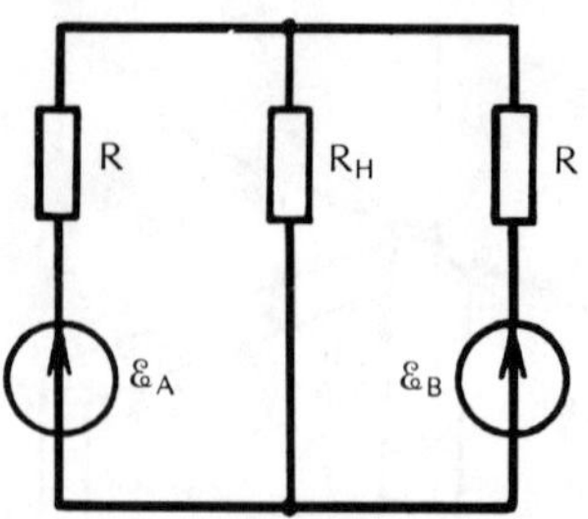

Figure 6.7

We assume that both dipoles have equal total impedances, matched with their transmission lines, and that the dipoles have no mutual coupling. The equivalent load power diagram is shown in Figure 6.7, where resistors R function as internal oscillator impedances.

The field components of the propagating wave at the point $z = z_0$ on the $z0x'$ and $z0y'$ axes will be

$$E = i_{x'}\left\{ \exp\left[j\left(\omega_0 t - \frac{2\pi z_0}{\lambda}\right)\right]\cos\beta \right.$$

$$\left. + \exp\left[j\left(\omega_0 t - \frac{2\pi z_0}{\lambda} + \Delta\phi\right)\right]\sin\beta \right\} E_0$$

$$+ i_{y'}\left\{ - \exp\ j\left(\omega_0 t - \frac{2\pi z_0}{\lambda}\right)\ \sin\beta \right.$$

$$\left. + \exp\left[j\left(\omega_0 t - \frac{2\pi z_0}{\lambda} + \Delta\phi\right)\right]\cos\beta \right\} E_0. \tag{6.2.8}$$

After trigonometric conversions we obtain

$$E = i_{x'}(1 + \cos\Delta\phi\,\sin 2\beta)^{0.5}\,\exp\left\{ j\left[\omega_0 t + \text{arctg}\left(\sin\Delta\phi\,\frac{\sin\beta}{\cos\rho}\right.\right.\right.$$

$$\left.\left.\left. + \cos\Delta\phi\,\sin\beta\right) - \frac{2\pi z_0}{\lambda}\right]\right\}E_0 + i_{y'}(1 - \cos\Delta\phi\,\sin 2\beta)^{0.5}$$

$$\times \exp\left\{ j\left[\omega_0 t + \text{arctg}\ \left(-\sin\Delta\phi\,\frac{\cos\beta}{\sin\beta}\right.\right.\right.$$

$$\left.\left.\left. + \cos\Delta\phi\,\cos\beta\right) - \frac{2\pi z_0}{\lambda}\right]\right\}E_0\ , \tag{6.2.9}$$

$$\mathbf{E} = \mathbf{i}_{x'}\, E_A + \mathbf{i}_{y'}\, E_B \, .$$

The emf of the two generators that power the load are determined by the field components at the point of reception and are

$$\dot{\epsilon}_A = k\dot{E}_A \exp\left[-j\,\frac{2\pi}{\lambda}\,(\ell \pm \Delta\ell)\right], \tag{6.2.10}$$

$$\dot{\epsilon}_B = k\dot{E}_B \exp\left(-j\,\frac{2\pi}{\lambda}\,\ell\right).$$

According to Thevenin's theorem the load resistance will be

$$U_0 = \frac{\dot{\epsilon}_A + \dot{\epsilon}_B}{2RR_o + R^2}\,RR_o = \frac{\dot{\epsilon}_A + \dot{\epsilon}_B}{2 + (R/R_o)} \tag{6.2.11}$$

Substituting (6.2.9) and (6.2.10) into (6.2.11), we obtain

$$\dot{U}_o = \frac{k}{2 + (R/R_o)}\left[\exp\, j\left(\omega_0 t - \frac{2\pi z_0}{\lambda} - \frac{2\pi\ell}{\lambda}\right)\right]\left[(1 + \cos\Delta\phi \sin 2\beta)^{0.5}\right.$$

$$\exp\left\{j\left[\mathrm{arctg}\left(\sin\Delta\phi\,\frac{\sin\beta}{\cos\beta}\right.\right.\right.$$

$$\left.\left.\left.+ \cos\Delta\phi\,\sin\beta\right) \mp \frac{2\pi\Delta\ell}{\lambda}\right]\right\} + (1 - \cos\Delta\phi\,\sin 2\beta)^{0.5}$$

$$\left.\times \exp\left\{j\left[\mathrm{arctg}\left(-\sin\Delta\phi\,\frac{\cos\beta}{\sin\beta} + \cos\Delta\phi\,\cos\beta\right)\right]\right\}\right].$$

The modulus of this expression is

$$|\dot{U}_o| = \frac{k}{2 + (R/R_o)}\left[1 \mp \sin\left(\frac{2\pi\Delta\ell}{\lambda}\right)\sin\Delta\phi\right.$$

$$\left.+ \cos\left(\frac{2\pi\Delta\ell}{\lambda}\right)\cos\Delta\phi\,\cos 2\beta\right]^{0.5}.$$

The values $\Delta\phi$ and $\Delta\psi = 2\pi\Delta\ell/\lambda$ describe the polarization of the received signal and of the receiving antenna, respectively. Returning to formulas (6.2.4), which establish the relation between $\sin\Delta\phi$ and $\cos\Delta\phi$ and the ellipticity factor, we may write

$$|U_o| = \frac{k}{2 + (R/R_o)}\left[1 \mp \frac{2k_s}{1 + k_s^2}\,\frac{2k_a}{1 + k_a^2}\right.$$

$$\left.+ \frac{(1 - k_s^2)}{(1 + k_s^2)}\,\frac{(1 - k_a^2)}{(1 + k_a^2)}\,\cos 2\beta\right]^{0.5} \tag{6.2.12}$$

Here k_s and k_a are the ellipticity factors of the irradiating wave and of the receiving antenna; β is the angle between the major axes of the polarization ellipses of the wave and antenna; the symbol "+" is used when the rotation directions of the received wave and of the wave radiated by the antenna in the transmission mode coincide, and the "−" sign denotes different rotation directions.

The maximum modulus of the load voltage will occur when the polarizations of the signal and antenna are completely matched. It is easy to see that when $k_a = k_s$, $\beta = 0$ and the rotation directions are identical

$$|\dot{U}_o|_{max} = \frac{\sqrt{2}\ k}{2 + (R/R_o)}\ .$$

Since

$$\gamma_p = S_r/S_{rmax} = |\dot{U}_o|^2/|\dot{U}_o|^2_{max}$$

then

$$\gamma_p = \frac{1}{2}\left[1 \mp \frac{4k_s k_a}{(1 + k_s^2)(1 + k_a^2)} + \frac{(1 - k_c^2)(1 - k_a^2)}{(1 + k_c^2)(1 + k_a^2)}\cos 2\beta\right].\quad (6.2.13)$$

Formula (6.2.13) enables us rather easily to estimate the effectiveness of polarization selection. Suppose, for example, that a linearly polarized antenna receives a signal with orthogonal linear polarization. In this case $k_s = k_a = 0$ and $\beta = \pi/2$, and therefore $\gamma_p = 0$

When a circularly polarized wave is received by a circularly polarized antenna with the identical rotation direction the equalities $k_s = k_a = 1$, $\beta = 0$ and $\gamma_p = 1$ are satisfied. If the rotation directions are different, then $\gamma_p = 0$.

When the wave and antenna have elliptical polarization with different ellipticity factors ($k_s = k_a = k_{el}$) and have the same rotation direction, the reception polarization factor is determined by the relation

$$\gamma_p = \frac{(1 + k_{el}^2)^2 + 4k_{el}^2 + (1 - k_{el}^2)^2\cos 2\beta}{2(1 + k_{el}^2)^2}\ .\quad (6.2.14)$$

The graphs in Figure 6.8 are used to determine the values of γ_p as a function of k_{el} and β when the E vector has the identical direction of rotation.

When the rotation direction is different

$$\gamma_p = \frac{(1 + k_{el}^2)^2 - 4k_{el}^2 + (1 - k_{el}^2)^2\cos 2\beta}{2(1 + k_{el}^2)^2}\ ;\quad (6.2.15)$$

the corresponding graphs are given in Figure 6.9.

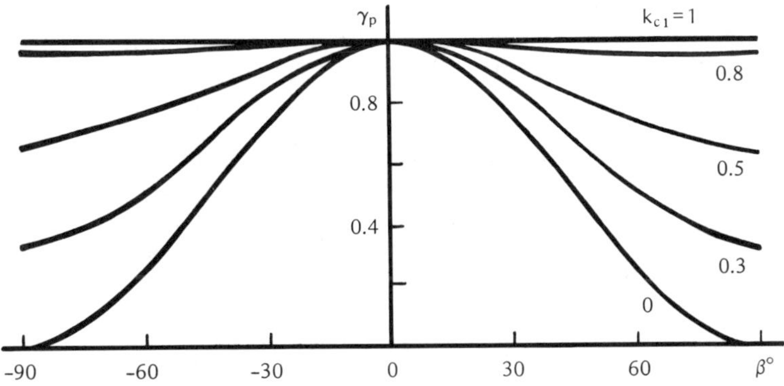

Figure 6.8

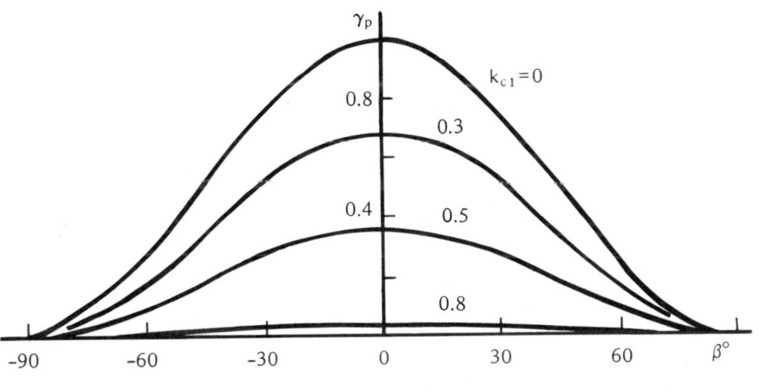

Figure 6.9

If the polarization characteristics of the wave and antenna differ only in terms of the ellipticity factor it is helpful to refer to ellipticity angle α to clarify the results of selection. When the ellipticity coefficients are replaced by ellipticity angles formula (6.2.13) is rewritten as

$$\gamma_p = \frac{1}{2}(1 + \sin 2\alpha_s \sin 2\alpha_a + \cos 2\alpha_s \cos 2\alpha_a \cos 2\beta). \qquad (6.2.16)$$

The rotation directions are determined here by the signs of angles α_s and α_a. When $\beta = 0$ and the electrical field intensity vectors have the same rotation direction (6.2.16) acquires the form

$$\gamma_p = \cos^2 (\alpha_s - \alpha_a). \qquad (6.2.17)$$

Formula (6.2.17) shows that when $\alpha_s - \alpha_a = \pm\pi/2$ the output power of the signal after the selector becomes equal to zero. The physical explanation of this result is that when the polarization angles differ by $\pm\pi/2$ the wave and antenna have orthogonal polarization.

Assuming that the receiving antenna channel is a linear system, which is precisely matched with the useful signal in terms of polarization, we may use formulas (6.2.13)–(6.2.17) to analyze the output power ratio of the signal and jamming S and J of the polarization selector (if the power densities of the antenna input jamming and signal are equal).

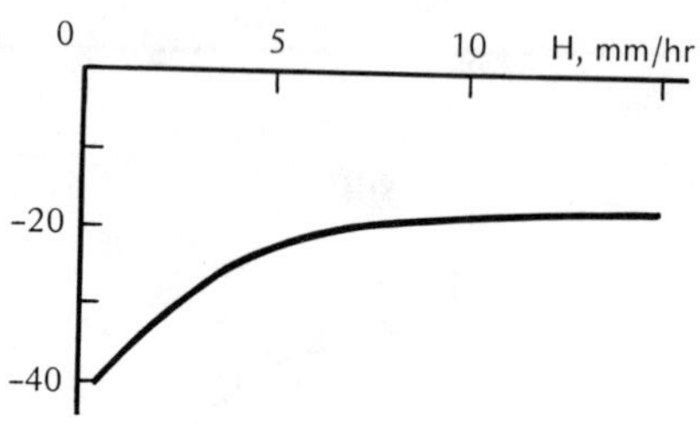

Figure 6.10

For example, by using polarization selection it is possible to greatly weaken precipitation echos and to improve sharply the receiver input signal-to-noise ratio. If raindrops were perfect spheres, then the wave reflected when they are irradiated with a circularly polarized wave would also be circularly polarized, but it would have the opposite rotation direction. In this case reflections from precipitation could be completely suppressed with a polarization selector. Raindrops actually are somewhat oblate and the reflected wave has elliptical polarization. The oblateness of raindrops increases with the rate of precipitation. For example, when the rate of precipitation is 15 mm/hr the ellipticity factor of a reflected wave is 0.88, but in a drizzle at a rate of 1-2 mm/hr $k_{el} \geqslant 0.98$. A curve, depicting the degree of suppression of precipitation reflections as a function of precipitation intensity H, in mm/hr, is given in Figure 6.10. Reflections from low-intensity precipitation are attenuated by 35-40 dB [74].

For example, a signal reflected by an airplane has, first elliptical polarization when the irradiating wave is circularly polarized and, second, the polarization parameters change continuously. Consequently, the reception polarization factor will be substantially smaller than unity for useful signal reception. According to experimental data, when circular polarization is used the average receiver input signal power is 6-8 dB lower in comparison with the case when linear polarization is used (under otherwise identical conditions). Consequently, the use of circular polarization in radars increases the signal-to-noise ratio for gentle rain, for example, not by 35-40 dB, but by 27-34 dB [74].

In order to assess how a useful signal passes through a polarization selector it is necessary to average the result of calculation by formulas (6.2.13)– (6.2.17) for different polarization parameter discrepancies between the incident wave and antenna. The probability of the appearance of signals with different polarization parameters must be taken into consideration. In view of the above statements the analysis is extremely difficult and the answer is usually obtained experimentally.

6.3. Change of Operating Frequency of Radio Systems

When people talk about a change of frequency they usually mean a change of the high-frequency carrier generated by radio systems. In application to radar, however, this may also mean a change of the pulse repetition frequency F_p and of the beam scanning frequency F_{sc}.

Under certain conditions changing frequency is an effective means of enhancing the jamming immunity of systems and simultaneously substantially improves some of their basic performance.

1. Effectiveness of Change of Carrier Frequency

There are several ways of changing the carrier frequency of radio signals. The simplest is to use two transceiver channels, tuned to different frequencies f_1 and f_2. The channels operate alternately. They are switched manually by the operator or automatically by special analysis systems, which establish the presence of jamming in the receiving channel.

Another technique is to change continuously and relatively slowly the working frequency of the system in accordance with a prescribed law, for example $f_s = f_0(1 + k \sin \Omega t)$, where Ω is low frequency and k is a coefficient that determines the deviation of frequency f_s.

The third method differs in that the frequency is changed step by step from one to another, but the system operates on each of the selected frequencies for a rather long time, for example during many radar pulse repetition periods.

Although these methods of changing frequency do find practical application [166], they must be regarded as unpromising. They do not enhance the basic properties of electronic systems, and in view of the state of the art of electronic instruments and electronic countermeasure techniques [5, 24], these methods for practical purposes also do not improve the protection of systems from jamming.

A rapid, random change of the frequency of a system, for example a change of frequency from pulse to pulse employed in radar, is fundamentally different [104, 131]. Changing the carrier frequency of radar signals accomplishes the following:

- improves target detection characteristics;

- reduces target angular coordinate measurement errors;

- increases the range and accuracy of target tracking in a background of clutter from the earth and water surfaces;

- substantially improves the protection of radar from jamming and mutual interference.

The improvement of the basic radar performance is related to the averaging of the cross section of the target σ_t for a rapid change of the frequency of the transmitted signals over a rather wide range. The power of a signal reflected by an actual target, as is known, experiences great changes, depending on the aspect angle of the target. Reflected signal power and σ_t are proportional to each other for a given direction. As the target moves relative to the radar, or as its echoes fluctuate randomly, the signal fluctuates slowly in accordance with the reflection pattern.

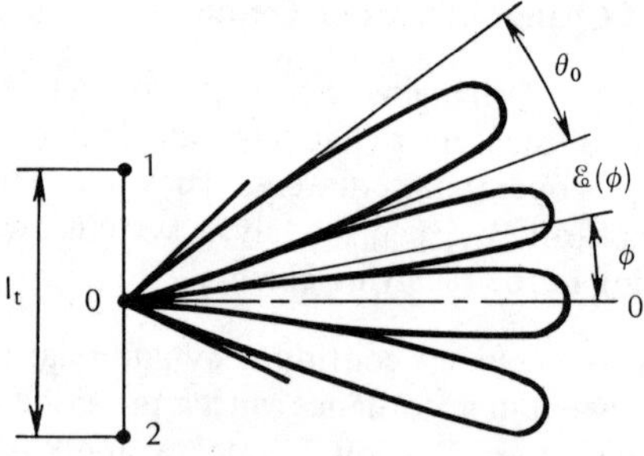

Figure 6.11

To explain the physical essence of the process by which the values of σ_t are averaged we shall examine the simplest model of a compound target, representing two isotropically radiating points 1 and 2 (Figure 6.11) with the identical reflection power, separated from each other by distance l_t (the linear size of the target). The field reradiation pattern in this case is described by the formula [134]

$$F(\phi) \sim \cos \left| \frac{\pi l_t \sin \phi}{\lambda} \right|, \qquad (6.3.1)$$

where ϕ is the angle to the receiver. The width of an individual lobe of the radiation pattern at the zero level is (for $\phi \leqslant 10\text{-}15°$)

$$\theta_0 = \lambda/4 l_t. \qquad (6.3.2)$$

As wavelength λ changes the reflection pattern is deformed due to a change of lobe width; only the position of the axis of the lobe that is symmetrical with respect to line 00 will remain unchanged.

For example, when $\lambda = 3$ cm and $\ell_t = 10$ m the lobe width is $\theta_0 = 0.043°$. If wavelength changes by 1%, then the spatial position of the 50th lobe of the reflection pattern (counted from line 00 and angle $\phi \approx 2°$) changes by $0.5\theta_0$. As a result maximum radiation will occur as λ changes for the direction that was previously characterized by zero reflection. Consequently the values of σ_t will change from zero to maximum for angles $\phi \geqslant 2°$ during the time of illumination of the target for the stated change of frequency. When the burst of reflected signals is added together the resulting effect (burst energy) will be proportional to the average target cross section.

If the carrier frequency of the signals changes randomly from pulse to pulse within the band B_t, the frequency distribution density is constant and distribution of average intervals between frequency jumps is uniform, then the average of the modulus of the difference of two adjacent frequencies is [129]

$$M_{|f_i - f_{i+1}|} = B_t/2 \qquad (6.3.3)$$

Therefore, when $B_t = 200$ MHz and $\lambda = 3$ cm, the averaging of σ_t from zero to maximum during the burst of pulses received each time the target is illuminated, should be expected to have a probability equal to unity. Thus, the possibility that the reflected signals will fade strongly due to the fact that the line of sight is close to one of the zeros of the target reflectivity pattern, is precluded.

Analysis [131] indicates that the correlation between successive echo signals will disappear if the frequency shift from one radiated pulse to the next is not less than

$$\Delta f \geqslant c/\ell_t, \qquad (6.3.4)$$

where c is the speed of light. For $\ell_t = 10$ m we obtain $\Delta f \geqslant 30$ MHz.

This effect results in a substantial improvement of target detection characteristics. These characteristics represent the dependence of the detection probability P_d of signals on the radar receiver input signal-to-noise ratio for a constant false alarm probability P_{fa}.

Three detection characteristics are presented in Figure 6.12 [132] for comparison: a) detection of a nonfluctuating target (ideal case); b) detection of a real target, characterized by slow fluctuations of σ_t; c) detection of fluctuating target with a fast, abrupt change of radar frequency. The false alarm probability is identical in all cases ($P_{fa} = 10^{-5}$), and so is the number of integrated pulses ($n_p - 20$). Under the stated conditions, as can be seen, the use of

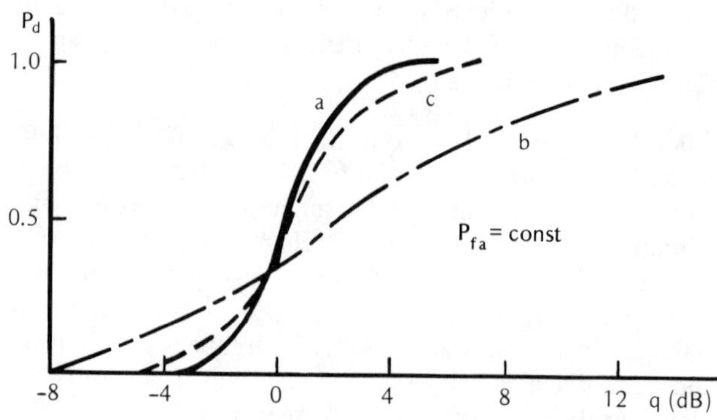

Figure 6.12

frequency agility reduces the required signal-to-noise ratio by approximately 7 dB (i.e., by a factor of about 6) for $P_d = 0.9$ under otherwise identical conditions. This corresponds to an increase of radar range by a factor of 1.6 for the same power consumption.

Fast averaging of σ_t results in a substantial reduction of angle tracking errors related to the amplitude fluctuation of the echo signals. It also reduces angular coordinate measurement errors caused by fluctuations of the position of the effective reflection center of a target. It is assumed in both cases that the effect of averaging is achieved in time T_{av} that is substantially shorter than the effective time constant t_o of an angle tracking system. If t_o is assumed equal to 1 s, then, for $T_{av} = 0.1$ s, these effects will occur.

The abrupt random changing of radar frequency is an extremely effective means of combatting active jamming, especially the most common and dangerous kinds of noise jamming. When this technique is used to change radar frequency, spot jamming (i.e., jamming with a spectral width close to the passband of the radar receiver) conceals from radar observation only the space behind the jamming vehicle, and the space in front of the jamming vehicle, in the case of spot jamming in principle cannot be concealed. The reason for this is that during each radar operating cycle jamming begins to be radiated on the radar frequency only after the next transmitted pulse is received by the reconnaissance receiver aboard the jamming vehicle. Consequently it is possible to determine the precise range of the vehicle. Also, cooperative target protection, for example by jamming from two points in space, is made considerably more difficult. In this jamming method [24] two targets, the angular coordinates of which are not resolved by the radar, simultaneously transmit spot noise jamming and thereby preclude the possibility of their detection and range resolution when a constant radar frequency is used. In such a situation missiles fired at the paired target miss badly, since by far the largest part of the trajectory of a missile is directed toward a point lying half way between the targets.

Changing the radar frequency from pulse to pulse in the presence of spot jamming fundamentally changes the situation, since the position of the target closest to the radar becomes known. Under actual conditions the distances of two targets (airplanes for example) from a radar will never be identical. And when a distance difference Δr exists it is possible, by gating the beginning of the section concealed by jamming, to identify the signal of only one target. It is assumed here that the length of the gated section Δr_g is less than Δr. If, for example, Δr_g = 20-30 m, then the condition $\Delta r_g < \Delta r$ will be satisfied with a probability close to unity.

When barrage jamming, the spectral width of which is not less than the radar frequency tuning band ($B_j \geqslant B_t$), is used, the jamming conceals the entire range of ranges within some sector θ_j. In this case the distance to the jamming vehicles cannot be resolved; jamming from two points in space becomes feasible. However, such an increase of the effectiveness of jamming is bought at a very dear price, namely a sharp increase of the radiation power of the jamming transmitter. It was shown in Chapter 2 (formula (2.1.9)) that in order to suppress a radar it is necessary to make the receiver input signal to noise ratio J/S (within the receiver passband) at least as large as the required jamming coefficient k_j. Accordingly the total power of barrage jamming at the receiver input should be

$$P_{jb} = J \, \frac{B_t}{B_r} = k_j S \, \frac{B_t}{B_r} \,, \qquad (6.3.5)$$

i.e., an extremely large increase of noise power is required in comparison with the case of spot jamming.

According to published data [131, 132, 133], fast random tuning of a radar can be accomplished in a range of from units (3-5) of percent of the center frequency f_0 to a prospective limit of (0.1-0.15)f_0. For radars in the 10 cm band (f_0 = 3,000 MHz), for example, a 5% change from f_0 is B_t = 150 MHz, and when the radar receiver passband is B_r = 1.5 MHz the ratio (B_t/B_r) = 100, i.e., the barrage jamming transmitter power under otherwise identical conditions is 100 times higher than that of a spot jamming transmitter.

Fast retuning within wide limits is a means of combatting the mutual interference of stations in the same frequency band. When N_{RS} radar stations, operating on fixed frequencies in the same band, are located in the mutual interference zone, there is always a high probability that the frequencies of the individual radars will be close together and that there will be mutual interference. By changing frequency from pulse to pulse within a wide range, mutual interference can be eliminated for practical purposes. If the effective radar receiver passband is B_r, the effective spectral width of the radiated signals is B_O and the tuning band is B_t (here $B_t \gg N_{RS} B_O$ and $B_t \gg N_{RS} B_O$), then the probability p_i that the signals of neighboring RS will enter the receiving channel of each of them is

$$p_i = \frac{B_r + B_O}{B_t} (N_{RS} - 1) \tag{6.3.6}$$

It is assumed here that only the extreme frequencies of the spectrum of the i-th signal, with spectral width B_O, and the passband of the i-th receiver (see Figure 6.13) need to coincide in order for mutual interference to occur. When $B_r = 2$ MHz, $B_O = 2B_o = 4$ MHz, $B_t = 500$ MHz (5% of $f_o = 10,000$ MHz) and $N_{RS} = 10$ we obtain $p_i \approx 0.1$. This means that 10% of the operating frequencies of each radar, on the average, will be accompanied by mutual interference. This is not a dangerous situation. It is possible, for example, to lock out automatically the receivers when they are acted upon by mutual interference; the loss of 10% of the radiated signals has virtually no effect on the basic performance of radars, and the influence of mutual interference will be eliminated.

A rapid change of radio frequency is accomplished by two fundamentally different means [104, 131, 132]. The first consists in using a high-power self-excited oscillator (a magnetron for example), the frequency of which is changed sufficiently rapidly (for example from pulse to pulse) by changing the frequency of the resonator. In this case a radar receiver uses as the LO, for example, a backward wave tube, tuned to the frequency of the transmitter by a fast AFC system. The receiver tuning time here is units of microseconds.

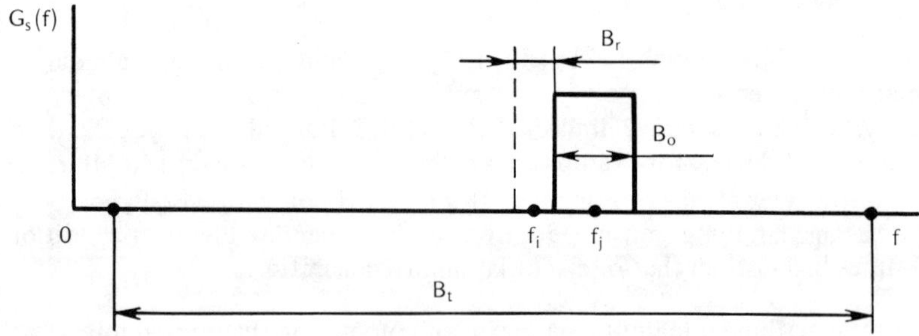

Figure 6.13

The second method is to use a set of switched low-power crystal oscillators
with subsequent frequency multiplication and power amplification; the
voltage of the same master oscillator, but with a different frequency multiplication coefficient, is used to ensure radar signal reception.

The drawbacks of radars in which frequency is changed from pulse-to-pulse
in rather wide ranges include problems of ensuring coherent processing of the
received signals and, in particular, moving target indication.

2. Effectiveness of Changing Radar Pulse Repetition Frequency

The "blind" speeds in pulsed coherent radar can be changed, interfering
reflections from distant targets can be eliminated, and the jamming immunity
of pulsed radar in relation to multiple pulse responsive jamming [24, 134,
166, 167] can be improved by changing the radar pulse repetition frequency.

As is known, pulse-to-pulse cancellation systems are used in coherent-pulse
radar for extracting moving target signals in a background of fixed or nearly
fixed clutter. The operation of these systems was examined in Chapter 5.

Signals with a Doppler frequency that is multiple of the radar pulse repetition
frequency are completely suppressed by interperiod cancellation systems. The
target speed relative to the radar, corresponding to such a Doppler frequency,
is called the blind speed.

One way to combat blind speeds is to change (stagger) the radar pulse repitition
period. Suppose, for example, that the radar pulse repetition period alternates
between two values T_1 and T_2.

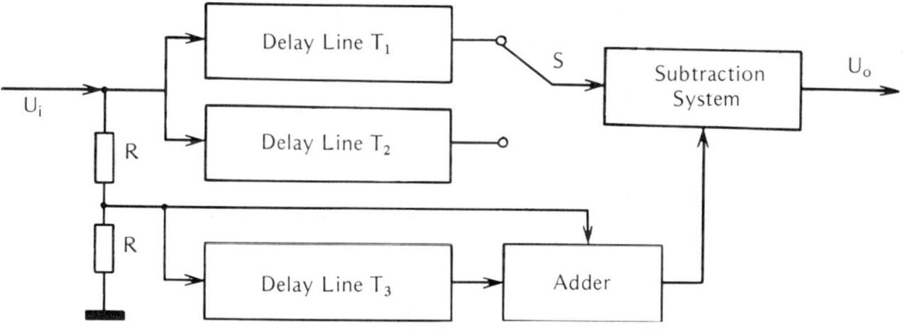

Figure 6.14

In this case a dual canceller system [167], illustrated in Figure 6.14, must be
used in the radar. In this system the sum of half the voltage $U_i(t)$ and of the
same half, delayed by time $T_1 + T_2$, is supplied to one of the inputs of the

subtraction system. Total voltage $U_i(t)$, delayed by time T_1 or T_2, is supplied to the other input of the subtraction system. Here the signals arriving during time interval T_2 are delayed by time T_1, and signals arriving during time interval T_1 are delayed by time T_2. The output voltage of the subtraction system is

$$U_0(t) = \frac{1}{2} U_i(t) + \frac{1}{2} U_i(t - T_1 - T_2) - U_i(t - T_1)$$

when switch s in is the top position and

$$U_0(t) = \frac{1}{2} U_i(t) + \frac{1}{2} U_i(t - T_1 - T_2) - U_i(t - T_2)$$

when switch s is in the bottom position and

The frequency response of such a cancellation system, regardless of the position of the switch, is given by the expression

$$|\phi_2(f)| = 2|[1.5 - \cos 2\pi f T_1 - \cos 2\pi f T_2$$
$$+ 0.5 \cos 2\pi f(T_1 + T_2)]^{0.5}|. \tag{6.3.7}$$

When $T_1 \neq T_2$ this characteristic differs substantially from the frequency response $|\phi_1(f)|$ of an interperiod cancellation system for a constant radar pulse repetition period (see formula (5.3.15)).

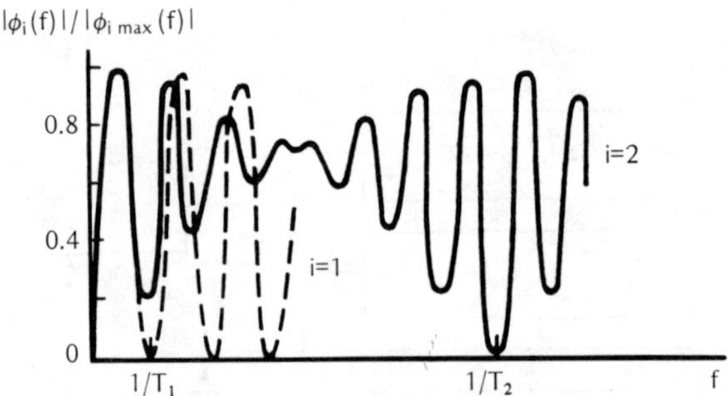

Figure 6.15

Standard graphs of the function $|\phi_i(f)|/|\phi_{i\,max}|$ for $i = 1$ and $i = 2$ are given in Figure 6.15 for comparison. The ratio T_1/T_2 for the function $|\phi_2(f)|$ is assumed to be 6/7. As can be seen, when $T_1 \neq T_2$ the number of blind speeds decreases substantially and the distance between the nulls of the frequency response on the frequency axis increases correspondingly.

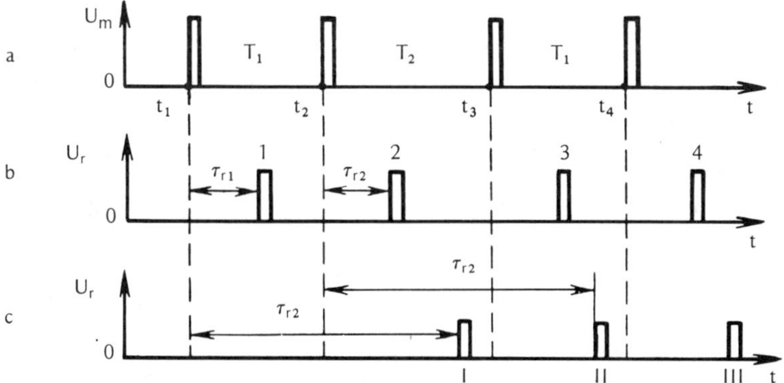

Figure 6.16

Target signals with a delay exceeding the repetition period can be eliminated by changing the radar pulse repetition period. If T_p = const and the delay time of signals from a distant target $\tau_r = (2r/c) > T_p$, then such signals will occupy a stable position on the radar display at range $\Delta r = (c/2) \times (\tau_r - T_p)$; a false blip will appear.

The radar pulse repetition period may be staggered to avoid the possibility of the appearance of false blips. Radar modulator pulses U_m, with two repetition periods T_1 and T_2, are shown in Figure 6.16a. Figure 6.16b shows receiver output signals U_r. A target located within the range of a radar, for which the signal delay time is smaller than the repetition period, $\tau_{r1} < T_1 < T_2$, is characterized in Figure 6.16b and c. The target whose signals are shown in Figure 6.16c is located at a great range, beyond the effective radar range. The signal delay time of the second target exceeds the pulse repetition period: $\tau_{r2} > T_2 > T_1$.

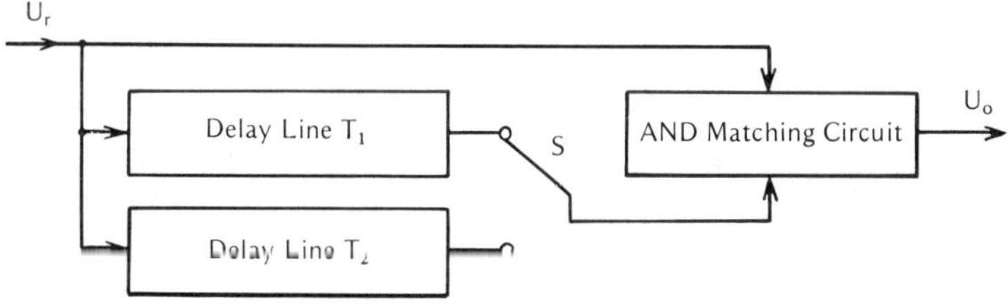

Figure 6.17

By using the video pulse processing system illustrated in Figure 6.17 [167], it is possible to pass only the signals of the first target to the radar display system. This system includes an interperiod pulse matching system (AND circuit) and two signal delay lines T_1 and T_2. Video pulses are supplied to the AND circuit directly from the receiver output and through one of the delay lines. Delay line T_1 is activated when the interval between transmitted pulses is also T_1. If the time between the radiated pulses is T_2 delay line T_2 is used. When these conditions are satisfied signal 1 (Figure 6.16b), with delay time τ_{r_1}, delayed by time T_1, coincides with undelayed signal 2, which has the same delay time: $t_1 + \tau_{r_1} + T_1 = t_2 + \tau_{r_2}$, since $t_1 + T_1 = t_2$. Consequently the signals from the near target pass through the processing system. Signal I, delayed by time T_2, will not coincide with undelayed signal II: $t_1 + \tau_{r_2} + T_2 \neq t_2 + \tau_{r_2}$, since $t_1 + T_2 \neq t_2$ and signal II, delayed by time T_1, does not coincide with signal III, etc.

If the delay time of the signals from the distant target exceeds $T_1 + T_2$, then three pulse repetition periods T_1, T_2, T_3 and three delay lines, switched in accordance with the change of the pulse repetition period, may be used.

3. Changing the Radar Pulse Repetition Period is a Means of Combatting Jamming.

By using this measure it is possible, for example, to reduce the effectiveness of multiple pulsed responsive jamming against radars. If the radar pulse repetition period is strictly constant, then multiple responsive jamming creates stable blips across the entire range scale of the radar display and the position of the jamming vehicle will be camouflaged. But if the repetition period is changed (for example is random), then only the space beyond the jamming vehicle can be concealed with multiple pulsed jamming. The exact position of the jammer will be seen on the radar display (it is the blip closest to the center of the radar scope).

The diagrams in Figure 6.18 explain what we have said above. Figure 6.18a shows pulses transmitted by the radar at moments of time t_1, t_2, t_3, which correspond to the beginning of range scanning on the radar scope. High-frequency pulses, which illuminate the target on which the jamming transmitter is installed, are shown in Figure 6.18b. The diagrams in Figure 6.18c characterize the signal reflected by the target and jamming signals; Figure 6.18d shows the same signals at the radar receiver input.

Let us assume that the radar pulse repetition period T_p is constant and the length of the sequence of jamming signals is T_p. Then, as can be seen by the diagrams in Figure 6.18, the burst of jamming pulses corresponding to illuminating signal 1 will produce stable blips in two sections of the range scale: in the section from r to r_{max} in the first radar operating cycle (the time from t_1 to t_2), and from zero to r in the second cycle (after time t_2). But if the radar pulse repetition period is random the position of the jamming signals on the time axis after time t_2 will also change randomly from period to period. Consequently the jamming signals do not make bright stable blips in the section of the range scale between zero and r.

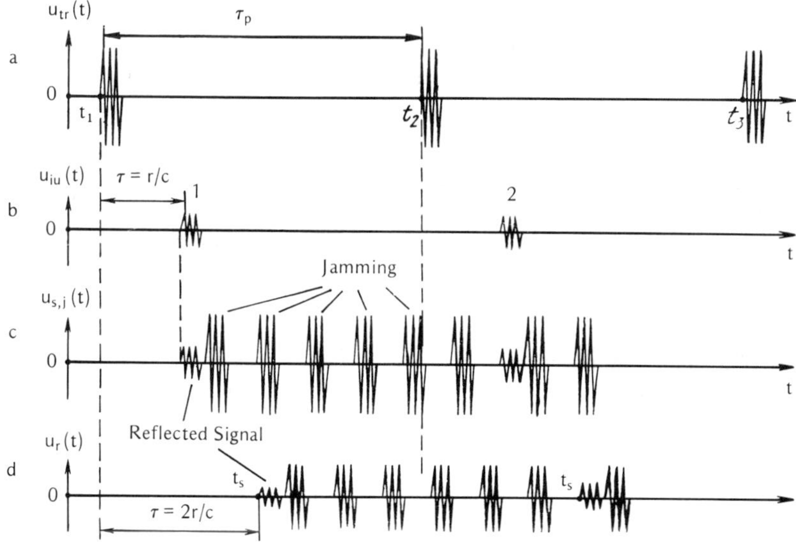

Figure 6.18

6.4. Frequency Diversity Radar

A frequency diversity radar transmits, receives and processes n ($n \geqslant 2$) radio signals with different carrier frequencies. To be able to do this such a radar station has n transmitters and receivers with the corresponding antenna system, and a multichannel signal processing system. A frequency diversity radar has the following advantages over a single-frequency system: the overall transmitted power is increased, target detection range is increased for given transmitted power, and the reliability and noise immunity of the system in relation to jamming and natural interference are improved.

There are two different ways to use several operating (carrier) frequencies in one radar. The first way is to generate a separate antenna beam for each operating frequency; the second way is to radiate signals on several frequencies simultaneously in the same beam.

Diagrams of the first type of radar are shown in Figure 6.19 for the case when three operating frequencies are used. A single point target in this case will be illuminated by radio waves on the same frequency; targets located at the inter-section of two adjacent beams patterns are the exception. Such radars essentially amount to the replacement of a single-frequency system by several independent systems, operating on different frequencies and observing different regions of space. The utilization of such a multifrequency system substantially improves the performance of the radar system as a whole. When n frequencies are used the total transmitted power of the radar increases by the same factor.

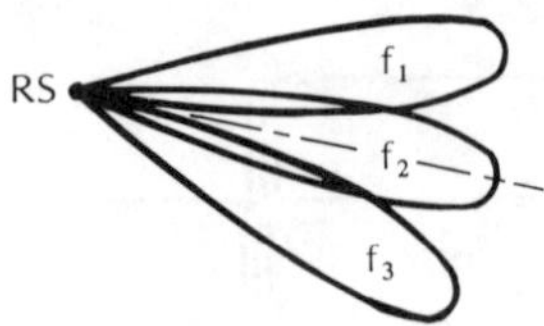

Figure 6.19

The maximum power of radar transmitters is limited by a number of factors:
breakdown in the high-frequency lines, cooling problems, high modulating
voltage, etc. In view of these power limitations of an individual transmitter
the use of frequency diversity makes it possible to increase the total transmitted
power of a radar. Consequently the effective radiated power (ERP) of a system
(the product of radiated power P_t and antenna gain G) is increased, with the
result that the range of the radar is increased and jamming immumity is improved.

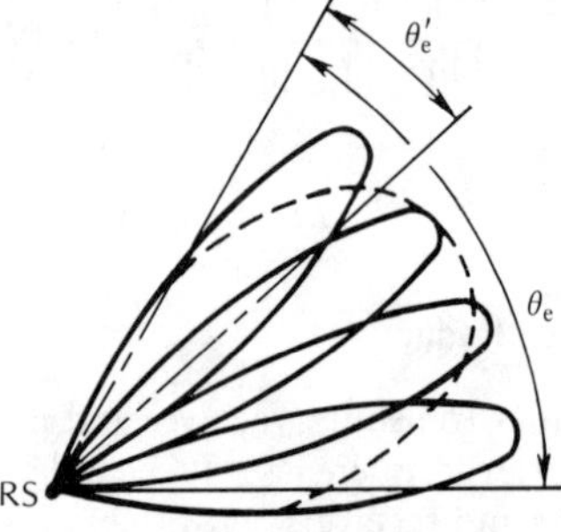

Figure 6.20

By way of explanation let us assume, for example, that a radar is designed to
scan a region of space, the boundaries of which are determined in angular
coordinates in the vertical plane by sector θ_1 (Figure 6.20), and in the horizontal plane by a sector of $360°$. We assume that in a single-frequency radar, in
order to observe the scanned zone for a given time T_S with illumination of a
single target T_i, it is necessary to use an antenna with a fan beam of width of
θ_e in the vertical plane and θ_a in the horizontal plane. With an n-frequency
station the width of the radiation pattern of each channel in the vertical plane
can be made to equal $\theta_e' = \theta_e/n$, and sector θ_e can be covered with n beams.
The width of the antenna radiation pattern in the horizontal plane remains the
same (θ_a). When such a multichannel radar is used the time required for scanning
a given space and the point target illumination time remain constant, but the
antenna gain for each channel increases by a factor of n in comparison with a
single-frequency radar. Antenna gain can be estimated with the approximate
formula [155]

$$G = 25{,}000/\theta_e°\theta_a° \qquad\qquad (6.4.1)$$

where θ_e° and θ_a° are the widths of the antenna radiation pattern at one-half power for two mutually perpendicular planes, for example the vertical and horizontal.

If the powers of the individual transmitters of single-frequency and frequency diversity radar are assumed to be identical (i.e., the total radiated power of a multichannel radar increases by a factor of n), then the ERP of each channel of the multichannel radar will be n times greater than that of the single-frequency system. The radar ERP to a great extent determines active jamming immunity, since the ERP of a jamming transmitter must be increased proportionately. Consequently, in order to jam one of the channels of an n-frequency radar with n transmitters and spatially separate antenna beams it is necessary, under otherwise identical conditions, to increase the ERP of the jamming transmitter by a factor of n in comparison with conditions under which a single-frequency station is jammed. Here the jamming sector for frequency diversity radar will be n times smaller.

The jamming immunity of a frequency diversity radar in relation to clutter of manmade and natural origin also increases as a result of the narrowing of the antenna radiation pattern of each channel in comparison with a single-frequency radar. As is known [134], if particles that produce clutter (dipole reflectors, raindrops, etc.) are uniformly distributed within a space substantially larger than the radar resolution cell, then power C of clutter at the radar receiver input will be proportional to the resolution cell. For a pulsed radar this volume is determined by the product of the cross section of the antenna radiation pattern at given range r and radar pulse duration τ_p, expressed in units of range. Hence we have this relation [134]

$$C = k\pi \; \frac{\theta_e \theta_a}{4} \; r^2 \; \frac{c\tau_p}{2}, \qquad\qquad (6.4.2)$$

where c is the speed of light; k is the proportionality factor.

Formula (6.4.2) shows that a narrowing of the beam, for example in the vertical plane, by a factor of n results in a proportional reduction of the input power of clutter under otherwise identical conditions. The effective radar range is proportional to the square root of its antenna gain. Therefore, if the ranges of a single-frequency radar and of an individual channel of a frequency diversity system are compared (under otherwise identical conditions), then the range of the n-channel system will be $\sqrt{n}$ times greater (if the absorption of electromagnetic energy in the atmosphere is ignored).

Let us examine the second method of frequency diversity, when each target is illuminated by transmissions on several frequencies. If the ERP's of frequency diversity and single-frequency systems are assumed to be identical, then the target detection range and jamming immunity of the former will be greater. This is the distinguishing feature of frequency diversity radars

When the frequency of the waves that illuminate a target changes, the target reflection pattern is deformed, and consequently the amplitude of the reflected signals, received at a given aspect angle also changes (see §6.3 for further details). When one-time transmission on several frequencies is used, and the difference between these frequencies is selected in accordance with inequality (6.3.4), and the output signals of all channels are added, the resulting signals should be expected to have much smaller amplitude fluctuations for a moving target than in the case when a single-frequency radar is used. Reduction of the fluctuations of the reflected signals leads to an increase of target detection range (under otherwise identical conditions).

The increase of radar range can be estimated in the following manner. Suppose that signal processing in a frequency diversity system consists in comparing the receiver output voltage of each channel with the threshold and adding up of the voltages of the channels in which the threshold is crossed. Thus the total output voltage, supplied, for example, to the display, may be written as

$$U_o = \sum_{i=1}^{n} U_i \tag{6.4.3}$$

where U_i is the voltage that appears at the output of the threshold system of the i-th channel. The probability that the sum of the signal and noise in a given channel exceeds the threshold, i.e., $U_i > U_{th}$, is defined as the detection probability P_{di} of that channel. The the detection probability of an n-channel system $P_{d\Sigma}$ is the probability that the threshold level will be exceeded even only in one channel and for $P_{d_1} = P_{d_2} = \ldots = P_{di} = \ldots = P_{dn}$ is

$$P_{d\Sigma} = 1 - (1 - P_{di})^n \tag{6.4.4}$$

The false alarm probability of a system $p_{fa\Sigma}$ may be expressed analogously through the false alarm probability of an individual channel p_{fai}

$$p_{fa\Sigma} = 1 - (1 - p_{fai})^n \tag{6.4.5}$$

when $p_{fai} \ll 1$ the approximate equality

$$p_{fa\Sigma} \approx n p_{fai} \tag{6.4.6}$$

may be used. Two conditions must be statisfied when the detection characteristics of frequency diversity and single-frequency stations are compared. First, the total transmitted energy must be identical, i.e., $E_t = \sum_{i=1}^{n} E_{ti}$, where E_t and E_{ti} are the transmitted energies of a single-frequency and of the i-th channel of a frequency diversity radar, respectively. Second, the resulting detection probability $p_{d\Sigma}$ and false alarm probability $p_{fa\Sigma}$ of both types of radar must be identical.

In reception of signals, for example with an unknown initial phase and slowly fluctuating amplitude, the required signal-to-noise ratio for a single-channel system is connected to the detection and false alarm probabilities by the following Equation [52]:

$$q = \frac{2E_s}{N} = 2 \frac{\ln(p_{d\Sigma}/p_{fa\Sigma})}{\ln(1/p_{d\Sigma})} \; , \tag{6.4.7}$$

where E_s is the total energy of the received signals, used for a single detection; N is the spectral power density of the receiver noise.

Different detection and false alarm probabilities must be ensured in the individual channels of a multichannel system (p_{fai} and p_{di}). Therefore the signal-to-noise ratio must also be different. Using formulas (6.4.4) and (6.4.6) p_{di} and p_{fai} may be expressed as

$$p_{di} = \sqrt[n]{1 - p_{d\Sigma}} \quad \text{and} \quad p_{fai} = p_{fa\Sigma}/n.$$

Now returning to formula (6.4.7), we may calculate the input signal-to-noise ratio of one channel q_i of the receiver of a frequency diversity radar. For a given value of N, identical for both a single-channel radar and for any channel of an n-channel system, we may determine the required input signal energy E_{si} of an individual channel of a multichannel station $E_{si} = q_i N/2$.

The total input signal energy, distributed among the n channels, will be

$$E_{s\Sigma} = \sum_{i=1}^{n} E_{si} = nE_{si}.$$

If signal input energy nE_{si}, required in an n-channel system, is less than the required signal energy E_s of a single-channel radar, then, under otherwise identical conditions, the effective range of a frequency diversity radar (r_{fd}) will be greater than that of a single-frequency radar (r_s); the ratio of these ranges is

$$r_{fd}/r_s = \sqrt[4]{E_s/nE_{si}}$$

Since the energy ratio is q,

$$r_{fd}/r_s = \sqrt[4]{q/nq_i} \; . \tag{6.4.8}$$

The graphs in Figure 6.21 illustrate the results of such calculations for a three-frequency system. The solid curve in the figure corresponds to the detection characteristic of the three-frequency system, and the broken curve corresponds to single-frequency radar. The calculation results show, for example, that when $p_{fas} = p_{fa\Sigma} = 10^{-5}$ and $p_{d\Sigma} = 0.95$ the required signal-to-noise ratio for a single-frequency radar $q \approx 180$, and $3q_i = 60$ for a three-frequency system. Thus the increment of the effective range of the three-frequency system is $r_{fd}/r_s = 1.32$.

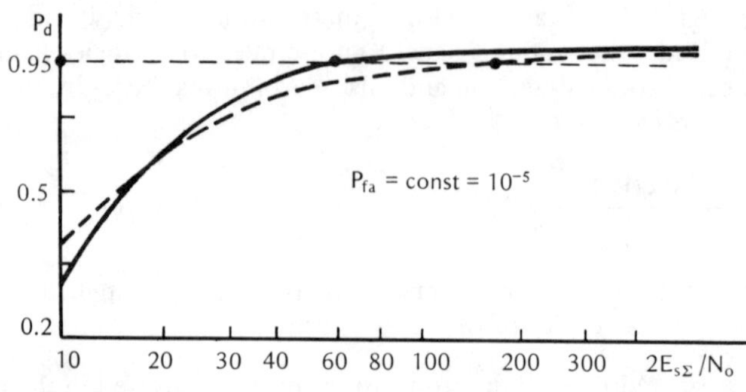

Figure 6.21

Various versions of the combined processing of the output signals of individual channels are employed to improve the jamming immunity of a frequency diversity. For example, the following processing variations are used in a three-frequency radar [43, 78] :

a) $U_O = k_1(U_1 + U_2 + U_3)$,

b) $U_O = k_2(U_1 U_2 + U_2 U_3 + U_1 U_3)$,

c) $U_O = k_3 U_1 U_2 U_3$.

In these variations the detection characteristics of radar deteriorate, but jamming immunity improves in the order in which they are listed.

The probabilities $p_{d\Sigma}$ and $p_{fa\Sigma}$ for a three-channel system using processing version a) are determined for given detection and false alarm probabilities (p_{di} and p_{fai}) of an individual channel by formulas (6.4.4) and (6.4.6)

If version b) is used, then

$$p_{d\Sigma} = 1 - (1 - p_{di}^2)^3 , \qquad\qquad (6.4.9)$$
$$p_{fa\Sigma} = 1 - (1 - p_{fai}^2)^3 \approx 3p_{fai}^2 \qquad\qquad (6.4.10)$$

Finally, we have for version c)

$$p_{d\Sigma} = p_{di}^3 \qquad\qquad (6.4.11)$$
$$p_{fa\Sigma} = p_{fai}^3 \qquad\qquad (6.4.12)$$

When probabilities $p_{d\Sigma}$ and $p_{fa\Sigma}$ are given we may use formulas (6.4.4), (6.4.6), (6.4.9)–(6.4.12) to calculate probabilities p_{di} and p_{fai}, and then formula (6.4.7) to determine the required input signal-to-noise ratio of one receiver. The results of such calculations are presented in the form of graphs in Figure 6.22, where triple the value of q_i corresponding to the total transmitted power of the radar is plotted on the abscissa. The results of the analysis show that version a) has the best detection characteristics and version c) the worst.

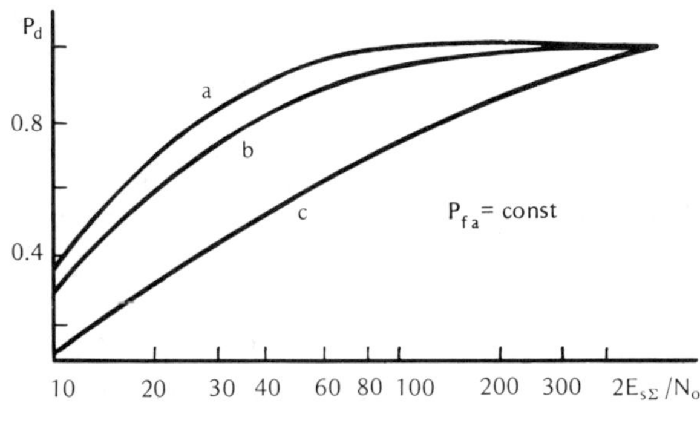

Figure 6.22

The jamming immunity of radars that use these versions of signal processing varies because in version a) it is sufficient to suppress only one frequency to jam the radar, in version b) at least two frequency channels must be jammed, and all three in version c).

Different pulse repetition frequencies may be used in different channels to improve the jamming immunity of a multichannel radar [43, 78]. The pulse repetition frequency in one channel, for example, may be selected such as to ensure the detection of fast moving targets in a background of stationary or slowly moving targets ($F_p \geqslant 2F_{dmax}$), and in another channel such as to ensure the unambiguous measurement of maximum range $r_{max}(F_p \leqslant c/2r_{max})$. When different repetition frequencies are used in radar channels the number of pulses that accumulate during target detection on different frequencies increases. Consequently the detection performance of the radar deteriorates. Exact analysis of the performance of such a system with $n > 2$ is frought with serious difficulties and cannot be done at the present time. Data obtained from preliminary studies [43] suggest that detection performance should not deteriorate significantly.

The improvement of the reliability of frequency diversity radar in comparison with a single-frequency system is related to the fact that the probability that all n channels will fail at the same time is vastly less than the probability of failure of the one channel of a single-frequency radar. At the same time a frequency diversity radar, in which only one channel continues to operate normally after all the others have failed, may be assumed to be functioning with some deterioration of performance. If the probability of trouble-free operation of a multichannel radar (p_{sn}) is defined as the probability of normal operation for a given period of time of just one of n channels, then p_{sn} is connected to the trouble-free operation probability of a single channel p_{si} by the trivial relation

$$p_{sn} = 1 - (1 - p_{si})^n. \qquad (6.4.13)$$

Calculations by formula (6.4.13) show that even for n = 3 the reliability of a system substantially exceeds that of a single channel. For example, when the trouble-free operation probability of one channel is 0.6 (low reliability), this probability for a three-channel system is 0.933.

Judging by published materials [37, 132, 195], it may be assumed that the frequency difference in a multichannel radar amounts to tens and hundreds of megahertz.

6.5. Application of Frequency and Phase Selection Systems

1.   Application of Frequency Stabilization and Automatic Frequency Tracking Systems

The required band of a radio receiver, i.e., its selective properties, in the final analysis, depends on the spectrum of the received signals, and also on the carrier instability of the radio transmitter and the tuning frequency of the receiver LO. Because of transmitter and LO frequency instabilities f_t and f_o, the band often must be made wider than optimum, which detracts from frequency selectivity and reception conditions in relation to the optimum.

As is known, there are several ways to improve the frequency stability of transmitters:

- use high-stability master oscillators;

- use crystal and various parametric frequency stabilization techniques [103];

- use automatic frequency stabilization systems.

The same techniques are also suitable for improving the frequency stability of receiver LO's. In most cases, however, receiver LO's must be tuned in a wide frequency band, and consequently the method based on the utilization of automatic frequency tracking systems acquires fundamental importance here.

Radio links may be divided into two groups from the standpoint of frequency selection:

- links in which band spreading is minimized by using a highly stable carrier oscillator and LO; in this case communications are conducted on prescribed (fixed) frequencies;

- links in which the receiver has an automatic frequency tracking system. Links of this type also include those in which an automatic frequency stabilization system is used in the transmitters.

In links of the first type the required band spreading is determined by the total frequency instability Δf_t and Δf_o of the radiated signal and LO. For definition we shall assume that intermediate frequency $f_{if} = f_t - f_o$, where f_t and f_o are the frequencies of the radiated signal and LO, respectively. Frequencies f_t and f_o, due to instability, change in time relative to their average values (mathematical expectations) f_t^* and f_o^*, which are assumed to be constant (the stationary case). Frequencies f_t^* and f_o^* differ from their nominal values f_{to} and f_{oo}, corresponding to the exact frequency settings of the transmitter and receiver. The difference $f_{to} - f_{oo} = f_{ifo}$ is the nominal intermediate frequency.

The required band spreading is determined both by the difference of the mathematical expectations from the nominal frequencies and by the instability of frequencies f_t and f_o. The frequency deviations $\Delta f_t = f_t^* = f_{t_o}$ and $\Delta f_o = f_o^* - f_{oo}$ are statistically independent random values. The required band spreading due to the initial frequency setting error Δf_i may be assumed equal to

$$2\Delta f_i = 2\sqrt{\sigma_t^2 + \sigma_o^2} , \tag{6.5.1}$$

where σ_t^2 and σ_o^2 are the variances of f_t and f_o, respectively.

Likewise, assuming the frequency drifts of the transmitter and LO to be stationary random functions of time with zero mathematical expectations and variances σ_{td}^2 and σ_{od}^2, we may assume the required spreading to be

$$2\Delta f_{dr} = 2\sqrt{\sigma_{td}^2 + \sigma_{od}^2} \tag{6.5.2}$$

The value

$$2\Delta f_{tot} = 2\sqrt{\Delta f_i^2 + \Delta f_{dr}^2} = 2[\sigma_t^2 + \sigma_o^2 + \sigma_{td}^2 + \sigma_{od}^2]^{0.5} \tag{6.5.3}$$

also determines the total band spreading of the receiver. If the band is not spread and the optimum band is used, then reception conditions will deteriorate increasingly as the ratio $\Delta f_{tot}/\Delta f_{opt}$ increases due to frequency errors.

In the second type of links such deterioration can be avoided by using automatic frequency tracking (AFT) systems. AFT systems are especially effective when the transmitter has a reference signal. These systems include the receivers of all radar stations located near radio transmitters (from which a reference signal can be taken) and certain radio links. They also may include, conditionally, the AFT systems of missiles with a semiactive homing system, in which the reference signal is picked up by the tail receiver of a missile. The problem of increasing the jamming immunity of AFT systems with a reference signal is not an urgent one, and consequently the design of automatic search and lock-on systems is simplified.

In AFT systems without a reference signal basic attention is devoted to ensuring the best search and lock-on conditions (minimum errors, lowest tracking interruption probability, etc.) for an AFT system. This requirement imposes constraints on the functional design of an AFT system and the choice of its parameters.

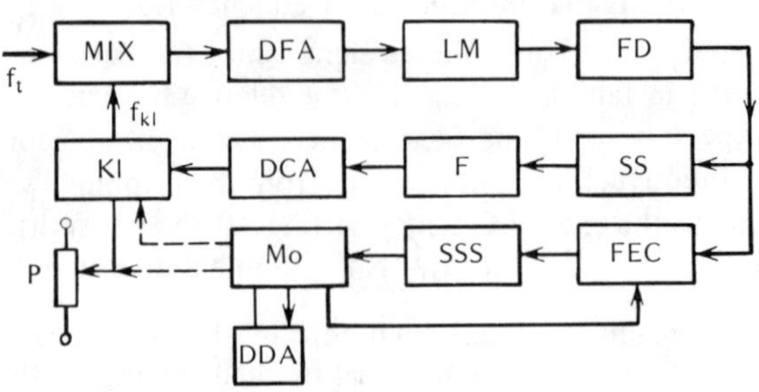

Figure 6.23

AFT System with Reference Signal.

One design modification of an **AFT** system of this type in application to a
klystron local oscillator is illustrated in Figure 6.23. As a result of offsetting
of the frequency f_t of the reference signal (continuous or pulsed) and of the
frequency f_{kl} of the klystron oscillator signal, a voltage of difference frequency
f is formed, which, after passing through the difference frequency amplifier
(DFA) and limiter (L), goes to a frequency detector (discriminator) (FD). If
lock-on does not occur, then the wideband search system is turned on: the
klystron is mechanically turned by a motor (Mo) in such a way that its frequency
changes linearly. When the end of the band is reached the motor reverses and
tunes the klystron in the opposite direction. At the same time the voltage on
the reflector is changed during the tuning process with a programmed potentio-
meter in such a way that maximum power is always maintained in the selected
klystron oscillating mode. The search stop (motor switch-off) system (SSS) is
activated at the instant the difference frequency f enters the operating band of
the frequency discriminator, but only if this is permitted by a frequency error
correction (FEC) logic system. The motor must be started and stopped in such
a way that after it is turned off the difference frequency will fall within the
lock-on band of the electronic loop of the AFT system. This loop is closed
through a filter (F), DC amplifier (DCA) and klystron reflector. These conditions are usually easy to satisfy. An extra search system (SS), which scans in a
narrow band and is subsequently switched to the tracking mode, may be included
in the electronic control loop (or parallel to it) to improve the reliability with
which the system is switched to the tracking mode.

After the motor stops the frequency error is corrected by the electronic loop,
the motor is disconnected from the tracking system and the search system
performs as an amplifier. The motor is turned on again only in the event of
prolonged and large drifts of frequency f. In the interest of design simplicity
a static **AFT** system (without an integrator) is usually selected.

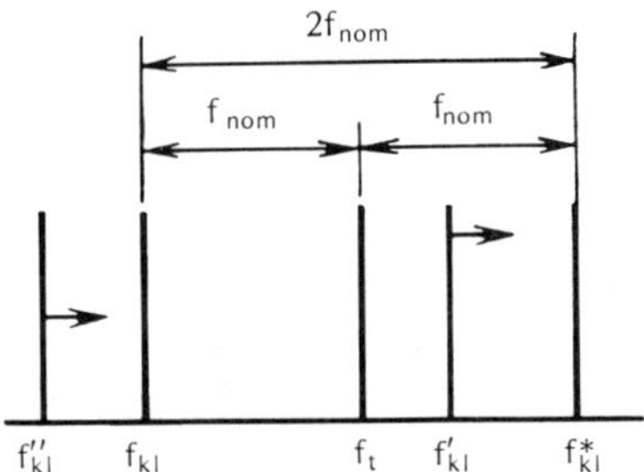

Figure 6.24

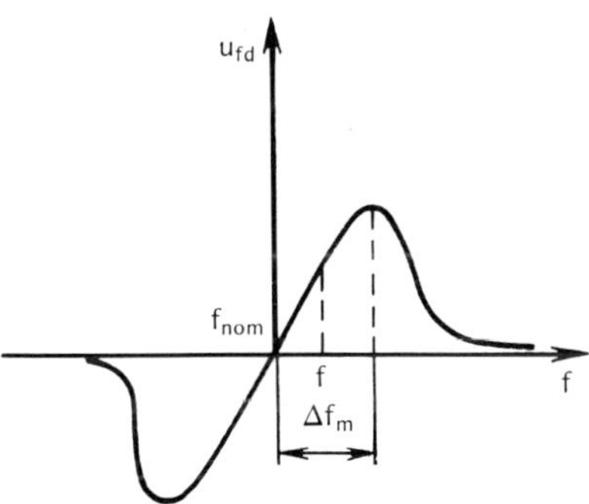

Figure 6.25

The frequency error correction (FEC) system operates in such a way that lock-on is precluded when difference frequency f appears as a result of inaccurate tuning of the klystron to frequency f^*_{kl}, differing from the required frequency by $\approx 2f_{nom}$, where f_{nom} is the nominal difference frequency, equal to the frequency to which FD is tuned. Let us assume that the klystron frequency begins to increase from f'_{kl} as soon as the system is turned on (Figure 6.24). As soon as the difference $f'_{kl} - f_t$ increases to the point where $f = f'_{kl} - f_t$ falls within the band of the amplifier of a DF without an FEC system, lock-on will take place on frequency f^*_{kl}, very different from frequency f_{kl}. To prevent that it is necessary to make the locking system sensitive to the polarity of the signal generated on the FD output when the difference frequency passes the aperture of the FD characteristic, which determines the dependence of the output voltage u_{fd} on f. For the examined direction of search and FD characteristic, illustrated in Figure 6.25, frequency f increases with f_{kl}, and consequently the FD output voltage will first have a negative, and then (after transient frequency f_{nom} passes through it) positive polarity. This reversal of polarities corresponds to mistuning, and the signal that travels from the FD to the FEC system prevents lock-on. Conversely, if search begins at low klystron frequencies, starting, for example, with f''_{kl} (Figure 6.24), then frequency f would decrease during the search process (we recall that $f = f_t - f_{kl}$) and entry into the aperture of the FD characteristic would occur from high frequencies $f > f_{nom}$, so that the FD output voltage would initially have positive polarity and the FEC circuit would stop searching. When the direction of search changes the order of polarity alternation and the logic of the FEC system are reversed. The necessary signal about the reversal of the direction of search is sent to the FEC system from the motor. When a signal is caught on frequency f^*_{kl}, corresponding to mistuning, the difference frequency will differ strongly from f_{nom}.

Because the AFT system has a reference signal, there are extensive opportunities to stabilize f with a high degree of accuracy relative to the nominal frequency f_{nom}, which coincides with the center frequency of the discriminator. It is important to note, however, that limitations imposed on the increase of the gain k of an AFT system are attributed to the fact that such a system contains components with inertia. If very high gains are excluded from the discussion, then we need only consider the comparatively small time constant T_{fd} of the FD filter in the electronic control loop and time constant T_f of the filter. The transfer function of a closed system in this case has the form [86]

$$\Phi(D) = \frac{k_e}{\tau_\zeta^2 D^2 + 2\zeta\tau_\zeta D + 1},$$

(6.5.4)

where

$$k_e = \frac{k}{k+1} \; ; \quad \tau_\zeta = \sqrt{\frac{T_{fd}T_f}{k+1}} \; ; \quad \zeta = \frac{T_{fd} + T_f}{2\sqrt{k+1}} \; .$$

(6.5.5)

Figure 6.26 shows a family of curves of the dependence of gain k on the time constant ratio $a = T_f/T_{fd}$, which provides the required damping factor ζ. The latter determines the transient response of a system. It follows from relations (6.5.5) that

$$k = \frac{(a+1)^2}{4a\zeta^2} - 1.$$

(6.5.6)

In order to obtain high values of k (of the order of 30-50) it is necessary (for $\zeta = 0.5-1.5$) that $a = 50-100$. The transient time is found from the relation

$$t_{tr} = \frac{2mT_{fd}T_f}{T_{fd} + T_f} \zeta,$$

(6.5.7)

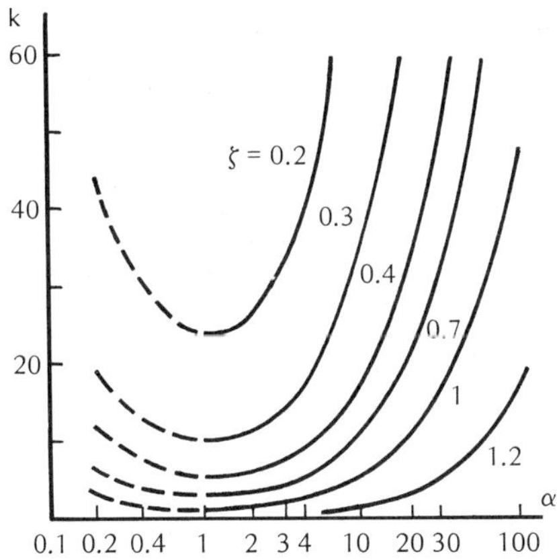

Figure 6.26

where m = 3–6. When $T_f \gg T_{fd}$ time t_{tr} for practical purposes does not depend on T_f and is determined for a given value of ζ by the smaller time constant T_{fd} and is

$$t_{tr} \approx 2mT_{fd}\zeta. \tag{6.5.8}$$

Consequently k can be increased for fixed values of ζ and T_{fd} by increasing the time constant of the filter, and for large k the transient time for practical purposes depends neither on T_f nor on k. However, all the above discussions do not apply when the values of k are very large and when extremely small inertias and delays in the DFA filters must be taken into account. Of course, the dynamic characteristics can be improved to some extent by including correcting circuits.

AFT Systems Tuned by Received Radio Signal.

A functional diagram of the system (Figure 6.27), which, in addition to the main control loop, consisting of a mixer (Mix), difference frequency amplifier (DFA), limiter (LM), frequency detector (FD), filter (F), amplifier (Amp) and LO, also contains additional components: search oscillator (SO), logic locking circuit (LLC) and locking relay (LR), which, unlike the one examined previously, is an essential element of the system, which switches the system from the search mode (S) to the lock mode (L). Because of the inclusion of this system the frequency of the oscillator in the search mode does not depend on the output voltage of frequency discriminator FD. Consequently interference that occurs at the FD output in this mode has no influence on the performance of the search oscillator and does not alter the pre-established logic of the switching of the system to the tracking mode.

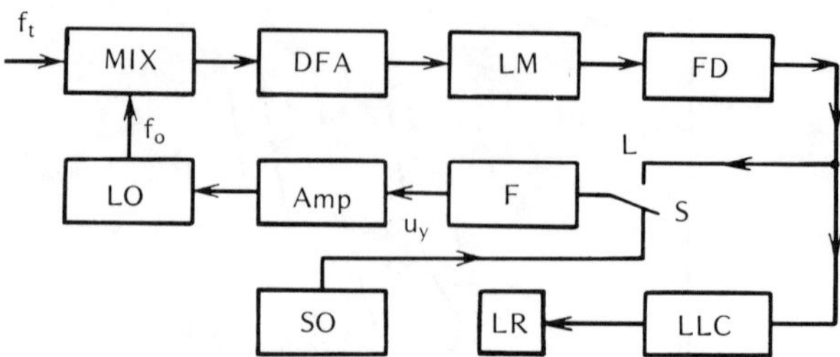

Figure 6.27

The locking process can be broken down into two operations. The first operation consists in the accumulation of voltage on the output of the linear stages
of the receiver or frequency discriminator for making a decision concerning
the presence of a difference frequency in the output voltage of a signal. This
is a typical detection problem. Search and decision making may be performed
variously. For example, frequency modulation of the LO may be induced by
the oscillations of an extra reference oscillator, followed by phase detection
of the voltage taken from the FD output; a two-step locking process may
occur here. In the first step, after voltage appears on the FD output, the
system stops searching but remains open. In the second step, after voltage is
accumulated on the output of the phase detector (PD), the final decision is
made concerning locking. If the PD output voltage does not reach a certain
level within a prescribed interval of time the system continues to search.

The second operation of the locking process consists in the correction of the
initial errors that occur in the system when the locking relay closes. Positional misalignment occurs in systems with a single integrator, and positional
misalignment and speed differences, attributed to the initial conditions on two
integrators, occur in systems with two integrators. The second operation is
more difficult from the standpoint of ensuring reliable locking, since errors
are large at the time of switching and interference passes through the feedback
channel. To reduce the time required for correcting initial errors the gain,
which is increased to the required level as soon as the initial errors are corrected
by the system, is often increased by the transient time. The gain should be
changed so as not to cause additional transient processes.

The specific design execution of the logic locking system depends on the
purpose of the AFT system, search speed requirements, acceptable storage time,
tolerable false alarm level, etc.

After the locking relay closes and initial errors are corrected for a comparatively
low noise level (when the system may be assumed linear), the structural dynamic
diagram illustrated in Figure 6.28 may be used. Interference voltage u_i at the
FD output is the result of conversion of the noise combined with the radio

signal at the receiver output. If the noise level is low and the limiting threshold of the FD is exceeded the one-sided spectral density of that voltage [77] is

$$G_i = \frac{G_0 k_{fd}^2}{U_1^2} \omega^2 ,$$ (6.5.9)

where G_0 is the one-sided spectral density of the FD input noise and U_1 is the FD input limiting threshold. Here $G_0 = k_{sa}^2 G$, where G is the spectral density of white noise or of wide band interference at the input of the selective difference frequency amplifier, which has gain k_{sa} and equivalent noise bandwidth B_n.

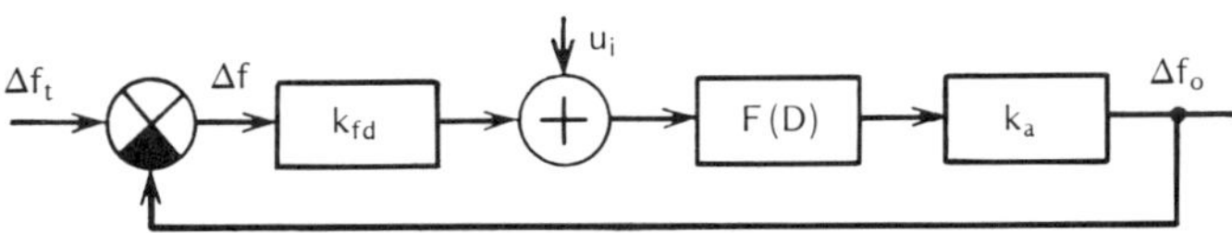

Figure 6.28

Proceeding from the structural dynamic diagram, we may find for the variance of the frequency error in an AFT system with a single-stage filter the expression [155, Vol. III]

$$\sigma_i^2 = \frac{G_0}{U_1^2} \frac{k^2}{T^2} B_n \left(1 - \frac{2\beta_n}{\pi B_n} \arctan \frac{\pi B_n}{2\beta_n} \right)$$ (6.5.10)

$$\approx \frac{4 G_0}{U_1^2} \beta_n^2 B_n$$

Here k and T are the gain of the system and the time constant of the filter; $\beta_n = (k + 1)/4T$ is the equivalent noise bandwidth of the closed AFT system. Approximate equation (6.5.10) is valid for $(\beta_n/B_n) \ll 1$.

From the standpoint of reducing the influence of interference it is advisable to narrow band β_n of the system by increasing time constant T of the filter.

High-level jamming can interrupt tracking, when the error $\Delta f = f - f_{nom}$ begins to substantially exceed the aperture of the FD characteristic and the system loses its capacity to follow the frequency of a signal. There are effective ways of determining tracking interruption conditions in radio automation systems [118]. However, the application of the results presented in [118] to AFT systems encounters problems, which are attributed to the complexity of the dynamic equivalent of the frequency discriminator. The basic laws of interruption are set forth in [13] for the simplest static AFT system with a single-stage filter. The transfer function of an open system in this case is

$$W(D) = \frac{k}{TD + 1}$$

where $k = k_{fd}k_a$; $T = RC$ is the time constant of the filter. The oscillator frequency deviation Δf_0 after a change of signal frequency by Δf_t (the case of operation in the straight line segment of the FD characteristic) is

$$\Delta f_0 = \phi (D) \Delta f_t,$$

where

$$\phi(D) = \frac{W}{1 + W} = \frac{k_{eq}}{\tau_e D + 1} . \tag{6.5.11}$$

Here $k_{eq} = k/(1 + k) \approx 1$ is the gain of a closed system; $\tau_{eq} = T/(k +1) \approx T/k$ is the equivalent time constant of an AFT system. Band B of a closed system at the 3 dB level is connected to the value τ_{eq} by the equation

$$B = 1/2\pi\tau_{eq}. \tag{6.5.12}$$

If the receiver input is acted upon by wideband noise, tracking interruption may occur, which in this case is defined as the final excursion of error Δf beyond the limits of the aperture of the FD characteristic. As the interruption characteristic we shall use, in accordance with [13], the probability p of transition to the indicated state during observation time t_{ob}. Probability p is proportional to time and depends both on the noise level and on the parameters of the system. Some quantitative data, obtained in [13] for comparatively high signal-to-noise ratios, are presented below. Interruption probability p for a given noise level depends very strongly on the ratio of the steady state frequency error Δf_0 to the abscissa B_m of the maximum of the FD frequency response. The probability of interruption is negligible for small ratios $\Delta f_0/B_m$[1]. As $\Delta f_0/B_m$ approaches unity the interruption probability increases with the noise level. Hence it follows that in order to reduce the interruption probability it is necessary to try to operate with small steady state errors.

---

[1] This is true in the case of a perfectly symmetrical FD response. In practice, however, interruption occurs even at $\Delta f_0 = 0$, which is caused by the asymmetry of the FD response.

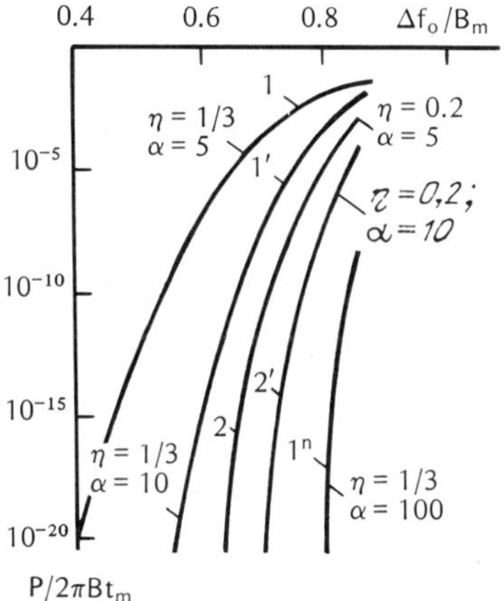

Figure 6.29

The graph in Figure 6.29 of the dependence of the relative interruption probability on the ratio of the steady state error Δf_0, caused by deviation of signal
frequency Δf_t, to the distance to the extremum B_m of the FD response, gives
an idea of the quantitative data on interruption. For static AFT systems,
operating in the linear mode,

$$\Delta f_0 = \Delta f_t/(1 + k) \approx \Delta f_t/k. \tag{6.5.13}$$

For an IF with good selectivity $B_m \approx B_{if}$. The graphs in Figure 6.29 were
plotted for the case of white noise at the receiver input; the individual curves
correspond to different values of α and η. Here $\alpha = B_{if}/\pi B$ is the ratio of the
IF[1] band to the band of a closed system, multiplied by 2π, and η is the ratio
of the effective FD input noise voltage (within the noise band of the IF or
FD circuits) to the effective FD input signal voltage. Hence it follows that
in order to provide more favorable conditions in relation to tracking interruption it is desirable to make band B of a closed system substantially smaller
than the IF band for a small steady state error (large k), which requires that
a filter with long time constant be used. It is important to note that for
Δf_0 close to zero and for a high noise level the above results should be regarded
as approximate.

---

[1] The curves in [13] correspond to different FD circuit band ratios and constants τ_e, but the
band of the circuits of an IF with high selectivity may be replaced by the band of the IF.

2. Application of Phase Locked Loops

Phase automatic frequency and phase tracking systems (phase locked loop — PLL) may be used as phase selection systems. The operation of such systems is based on the utilization of phase detectors PD for distinguishing the phases or frequencies of input signals.

Studies show that a PD may be viewed as a system that multiplies and smooths signals supplied to its inputs. If the frequencies of the signals do not differ very much, then only the low-frequency components of the difference frequencies need be taken into account in the output voltage; the high-frequency components (sum frequencies) are suppressed by the low-pass filter of the PD.

The reduction of the effect of interference by PLL systems is based not on narrowing of the IF band of receivers (as was done in the case of frequency AFT systems), but on the filtering action of the system and nearly coherent signal processing. A PLL system may be viewed as a system that performs (approximately) synchronous detection. To perform this operation a synchronous detector must have a reference signal, coinciding in phase with the input signal. The reference signal may be the voltage from the tracking oscillator of the PLL tracking system.

In synchronous detection there is no AM signal and noise is suppressed by a signal of virtually any level. Let us explain this by way of an example.

Suppose a PD is acted upon by a signal along with normal narrow band noise $n(t)$ (white noise, passing through the receiver IF):

$$u_i(t) = U_s(t) \cos \omega_s t + n(t). \tag{6.5.14}$$

We represent the noise as the sum of two random processes

$$n(t) = A(t) \cos \omega_s t + B(t) \sin \omega_s t,$$

where $A(t) = U_m(t) \cos \theta(t)$; $B(t) = U_m(t) \sin \theta(t)$ are normal noises with the same variance as $n(t)$, and $U_m(t)$ and $\theta(t)$ are slowly changing random processes.

With the reference signal $u_r = U_r \cos \omega_s t$ the PD output voltage is

$$u_{pd} = k_{pd} \left[U_s(t) \cos \omega_s t + n(t) \right] U_r \cos \omega_s t =$$

$$= \frac{1}{2} k_{pd} U_s(t) + \frac{1}{2} k_{pd} A(t) U_r.$$

Here components with frequencies of the order $2\omega_s$ are discarded because of the fact that the PD has a filter.

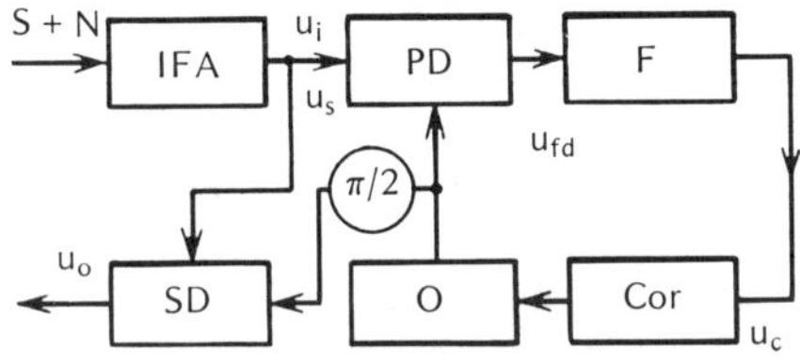

Figure 6.30

By virtue of synchronous detection the quadrature component B(t) in the output voltage vanishes and the output signal-to-noise ratio stays the same as at the input, since the variances of processes A(t) and n(t) are identical.

The above discussions, of course, do not take into account the additional filtering performed by the low-pass filter of the PD; i.e., it is assumed that the band of that filter is substantially wider than the IF band (and also wider than the energy spectrum of process B(t)).

The use of a synchronous detector with a PLL is illustrated in Figure 6.30. The PLL system includes phase detector (PD), low-pass filter (F), frequency control system (frequency corrector Cor) and tracking oscillator (O). In this case the tracking oscillator tracks the frequency of the signal $f_s = \omega_s/2$, which comes from the IF output of the receiver[1]. If ω_s is constant (or changes slowly in comparatively wide limits) and the initial oscillator frequency of the signal ω_{OO} (at zero control voltage u_c) coincides with the frequency of the signal $\omega_s = \omega_{SO}$, then a phase difference, close to $3\pi/2 \pm 2k\pi$, is established between the voltages of the signal and oscillator. After the phase is rotated by the angle $\pi/2$ the oscillator voltage is supplied as reference voltage to the synchronous detector (SD), the functions of which may be performed by a second phase detector. The IF output signal is supplied to the input of the synchronous detector.

We shall ignore noise at first. Then the PD output voltage may be written as

$$u_{fd} = k_d \cos \phi. \tag{6.5.15}$$

Here k_d is a coefficient, expressed in terms of voltage, and ϕ is the phase difference of the oscillator signal and IF output signal.

We try to make the coefficient k_d constant, for which purpose a high-gain AGC system is used in the IF, or an amplitude limiter is inserted in front of the PD. It is important to note that the cosinusoidal dependence in expression (6.5.15) was selected for definition, and this choice does not introduce significant changes in the final derivations.

---

[1] It is convenient to consider angular frequencies in the ensuing analysis.

For the filter output voltage we have

$$u_c = F(D) u_{fd} \qquad (6.5.16)$$

where $F(D)$ is the transfer function of the filter. Low-pass filters with transfer functions of the form $(TD + 1)^{-1}$ or $(\tau D + 1)(TD + 1)^{-1}$ are usually employed. A filter with an integrator is sometimes used.

The frequency control system is described by the expression

$$\Delta\omega_0 = \omega_0 - \omega_{00} = 2\pi k_c u_c \qquad (6.5.17)$$

Here k_c (V/Hz) is a coefficient that characterizes the relation between voltage u_c and the frequency deviation. For the phase difference we write

$$\phi = \int_0^t \omega(t) \, dt + \phi_0 = \omega/D. \qquad (6.5.18)$$

Here D is the differentiation symbol; ϕ_0 is the initial phase; ω is the difference frequency, equal to

$$\omega = \omega_s - \omega_0 = \omega_s - (\Delta\omega_0 + \omega_{00}) = \omega_s - 2\pi k_c u_c - \omega_{00} \qquad (6.5.19)$$

We introduce into the analysis the signal frequency deviation $\Delta\omega_d = \omega_s - \omega_{s0}$ and the initial deviation (frequency error) $\Delta\omega_s = \omega_s - \omega_{s0}$ Then we obtain from (6.5.15)–(6.5.19) the basic equation of the system:

$$\frac{d\phi}{dt} + 2\pi k_d k_c \, F(D) \cos \phi = \Delta\omega_0 + \Delta\omega_s \qquad (6.5.20)$$

We assume first that $\Delta\omega_s$ is a constant, so that the constant frequency difference

$$\Delta\omega_{init} = \Delta\omega_s + \Delta\omega_0 .$$

will be appear in the system. Then the equilibrium states of the system are determined by the equations:

$$F(D) = 1; \quad \cos \phi = \Delta\omega_{init}/\omega_m$$

where $\Delta\omega_m = 2\pi k_d k_c$ is the maximum possible angular frequency deviation of the oscillator and $2\Delta\omega_m$ is the pass band of the system.

It can be shown [199, 86] that the equilibrium states determined by the equation equation

$$\phi_{st} = -\phi_0 \pm 2k\pi \quad (k = 0, 1, 2, \ldots), \qquad (6.5.21)$$

where

$$\phi_0 = \arccos \frac{\Delta\omega_0 + \Delta\omega_s}{\Delta\omega_m} \tag{6.5.22}$$

is the main value of the inverse trigonometric function, correspond to the stable states.

When $\Delta\omega_0 = 0$ and $\Delta\omega_s = 0$, $\phi_{st} = 3\pi/2 \pm 2k\pi$, which corresponds to the statement made above.

The system can be linearized relative to the stable equilibrium states. To do this we assume that the signal frequency changed relative to $\Delta\omega_s$ by $\delta\omega_s$, so that the PD output voltage deviated from equilibrium by the small value δu_{pd}. Expanding function (6.5.15) into a series relative to the equilibrium state ϕ_{st} and retaining two terms of the expansion, we obtain for δu_{pd}

$$\delta u_{pd} = k_d k_0 \delta\phi, \tag{6.5.23}$$

where $k_0 = \sin\phi_0 = \sqrt{1 - \Delta\omega_{init}^2/\Delta\omega_m^2}$ is a coefficient that depends on the ratio of the initial frequency error to one-half of the pass band.

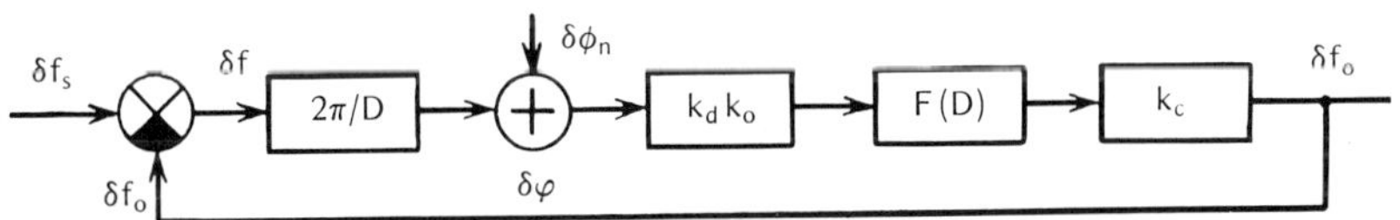

Figure 6.31

If the dependence of u_{pd} on ϕ were described by a sawtooth curve, then $k_0 = 1$. As the result of linearization we arrive at the structural dynamic diagram of the linearized system illustrated in Figure 6.31. Here δf_0 is the oscillator frequency derivation from equilibrium, δf_s is the signal frequency deviation from $\Delta f_s = \Delta\omega_s/2\pi$, $\delta\phi$ is the difference frequency deviation from ϕ_{st}. With this structural diagram we can analyze all dynamic characteristics of a system in linear approximation. In particular, if δf_s is constant, then $\delta f = 0$ in the steady state mode and $\delta\phi = \text{const}$.

Thus, when the signal frequency deviation in the steady state mode is constant, the oscillator frequency is set equal to the signal frequency; i.e., tracking is performed to the accuracy of phase. The structural diagram also takes into account the action of low-level noise, which can cause random phase fluctuations $\delta\phi_n$.

Actually, in accordance with the diagram in Figure 6.31, we may write for the instantaneous frequency and phase reproduction errors

$$\delta f_r = \delta f = \frac{1}{1 + W} \delta f_s - \frac{k_d k_o k_c \, F(D)}{1 + W} \delta \phi_n \tag{6.5.24}$$

$$\delta \phi_r = \delta \phi = \frac{1}{1 + W} \delta \phi_s - \frac{W}{1 + W} \delta \phi_n. \tag{6.5.25}$$

Here

$$W = \frac{2\pi}{D} k_d k_o k_c \, F(D) = \frac{2 \cdot 2\pi \, B_m}{2} k_o \, F(D) \tag{6.5.26}$$

is the transfer function of an open system; $2B_m = \Delta\omega_m/\pi$ is the pass band of the system; then $2\Delta\omega_m = 2\pi(2B_m) = 2\pi \cdot 2k_d k_c$.

The first terms of formulas (6.5.24) and (6.5.25) describe dynamic error, and the second the error resulting from interference. For the simplest case of a system with a single-stage filter $F(D) = (TD + 1)^{-1}$. It follows directly from the above relations that the variances of the phase and frequency reproduction errors σ_f^2 and σ_ϕ^2 due to the influence of interference are

$$\sigma_f^2 = \frac{G_0}{8\pi^2 U_s^2 T} \, \beta_n \tag{6.5.27}$$

$$\sigma_\phi^2 = \frac{G_0}{4U_0^2} \beta_n \tag{6.5.28}$$

Here $\beta_n = k_v/4$ is the equivalent noise band width of the system and $k_v = 2\pi k_d k_o k_c = 2\Delta\omega_m/2$ is the Q-factor (gain) of the system. Hence we arrive at the trivial conclusion: the noise immunity of a system improves as the band width of the closed loop gets narrower.

Analysis of the system in the presence of high-level noise is vastly more difficult. The basic results and procedure of the analysis are presented in the survey mono-graph [199], which includes a very large bibliography. We shall write the results of this analysis in general terms for the steady state mode.

When acted upon by high-level noise the system experiences sharp changes of the phase difference by angle $2k\pi$ (phase jumps or transitions). These are the result of the periodic nature of nonlinear function (6.5.15) and occur with in-creasing frequency as the signal-to-noise ratio decreases. Phase transitions may be ignored as long as the signal-to-noise ratio remains above 6 dB. As the noise level increases the probability of transitions increases, so that a discrepancy occurs between the signal frequency and the average oscillator frequency, and the sign of this discrepancy coincides with that of the initial error $\Delta\omega_{init}$. Finally, as the noise level continues to increase synchronism becomes completely disturbed and tracking interruption occurs. The results of quantitative analysis of these phenomena for typical systems are summarized in [199].

We should like to add that the most important results of analysis of the nonlinear modes of PLL systems are presented in works by Soviet researchers [171-173, 199].

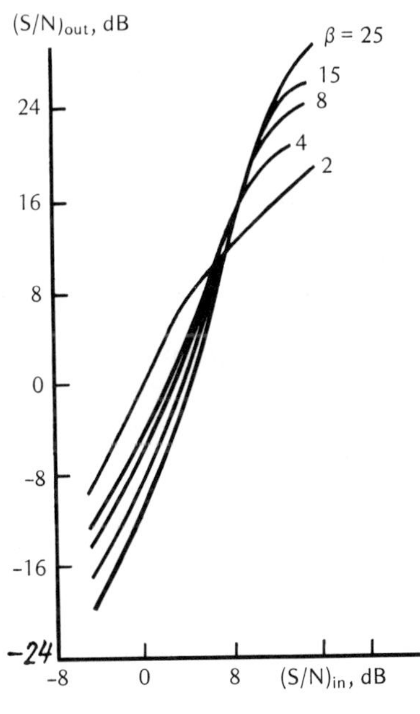

Figure 6.32

3. On Application of Tracking Receivers for Improving Noise Immunity of FM Receiver

As is known, threshold phenomena occur in the reception of FM signals in a noise background; a sharp deterioration of FM reception conditions occurs, beginning at some signal/noise ratio. Big noise spikes appear at the FM receiver output and the signal/noise ratio eventually begins to decrease rapidly, even in comparison with amplitude modulation. Threshold phenomena begin to occur in the vicinity of signal/noise amplitude ratios close to unity when noise is of the sinusoidal variety. Analysis of threshold phenomena for noise jamming is a difficult task. Analysis of these phenomena is described in special literature a survey of which can be found in [32, 77, 141]. Characteristic curves that illustrate the threshold phenomena in the presence of noise are shown in Figure 6.32. The power ratio of the carrier (signal) to the mean square noise at the receiver input (in the IF band) is plotted on the abscissa, and the same ratio at the FM receiver output is plotted on the ordinate. (Here the noise power is calculated in a certain frequency band, for example the IF band.) The curves were plotted for different modulation indices β. At the threshold there is a sharp reduction of the ratio $(S/N)_{out}$. As can be seen in the figure, the greater the frequency deviation, i.e., the larger the gain in the output signal/noise ratio in the region above the threshold, the lower the input noise levels at which threshold phenomena begin to occur (a high threshold).

Without going into an analysis of the threshold phenomenon, let us examine a simplified physical picture of why the threshold increases with the modulation index, in accordance with which threshold phenomena begin to occur when noise spikes, exceeding the carrier level, start to appear sufficiently frequently at the FD input[1]. Such a simplification is acceptable, because it qualitatively correctly describes the essence of the phenomenon.

When the modulation indices are large it is necessary to use a broadband IF. Therefore a sufficient number of noise spikes, comparable with the carrier level, is formed per unit time at a lower average noise level than in the case of small modulation indices, when the IF band may be considerably narrower. This also explains the earlier onset of the threshold for large β, corresponding to large values of $(S/N)_{in}$. By using tracking reception it is possible to substantially reduce (by 5-8 dB) the threshold signal/noise ratio and thus to improve reception noise immunity for large modulation indices. The physical reason for the resulting gain is explained below by way of example of the utilization of a frequency AFT system for receiving FM signals (Figure 6.27).

Quantitative treatments of questions related to gain are presented in the above-cited works [77, 141], and also in 69, 113]. The parameters of an AFT system in this case are so selected that the oscillator frequency changes will differ as little as possible from signal frequency changes. When an AFT system is used in this way it should have the highest possible speed and the LO frequency should track signal frequency changes as accurately as possible. Then voltage u_c will follow the law of frequency modulation and the entire AFT system may be viewed as a single-line frequency detector.

Because the oscillator signal is frequency-modulated, the maximum mixer output frequency deviation (at the IF input) will be substantially reduced, and consequently the IF band becomes much narrower. This is why the threshold decreases, since the probability of the formation of big noise spikes at the limiter input decreases. The limit on the narrowing of the IF bandwidth is attributed to the need to reproduce the spectrum of the modulating signals.

---

[1] We note that this simplified picture is similar to the one used in an analysis of threshold phenomena in the theoretical work [137].

It can be shown on the basis of the above discussions that the higher speed of a system (the wider its band B_m), the more strongly the threshold decreases. However, such is not the case, since the above discussions ignored the fact that the noise component of the LO frequency deviation, caused by the noise voltage at the FD output, increases as the band gets wider. This noise component causes the oscillator frequency noise, and consequently the noise component of the mixer output FM signal, to increase.

At some (sufficiently wide) band these noise components begin to predominate. This is also a threshold-like phenomenon (called the feedback phenomenon [69, 77]). Thus, as the band expands the forward threshold decreases and the feedback threshold increases. Therefore a tracking system has a certain optimum band, the determination of which is explained in [69]. This optimum falls near the point at which both thresholds are equal.

The feasibility of achieving the required speed was not examined in the above discussions. The time constant of the DFA, which functions as a smoothing filter [77], acquires great importance in such systems. Both frequency and phase AFT systems, as well as tracking filters, may be used for tracking reception. Features of the operation of the last two systems as threshold-lowering systems are examined in the literature [32, 141].

4. Frequency Selection in Automatic Frequency Tracking Systems

Frequency selection in AFT systems is the capacity of a system to select signals of a certain frequency in the presence of interfering signals on nearby frequencies. It is necessary here to stipulate the amplitude ratio of the selected and interfering signals. Frequency selection is also performed by band pass filters. The advantage of AFT systems is that they are able to perform selection in a wide frequency range without appreciable changes of selectivity characteristics. Tracking filters produce similar results [141, 32].

When frequency and phase AFT systems are used for selection the usual concepts of selection need to be refined because of the specific features of these systems.

Selection with Frequency AFT Systems. In tracking the frequency f_s of a signal, differing from the nominal frequency f_{s0} by a constant value Δf_s, the difference frequency f in an integrating loop system coincides with the nominal value f_{nom}. Interference will be suppressed if the frequency of the interference differs from f_s by a value exceeding $\pm\Delta f_{dfa}$, where $2\Delta f_{dfa}$ is the difference frequency amplifier passband. (The frequency response of the amplifier is assumed to have extremely steep edges.)

In a zero-order tracking system, operating under the same conditions, frequency error $\Delta f = \Delta f_s/(1 + k)$ occurs, where k is the gain of the system, so that difference

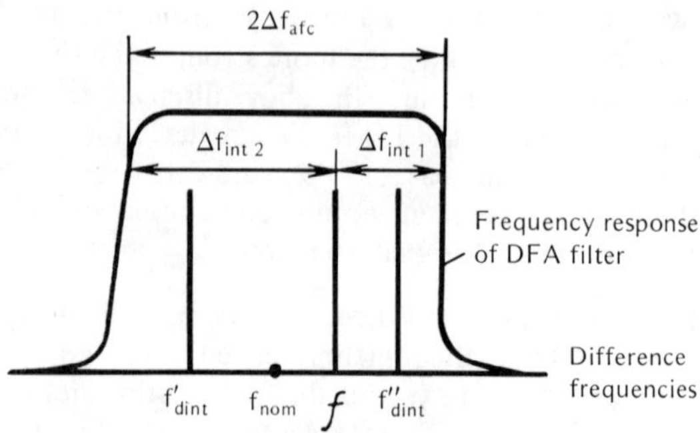

Figure 6.33

frequency f will differ from f_{nom} by that value (Figure 6.33). When the system
is acted upon by sinusoidal interference with frequency f_{int} difference frequency
$f_{dint} = f_{int} - f_o$ (f_o is the oscillator frequency) is formed, which may fall within
the DFA passband to the left (f'_{dint}) or to the right (f''_{dint}) of f. In the former
case the difference f_{int} and f_o may be smaller than Δf_{int2}, and in the latter case
smaller than Δf_{int1}, where Δf_{int1} and Δf_{int2} are the frequency intervals between
f and the boundaries of the frequency response of the DFA filter. Since $\Delta f_{int1} \neq$
$\neq \Delta f_{int2}$ (for $f_s > f_{so}$, $\Delta f_{int1} < \Delta f_{int2}$), then the interference is selected when its
frequency differs less from f_s for $f_{int} > f_s$ than for $f_{int} < f_s$. However, the overall band in which selection is performed remains equal to the band of the DFA
filter, i.e.,

$$2\,\Delta f_{dfa} = \Delta f_{int1} + \Delta f_{int2}\,.$$

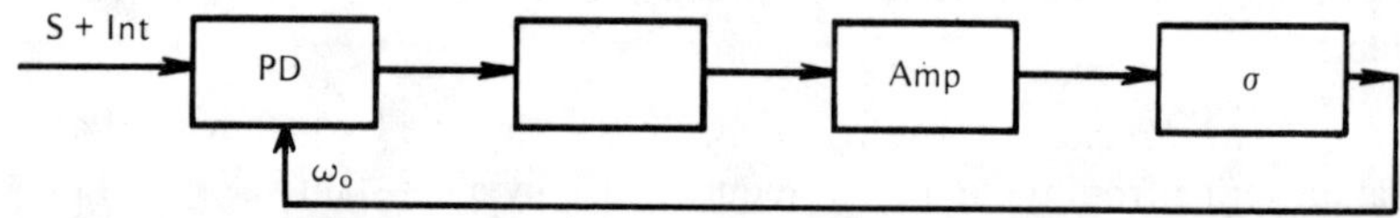

Figure 6.34

Selection with PLL Systems. The selective properties of the system, a functional
diagram of which is presented in Figure 6.34, in relation to interference with a
frequency different from f_s, are determined by the frequency selectivity of the
closed system, i.e., they depend on the filter passband and gain of the system.

The processes which take place in the system when it is acted upon by a signal and interference are exceedingly complex. Some conclusions concerning the selective properties of the system may be drawn from an examination of the effect of interference on typical systems [63, 189].

In the case of a sinusoidal interference signal the tracking oscillator is frequency-modulated; the first harmonic of the modulation frequency is equal to the difference frequency

$$\Omega = |\omega_s - \omega_{int}| = 2\pi \ |f_s - f_{int}| \ .$$

The selective properties of the system depend on the amplitude ratio of the interference and signal and on the ratio of the frequency difference $|f_s - f_{int}| = \Delta f$ to the capture band $2\Delta f_{cap}$. When the ratio $(\Delta f/2\Delta f_{cap}) < 1$ the action of the interference is sharply reduced; when $(\Delta f/2\Delta f_{cap}) > 1$ the effectiveness of the interference depends on the signal-to-noise amplitude ratio

$$\sqrt{q_{int}} = U_{int}/U_s$$

When the amplitude ratios of the interference and signal are small ($q_{int} < 0.25$) the mathematical expectation of the oscillator frequency (in the case of zero initial frequency error) stays equal to the signal frequency, i.e., the interference is "tuned out". For ratios $q_{int} > 4\text{-}6.25$ the average frequency of the tracking oscillator coincides with the interference frequency, and the signal will not be tracked.

When the input voltage amplitude ratios are close to unity and $\Delta f/2\Delta f_{at} < 1$, the frequency of the tracking oscillator is

$$\omega_o \approx (\omega_s + \omega_{int})/2.$$

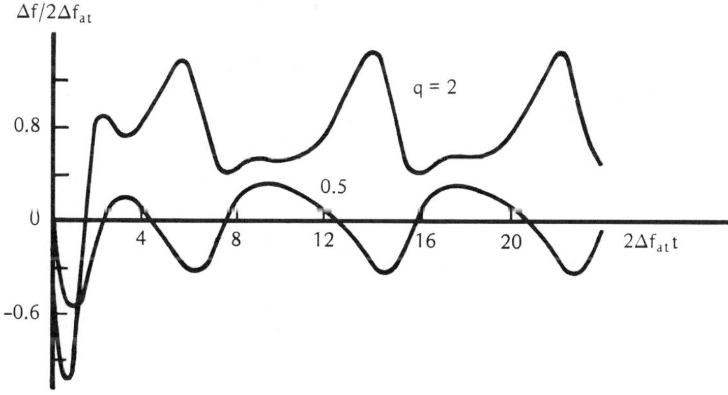

Figure 6.35

The relative difference frequency $\Delta f/2\Delta f_{at}$ of a tracking oscillator is depicted in Figure 6.35 as a function of the product of time multiplied by $2\Delta f_{at}$. This figure illustrates some of the conclusions made above [63]. The curves belong to a system with an RC smoothing filter.

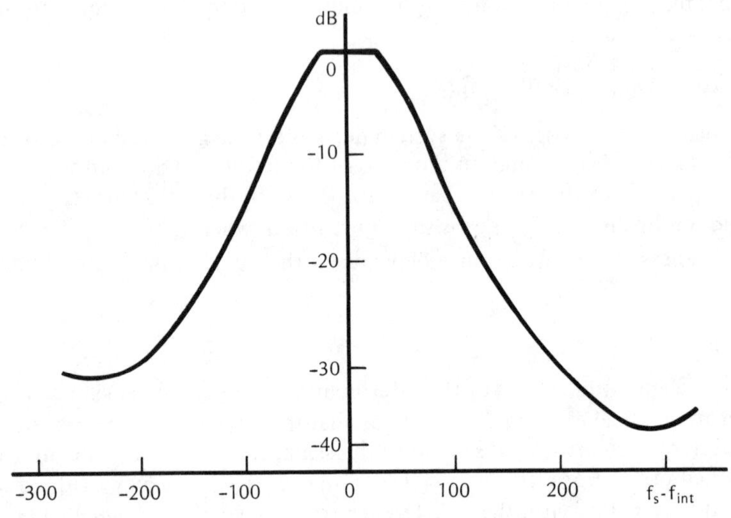

Figure 6.36

The above-described selectivity characteristics of PLL systems are not the only ones possible. In the case of radio communications systems, for example, selectivity may be characterized as the dependence of the attentuation of sinusoidal interference on frequency at a given level of distortions. An example of such a curve is shown in Figure 6.36 [44], where the frequency difference of the signal and interference is plotted on the abscissa axis and the number of decibels by which the interference must exceed the signal in order that the nonlinear distortion coefficient not exceed a certain level (5% here), taken with the opposite sign, is plotted on the ordinate axis. It is important to note here the absolute signal level (more accurately, by how much the signal level exceeds the limiting threshold) and the initial frequency error. In the example at hand the threshold is exceeded by 10 dB and there is no initial frequency error. The attenuation of interference outside of the ±200 kHz band reaches 30 dB while the band at the 0 dB level is equal to ±25 kHz. The curve in Figure 6.36 graphically demonstrates the feasibility of using PLL systems for combatting sinusoidal interference with a different frequency from that of the signal.

Chapter 7

Time and Amplitude Selection

7.1. Time Selection of Pulsed Signals

Time selection of pulsed signals in a jamming background is based on a difference between the selected pulses and jamming pulses in terms of time position (phase), repetition frequency and duration.

Random pulse jamming and noise jamming are regarded as sufficiently universal for pulsed radar. Protection of receivers using time selection techniques will be analyzed below.

1. Time Position Pulse Selection

Time position pulse selection is defined as the extraction of a nearly periodic sequence of pulses, displaced by some time interval relative to reference pulses. This time interval is a slowly changing function of time, so that it changes negligibly in the reference pulse repetition time T_p.

Time selection is accomplished with an automatic system that tracks the pulses of a selected sequence in terms of time position, i.e., an automatic time position selection system. There are two groups of such systems, one with reference pulses on the receiving end and one without. A typical example of a system of the first group is an automatic range tracking system of a pulse radar. Systems of the second group include, in particular, the automatic indicator of range-difference radio navigation systems [166].

The design features of automatic time selection systems are analyzed below, using as an example an automatic range tracker with reference pulses (gates). The functional diagram of such a system (Figure 7.1) consists of a clock (Clo), intermediate elements (IE) (low-pass filter, compensation circuits and integrator) and a time delay system (TDS), from which the clock receives tracking gates. The clock, connected to the receiver (Rec), generates u_{clo}, which depends on the time misalignment ξ between the center of the pulses P to be

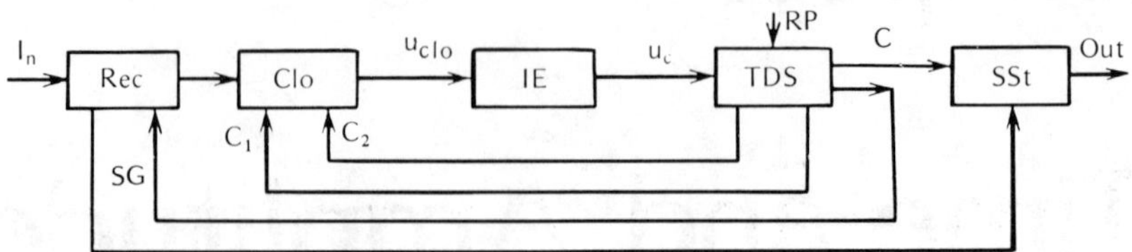

Figure 7.1

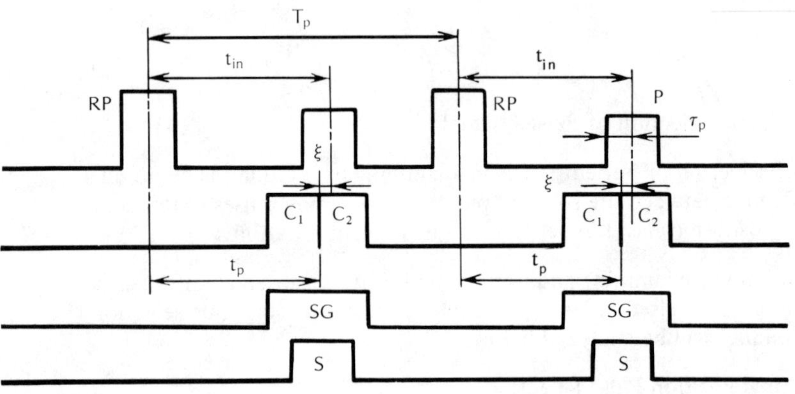

Figure 7.2

selected (Figure 7.2) and the center of tracking gates C_1, C_2. The latter are delayed relative to the reference pulses (RP) of a sequence by time t_p, proportional to control voltage u_c from the intermediate elements, i.e.,

$$t_p = t_{po} + k_c u_c. \tag{7.1.1}$$

Here t_{po} is the initial time lag corresponding to $u_c = 0$. It may be assumed without a loss of generality that $t_{po} = 0$.

Thus a TDS may be viewed as a converter of u_c and t_p with coefficient k_c. The clock, the comparison element of the system, operates for a short time (in comparison with T_p) and converts the time error $\xi = t_{in} - t_p$ to voltage u_{clo}. Clocks are called proportional and integrating, depending on the nature of this conversion. In the former case, for small ξ, voltage u_{clo}, after the passage of the n-th group of pulses, is proportional to the error ξ_{n-1} in the preceding $(n-1)$-st period, i.e.,

$$u_{clo}(n) = k_p \xi_{n-1}. \tag{7.1.2}$$

In the latter case the increment of the voltage that appears on the Clo output after the passage of the next n-th group of pulses, is proportional to error ξ_{n-1}:

$$\Delta u_{clo} (n - 1) = k_{clo}\xi_{n-1} \, , \tag{7.1.3}$$

so that

$$u_{clo}(n) = \sum_{k=0}^{n-1} \Delta u_{clo}(k) = k_{clo} \sum_{k=0}^{n-1} \xi_k. \tag{7.1.4}$$

The principles of the design and construction of Clo are widely known [86, 190, 155, 134].

In the intervals between gates, voltage u_{clo} either diminishes slowly (due to the natural leakage of storage capacitors), or remains constant. In the latter case voltage is cleared from a proportional Clo before the arrival of the next pulse group. However, an integrating Clo with ideal storage operates as a sample-and-hold circuit.

Either type of Clo is a system in which the linear part of the characteristics is comparatively short and does not exceed 2-3 pulse durations. Outside of the aperture of the characteristic, when the tracking and selected pulses become "uncoupled," $u_{clo} = 0$ (or $\Delta u_{clo} = 0$). This property of Clo is based on input pulse time position selection.

The tracking system functions in such a way that time error ξ is kept small and any change of ξ entails a change of u_{clo} (and accordingly of u_c) such that the tracking gates generated by the TDS are shifted toward a reduction of ξ.

A typical automatic range tracking system contains two integrators with a stabilizing circuit. An integrating Clo and an extra integrator in the IE are most commonly used; a proportional Clo and two IE integrators are used occasionally.

Time selection is accomplished by sending a special selector gate (SG) (Figure 7.2), generated in the TDS, to the receiver (Rec). The receiver is normally cut off and is opened only at the time of arrival of SG. The latter is synchronized with the tracking gates. Its duration is so selected as to not deform the discrimination characteristic of the Clo. The SG duration is usually close to the total duration of the tracking gates.

Signal pulses of constant amplitude must be sent to the Clo input of the system, since otherwise the gain of the system would not be constant. If the information carried by a sequence of selected pulses is contained in their amplitude (in conical scanning systems, for example) it is advisable to include one more selector stage (SSt), which receives from the TDS an enabling gate C, which is only a little longer than the selected pulses.

Pulses that do not fall within the aperture of the Clo characteristic are suppressed in this manner. It is important to note that this system, in addition to excluding asynchronous pulse jamming, provides a gain in relation to fluctuation noise.

The tracking system is a pulsed automatic control system. However, the parameters of most practical systems are such that a pulsed system can be replaced with the continuous dynamic equivalent with about the same characteristics, for practical purposes.

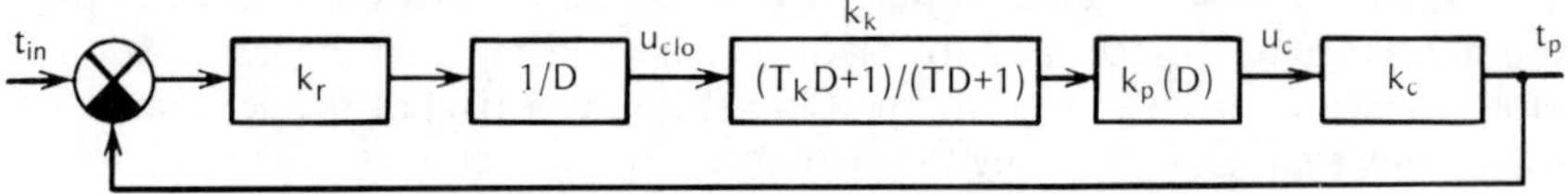

Figure 7.3

A structural dynamic diagram of an equivalent continuous system with two integrating elements is illustrated in Figure 7.3 [88, 190, 155]. Here k_r is the gain of a proportional Clo or dynamic equivalent of an integrating Clo, connected with the parameter k_{clo} of an integrating Clo by the relation $k_r = k_{clo}/T_p$. An equalizing circuit with the transfer function

$$W(D) = k_k(T_p D + 1)(TD + 1)^{-1}, \quad T_p > T.$$

is used in this case as the stabilizing circuit.

The second integrator (k_p/D) is connected after the equalizing circuit, directly in front of the TDS with the gain k_c.

The transfer function of a closed system, characterizing the relation between a given t_{in} and the time position t_p of the selected pulse, measured by the system, is

$$\phi(D) = \frac{k_a(T_k D + 1)}{TD^3 + D^2 + k_a T_k D + k_a}. \tag{7.1.5}$$

Here $k_a = k_r k_k k_p k_c$ is the gain of the system.

The basic dynamic features of the system are given below. It is a second-order servo system and, consequently, has no errors in the steady state mode if and only if t_{in} changes linearly or is constant. When t_{in} changes with constant acceleration a the steady state error is $\xi_{ss} = ak_a^{-1}$. The equivalent single-sided (noise) bandwidth of the system is

$$B_n = \frac{k_a T_k^2 + 1}{4(T_k - T)} \approx \frac{k_a T_k^2 + 1}{4T_k} \tag{7.1.6}$$

and is proportion to k_a and T_k (for $k_a T_k^2 \gg 1$). To obtain a favorable transient response it is helpful to select $T_k \sqrt{k_a} = 1.2$, which yields the relation

$$k_a \approx (2.5-4)\, B_n^2. \tag{7.1.7}$$

The variance of the time position error of the tracking gates, in relation to comparatively low-level wideband fluctuation noise (when ξ for practical purposes does not extend beyond the linear part of the Clo characteristic), if the selected Clo input pulses are not limited, is [88]

$$\sigma_t^2 = G_{\Delta t}(\omega)\, B_n, \tag{7.1.8}$$

where

$$G_{\Delta t} = (0.4-0.25)\frac{t_s^2 T_p}{U_p^2}\, \sigma_n^2 \tag{7.1.9}$$

Here U_p is the amplitude of the selected pulses; t_s is the tracking gate duration; σ_n^2 is the variance of the Clo input voltage.

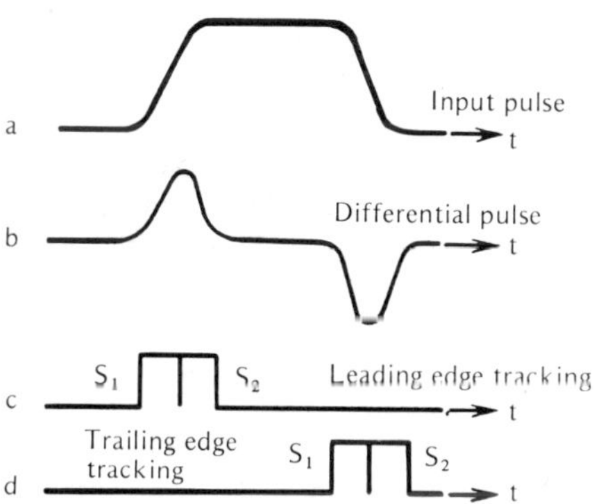

Figure 7.4

In the case of high-level noise, tracking interruption may occur in this system. The basic laws of interruption in range tracking systems are similar to those that hold true for frequency tracking systems. We note that because of second-order loop characteristics the probability that an appreciable steady-state constant error will occur in a ranging system is less than in a system with first-order characteristics, so that the probability of tracking interruption decreases.

Leading edge tracking or trailing edge tracking of a pulse (or combined leading and trailing edge tracking) is used to increase the noise immunity of a system in relation to chaff, and also for tracking formation targets [9] . For this purpose the selected pulses are first passed through a high-pass filter with a small time constant (in comparison with the duration of the leading or trailing edge) for example through a differentiating circuit. Then a system with two tracking gates is built, in which the tracking gates track the positive pulse (leading edge tracking) or negative pulse (trailing edge tracking) that is formed as a result of differentiation. This process is illustrated by the time diagrams in Figure 7.4.

2. Pulse Repetition Frequency Selection

This kind of selection is based on the coincidence of two streams of pulses. One of them comes from the receiver and the other (reference pulses) is formed in a selector. Only pulses that coincide in time with the generated reference pulses reach the output.

The most important practical case is the selection of an approximately periodic pulse train, in which the exact sequence of reference pulses cannot be assigned a priori. An example of this is the problem of the extraction of synchronization pulses in television receivers. In these cases a system is built which automatically tracks the phase of the received pulses. A functional diagram of such a system is illustrated in Figure 7.5.

The input (selected) pulses are fed to a clock or phase discriminator (Clo, PD), where their time positions are compared with the reference voltage (RV) from the tracking oscillator (TO). This voltage comes in the form of sinusoidal or sawtooth (in television) waves and may consist of two tracking gates, as was described above. As a result of phase comparison a mismatch voltage is generated, which passes through a filter (F), which also contains correcting circuits, for controlling the frequency tracking oscillator (TO). The result is a system which peiforms as the previously described PLL system. The only difference is that the input signal of this system is a pulse train, and not a sinusoidal signal. Thus the given system may be called a pulse-phase locked system. It is used extensively in TV as a smoothed synchronization system. It is distinguished from the automatic range tracking system in that the reference voltage is continuous and not pulsed.

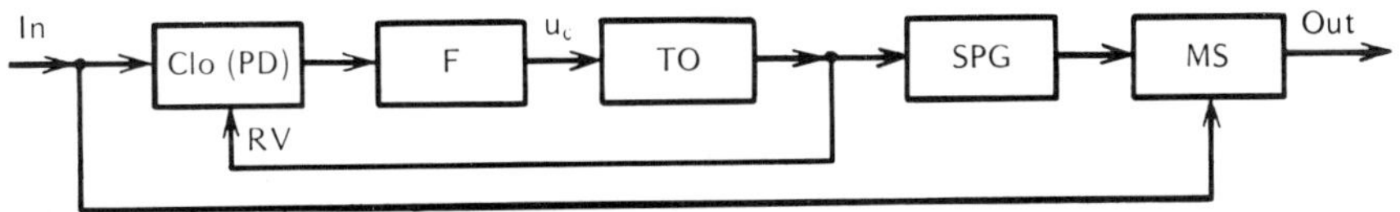

Figure 7.5

If the shape of the selected pulses must be preserved, then an extra matching stage (MS) is included, whose input receives the input signal and a selector pulse sequence, generated from the output of the tracking oscillator in the selector pulse generator (SPG).

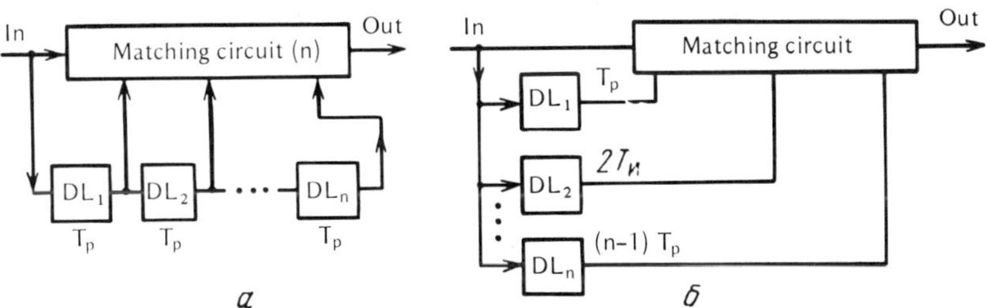

Figure 7.6

In addition to the system described above, with an exactly known (and nearly constant) pulse repetition period T_p, it is also possible to use a system with coincidence stages, operating open loop. There are two versions of such systems (Figure 7.6). In the first the coincidence circuit receives n pulses from series-connected delay lines DL_1, DL_2, . . . , each of which as a delay time equal to the pulse repetition period T_p. In the second, parallel-connected delays, each with delay time $t_{del} = T_p, 2T_p, \ldots, (n-1)T_p$, are used. The difference between these systems is basically structural (the first one is simpler in design). The output signal of the coincidence circuit appears only when the input pulse repetition frequency is equal to (or is a multiple of) the delay times. A diagram explaining the operation of a system for a three-pulse coincidence circuit is shown in Figure 7.7, in which it can be seen that only pulses with period $T_p = t_{del}$ reach the output, and random jamming pulses (not hatched in Figure 7.7) are eliminated. Since period T_p is not strictly constant and the delay time is not strictly stable, the selector pulses from the delay lines should have a somewhat longer duration than the pulses of the main sequence (for the sake of simplicity the selector pulse generating stages are not shown in Figure 7.6).

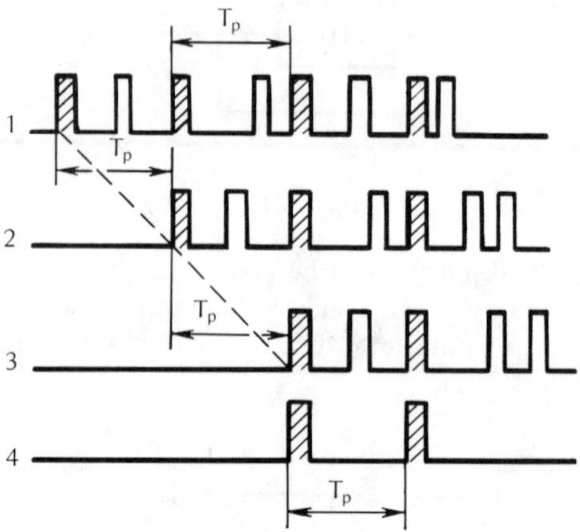

Figure 7.7

Pulses generated as the result of false combinations, formed from input noise pulses, may penetrate to the output of a coincidence stage. The number of penetrating pulses is proportional to the average number M_p of jamming pulses per unit time and depends on the ratio between pulse duration τ_p (the durations of jamming pulses are also assumed to be τ_p) and selector pulse duration $\tau_{S1}, \tau_{S2}, \ldots$. It may be shown that the number N_{fs} of false combinations per unit time for a system with two delay lines is [145]:

$$N_{fs} = M_p \left[1 - \frac{\tau_S}{(\tau_{S1} + \tau_p)(\tau_{S2} + \tau_p)} \right] \left\{ 1 - \exp \left[-(\tau_S + \tau_p) M_p \right] \right\}$$

$$\times \left\{ 1 - \exp \left[-(\tau_{S2} + \tau_p) M_p \right] \right\}. \tag{7.1.10}$$

More general characteristics of output pulses are presented in [145].

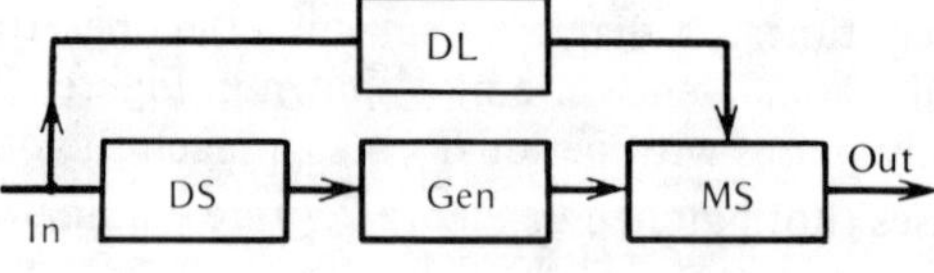

Figure 7.8

3. Pulse Duration Selection

There is a large variety of systems capable of selecting pulses, the duration of which is less than, exceeds or is equal to a given value. Of greatest importance from the standpoint of jamming immunity are selectors of the third group. Their functional diagram is shown in Figure 7.8.

The input pulse train goes to a duration selector (DS) which passes pulses with the selected duration to the output. From these pulses a generator (Gen) forms a standard pulse, the duration of which corresponds to the duration assigned for selection. The generator is followed by a matching stage (MS), which passes the selected pulse to the output. The delay line (DL) delays the output pulses for a time equal to the time delay in the generator (Gen). We note that this system provides distortion-free transmission of the selected pulses. If pulse shape need not be preserved the system is simplified and in this case will consist of two stages: DS and Gen.

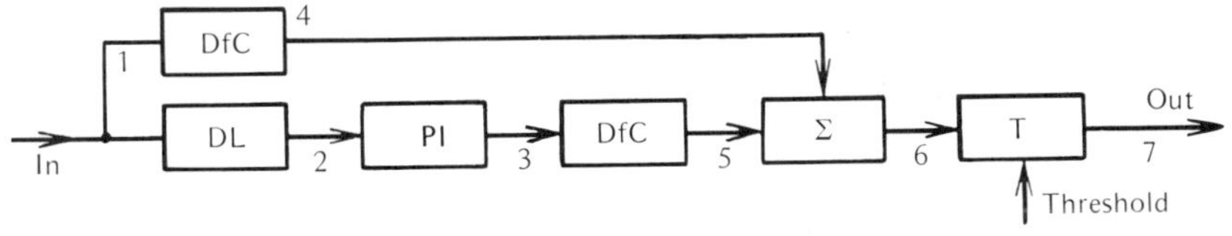

Figure 7.9

The functional diagram of a duration selector may be represented as delay line DL, the delay line of which is the duration of the selected pulse, a pulse polarity reversing stage (phase inverter — PI) in an adding system (Σ), differentiating circuits (DifC) and threshold (T) (Figure 7.9). As can be seen in Figure 7.10, which illustrates the operation of this system, a pulse

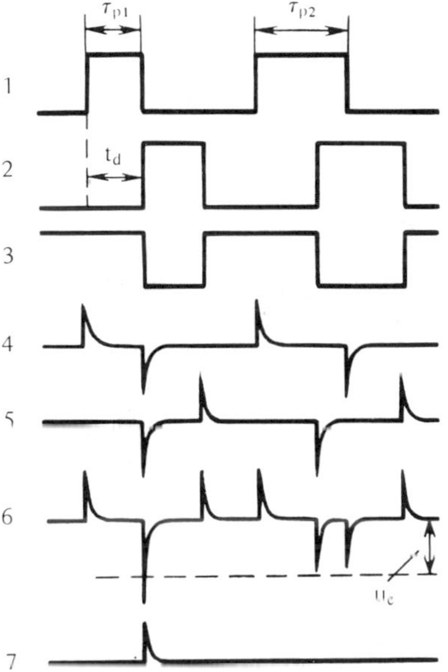

Figure 7.10

will appear on the threshold output only when the duration τ_{p1} of the pulses is equal to delay time t_d in the delay line (the left column of pulses). Otherwise the threshold will not be exceeded.

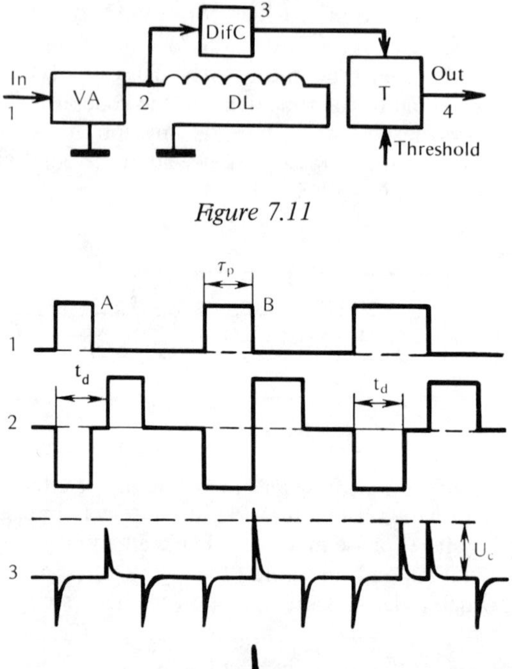

Figure 7.11

Figure 7.12

Practical selection systems can be very simple [170]. For example, the functional diagram of a selector with an end-shorted line is illustrated in Figure 7.11. When voltage waves are reflected from the end of the line they experience a polarity reversal. If the duration of the input pulse is double the delay time, then a drop, equal to twice the pulse amplitude (the middle diagram in Figure 7.12) will occur on the input of the line in time τ_p.

Therefore a differentiated pulse has maximum amplitude, exceeding the threshold and a pulse is generated at the output of the system. Output pulses will not appear when the relations between τ_p and t_d are different (the left hand and right hand diagrams in Figure 7.12).

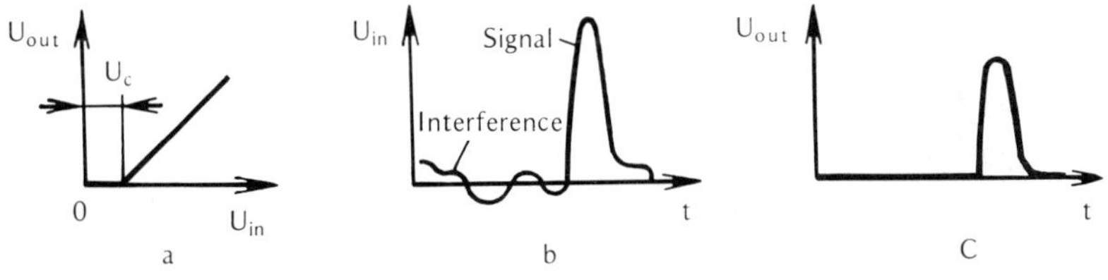

Figure 7.13

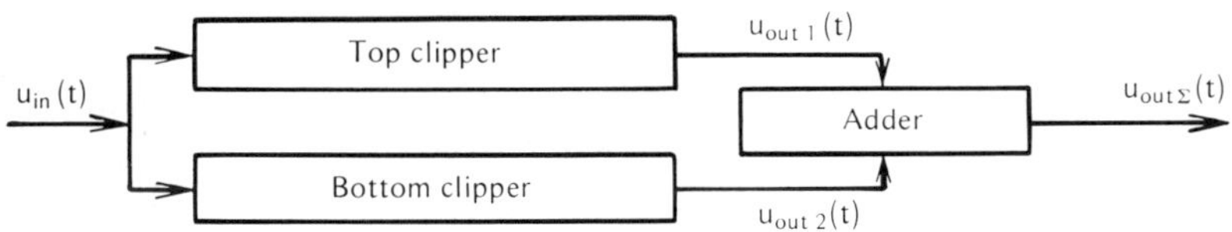

Figure 7.14

7.2 Amplitude Selection

1. Selection of Bottom-Clipped Signals

This kind of selection is used in cases when the amplitude of the useful
signal substantially exceeds the amplitude of jamming. Selection is performed
by an amplitude selector, which is a bottom clipper or slave pulse generator.
When a bottom clipper is used, voltage U_{out} at the output appears only
when the input voltage U_{in} exceeds the clipping threshold U_{cl}.

The characteristic of the clipper $U_{out} = f(U_{in})$ and its input and output
voltages $U_{in}(t)$ and $U_{out}(t)$, respectively, are illustrated in Figure 7.13a,
b and c.

In application to high-frequency signals such selection may be called minimum
clipping selection. By using separate bottom and top clipping of high-frequency
signals and then adding the output voltages of the clippers it is possible to
eliminate all input voltages characterized by the relation

$$|u_{in}(t)| < U_{cl} .$$

The corresponding functional diagram of the selector is shown in Figure 7.14.
The input voltage $u_{in}(t)$, the characteristic of the top clipper $u_{out_1} = f_1(u_{in})$,
the output voltage of that clipper $u_{out_1}(t)$, the characteristic of bottom
clipper $u_{out_2} = f_2(u_{in})$, the output voltage of the bottom clipper $u_{out_2}(t)$
and, finally, the variable component of the sum voltage $u_{out\Sigma}(t)$ at the output
of the selection system are shown in positions a, b, c, d, e, f in Figure 7.15.

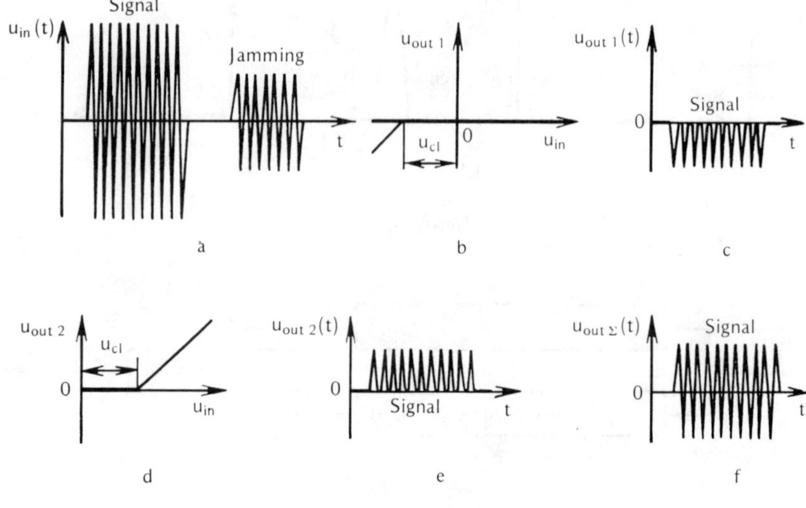

Figure 7.15

When a slave pulse generator, for example a flip-flop or blocking oscillator, is used as the selector the threshold level is a blanking voltage (bias voltage). When the input voltage exceeds the threshold a standard voltage pulse with a given amplitude and duration is generated.

2. Pulse Level Selection

This kind of selection is advantageous in cases when the amplitude of the useful signals is substantially less than the amplitude of jamming, or the amplitude of the useful signals fluctuates near some level.

Voltage pulses, the amplitude U_{in} of which is less than the given threshold level U_{cl}, may be selected; signals for which the condition $U_{in} < U_{cl}$ is valid pass to the selector output.

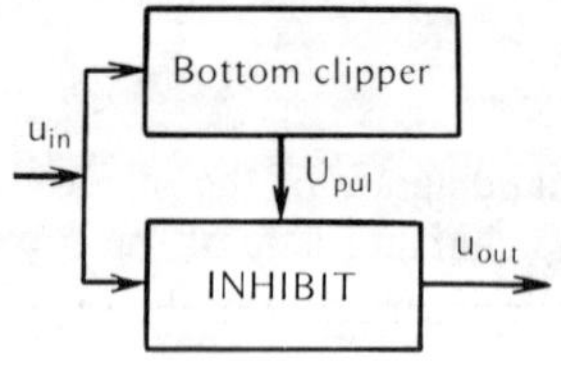

Figure 7.16

A bottom clipper and logic AND NOT (INHIBITORY)-gate are used jointly for low-amplitude pulse selection. A functional diagram of such a selection system is shown in Figure 7.16. Pulses whose amplitude exceeds the threshold pass through the bottom clipper. Voltage pulses with amplitude U_{pu1} go from the clipper output to the inhibiting input of the INHIBITORY-gate. To the second (information) input of that circuit is supplied the input voltage. The pulse voltage is transmitted from the information input to the output of the INHIBITORY-gate only when there is no voltage on its inhibiting input. Consequently the output of the INHIBITORY-gate in a selector such as the one illustrated in Figure 7.16, receives only pulse signals for which the relation $U_{in} < U_{cl}$ is valid.

A selecting system which passes only pulsed signals, the amplitude U_{in} of which falls with a given range: $U_{cl_1} < U_{in} < U_{cl_2}$, may be used.

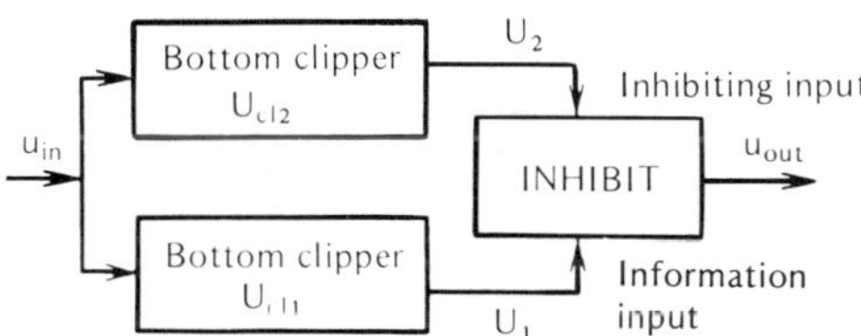

Figure 7.17

A functional diagram of such a selector is shown in Figure 7.17. The input voltage pulses pass through two bottom clippers with clipping levels U_{cl1} and U_{cl2}. Voltages with amplitudes U_1 and U_2 from the outputs of the first and second clippers are supplied to the information and inhibiting inputs of the INHIBITORY-gate. If $U_{in} < U_{cl1} < U_{cl2}$, then the pulses do not pass through the selector, since there is no voltage on the information input of the INHIBITORY-gate. When $U_{in} > U_{cl2}$ pulses also are not passed, since there is voltage on the inhibiting input. Only when $U_{cl1} < U_{in} < U_{cl2}$ does voltage appear on the information input of the INHIBITORY-gate and disappear on its inhibiting group.

3. Storage

The storage method is a special case of matched filtering, which provides the maximum signal-to-noise power ratio if the jamming is white noise. In matched filtering, the transfer function of the filter is, to the accuracy of a constant factor and constant delay, the complex-conjugate of the spectrum of the useful signal. At the same time the storage method is performed with an adder or integrator, called storage elements, regardless of the form of useful signal.

The storage method with an adder is based on the following principle. A predetermined number of readings is taken in mixture $u_{mix}(t)$ of signal and jamming within a prescribed time T_{st}. The values $u_{mix}(t)$ at the reading points are added and then, on the basis of the total signal, a resolving (threshold) system makes a decision concerning the presence or absence of a useful signal in mixture $u_{mix}(t)$. When an integrator is used the resolving system receives a signal proportional to

$$\frac{1}{T_p} \int_0^{\tau_p} u_{mix}(t) \, dt.$$

The function of the storage element is to improve useful signal detection conditions at the receiver input in the presence of jamming. Storage is advantageous in cases when the jamming is additive wide-band noise and may be used in all kinds of electronic systems.

To illustrate the capabilities of these storage methods we shall examine first of all the problem of the detection of video pulses with constant amplitude U_s, received along with white noise $\xi(t)$. It is assumed here that the spaces between adjacent periodic pulses exceed the jamming correlation time and that an adder is used in the storage element.

If the storage element reacts to n output pulses, then the adder output signal u_Σ is

$$u_\Sigma = \sum_{i=1}^n (U_s + \xi_i) = nU_s + \sum_{i=1}^n \xi_i \,,$$

where ξ_1 is the value of function $\xi(t)$ at moment time t_i ($i = 1, 2, \ldots, n$), corresponding to the measurement of the amplitude of the i-th pulse. Voltages nU_s and $\sum_{i=1}^n \xi_i$ characterize the useful signal and jamming during decision making as to presence of a signal in the received mixture $u_{mix}(t)$.

Since random values $\xi_1, \xi_2, \ldots, \xi_n$ are not cross-correlated, then, when the function $\xi(t)$, which is assumed to be stationary, has zero mathematical expectation, the variance σ_j^2 of jamming $\sum_{i=1}^n \xi_i$ is determined by the following formula:

$$\sigma_j^2 = n\sigma_\xi^2 \,.$$

Here σ_ξ^2 is the variance of the random function $\xi(t)$. Then the ratio q_{st} of the square of amplitude $n^2 U_s^2$, of the useful signal, to variance σ_j^2, is

$$q_{st} = nU_s^2/\sigma_\xi^2 .$$

If there were no storage element the ratio q_1 of the square of amplitude U_s^2 to variance σ_ξ^2 of jamming $\xi(t)$ would be

$$q_1 = U_s^2/\sigma_\xi^2 .$$

$$q_{st} = nq_1 . \tag{7.2.1}$$

Hence it follows that storage improves the signal-to-noise ratio by a factor of n and accordingly increases the probability that a useful signal will be detected.

The fact that $\xi_1, \xi_2, \ldots, \xi_n$ are cross-correlated makes the ratio q_{st}/q_1 equal not to n, but to some smaller value. Actually, in this case dispersion σ_j^2 is

$$\sigma_j^2 = M\left(\sum_{i=1}^n \xi_i\right)^2 = M\left\{\sum_{i=1}^n \xi_i^2\right\} + 2 \sum_{i=1}^n \sum_{\ell=1}^n M\left\{\xi_i \xi_i + i\right\},$$

where the symbol $M\left\{\cdot\right\}$ signifies statistical averaging.

If through $\xi\xi_\ell$ we denote the product of any pair of values of the function $\xi(t)$, measured in ℓ periods T_p, then we obtain

$$\sigma_j^2 = n\sigma_\xi^2 + 2 \sum_{\ell=1}^{n-1} (n - \ell) R(\ell).$$

Here $R(\ell) = M\left\{\xi\xi_\ell\right\}$ is the second mixed central moment for random values ξ and ξ_ℓ. Therefore

$$q_{st} = \frac{n^2 U_s^2}{\sigma_j^2} = \frac{nU_s^2}{\sigma_\xi^2\left[1 + 2 \sum_{\ell=1}^{n-1} \frac{n-1}{n} r(\ell)\right]},$$

where $r(\ell) = R(\ell)/\sigma_\xi^2$ is the correlation coefficient of random values ξ and ξ_ℓ. Because $q_1 = U_s^2/\sigma_\xi^2$, we have

$$q_{st} = \frac{n}{1 + v} q_1 .$$

From this formula, where $v = \dfrac{2}{n} \sum\limits_{\ell=1}^{n-1} (n-\ell)\, r\,(\ell)$, it follows that when the values

of the function $\xi(t)$ are correlated at the times of its reading, when $v \neq 0$, ratio q_{st} is smaller than in the absence of correlation. If, for example, the correlation relation need be considered only during time T_p, then $r(1) \neq 0$, and

$r(2) = r(3) = \ldots = 0$. Under these conditions $v = \dfrac{2(n-1)}{n}\, r(1)$, and when

n is sufficiently large the approximate equality $v = 2r(1)$ is valid, so that

$$q_{st} \approx \frac{n}{1 + 2r\,(1)}\, q_1 .$$

The above relation shows that as $r(1)$ increases ratio q_{st} decreases and when $r(1) = 1$ becomes one-third of what it is when $r(1) = 0$.

If the useful signal is characterized by the continuous function of time $u_s(t)$, then the storage element includes an integrator in addition to an adder. Let $u_s(t) = u = \text{const}$, and let the integrator compute the average value of the function $u_{mix}(t) = u + \xi(t)$ in time T_{st}. The output voltage $u_i(t)$ of this integrator is

$$u_i(t) = u + \frac{1}{T_{st}} \int_0^{T_{st}} \xi(t)\, dt .$$

The variance σ_j^2 of the jamming component

$$u_j(t) = \frac{1}{T_{st}} \int_0^{T_{st}} \xi(t)\, dt$$

with zero mathematical expectation, is determined by the following formula:

$$\sigma_j^2 = M \left\{ \frac{1}{T_{st}} \int_0^{T_{st}} \xi(t)\, dt \right\}^2 = M \left\{ \frac{1}{T_{st}^2} \int_0^{T_{st}} \int_0^{T_{st}} \xi(t)\xi(t_1)\, dt\, dt_1 \right\}, \quad (7.2.2)$$

but $M\,\xi(t)\xi(t_1) = R(t - t_1)$ is the correlation function of the interference.

Introducing the integration variable $t - t_1 = \tau$, we obtain

$$\int_0^{T_{st}} R(t-t_1)\, dt_1 = \int_{t-T_{st}}^{t} R(\tau)\, d\tau = C .$$

Function $R(\tau)$ is even and $R(0) = \sigma_\xi^2 > R(\tau)$. Therefore C will acquire its maximum value when $t = 0.5T_{st}$. Consequently

$$C \leqslant \int_{-0.5T_{st}}^{0.5T_{st}} R(\tau)\, d\tau.$$

For a comparatively large interval T_{st}, when the function $R(\tau)$ on the edges of this interval is small in comparison with σ_ξ^2, the integration limits $-0.5T_{st}$ and $0.5T_{st}$ may be replaced with $-\infty$ and ∞, respectively. Then

$$C \leqslant \int_{-\infty}^{\infty} R(\tau)\, d\tau = \sigma_\xi^2 \tau_c$$

where $\tau_c = \dfrac{1}{\sigma_\xi^2} \displaystyle\int_{-\infty}^{\infty} R(\tau)\, d\tau$ is the correlation interval.

Returning to relation (7.2.2), we obtain

$$\sigma_j^2 \leqslant \frac{\tau_c}{T_{st}}\, \sigma_\xi^2 .$$

Since the square of the useful voltage is u^2, then

$$q_{st} \geqslant T_{st} u^2 / \tau_c \sigma_\xi^2 .$$

If there were no storage element, then

$$q_1 = u^2 / \sigma_\xi^2 ,$$

$$q_{st} \geqslant T_{st}\, q_1 / \tau_c .$$

$$(7.2.3)$$

Comparing relations (7.2.1) and (7.2.3), we see that in (7.2.1) the role of n is played by the number T_{st}/τ_c of uncorrelated values of jamming in interval T_{st}. This means that the adder and integrator provide the identical attenuation of jamming. However, integration often is technically easier to perform than addition.

Code pulse groups are often used in addition to single periodic pulses and continuous signals.

The storage elements also may be used in electronic systems with the latter kind of signals. For example, there are systems with storage elements that make decisions on the basis of the majority principle. The essence of this principle consists in the fact that a code combination is duplicated, i.e.,

is transmitted n times. The receiver separates the elementary signals of the
code combination in separate circuits and the number of zeros and ones
for a given position of the code is counted in each circuit. If the number
of zeros is larger than the number of ones, the decision is made that a zero
was transmitted in the given position of the code. When the number of ones
exceeds the number of zeros a one is written. Analysis indicates that decision
making as to the presence or absence of a signal on the basis of the majority
principle reduces the error probability in the presence of strong pulse
jamming.

In communications with signals corresponding to various code combinations
both coherent (predetector) and incoherent (postdetector) storage of pulses
characterizing each code position also are possible. Under such conditions
the use of an n_z-value code requires the use in the receiver of n_z storage
elements, and the noise immunity gain is determined by formulas (7.2.1) or
(7.2.3), depending on whether adders or integrators are used.

Voltage u_Σ, generated by the adder, can be obtained not only by adding the
signals that appear or are counted at different moments of time, but also by
adding the voltages generated simultaneously by independent channels of
one multichannel electronic system or by several analogous electronic systems,
intended for the transmission of the same message.

Multichannel systems capable of the simultaneous transmission of identical
messages are designed in accordance with the frequency- or code-selection
principle.

At the same time periodic pulses are added in pulse radar and multichannel
electronic systems using time- or code-selection, in which the same message
is transmitted through each channel. In this case the individual code combinations must be transmitted one at a time in code-selection systems.

It follows from the above discussion that the signal-to-noise ratio can be
increased in the storage method by increasing the time over which a decision
is made concerning the presence of a useful signal, or by increasing the
bandwidth of the electronic system. The bandwidth must be increased, in
particular, when frequency-division multichannel radio links are used.

4. Angle Gating

Angle gating increases the resolving power and jamming immunity of angle
measuring systems by controlling their parameters. The main lobe width
$\Delta\theta_0$, the level and shape of the side lobes of the angle discriminator and
angle $\Delta\theta_e$ between the nulls of the main lobe of the angle discriminator
are most often subjected to control. For this purpose the parameters of
systems that participate in the formation of the angle discriminator, for
example the antenna, AGC system, signal processing elements, etc., are
usually varied.

There are three known angle gating methods [2, 101, 124, 161, 187, 214]. The first is based on the selection of the desirable parameters of the antenna system. In the second method the side lobes of the angle discriminator are eliminated, for which purpose familiar side lobe cancellers are used (see $5.2). The third method offers an opportunity to increase the resolving power by including different nonlinear circuits in the structural diagram of the angle tracking system and by using logic operations of the switching type for "cutting off" undesired angle information.

Choice of Antenna Parameters. The classical method of increasing the angle resolving power is to increase antenna size L or to shorten the wavelength (in the general case it is necessary to reduce λ/L). For given aperture L, however, the parameters $\Delta\theta_0$ and $\Delta\theta_e$ may be changed by selecting the angle θ_0 between the tracking axis and the peaks of the radiation pattern in angle trackers systems that use the amplitude method of signal comparison for direction finding (see Figure 2.13).

The influence of antenna parameters on the width of the angle discriminator curve is evaluated by analyzing a family of discriminator curves.

The (normalized difference patterns) of amplitude monopulse systems are described by the formula [24]

$$u_{df} = \frac{k\mu^2 \left[F^2(\theta_0-\theta) - F^2(\theta_0+\theta)\right]}{\left\{1 + \mu \left[F(\theta_0-\theta) + F(\theta_0+\theta)\right]\right\}^2}, \tag{7.2.4}$$

where μ is the equivalent gain of the AGC system [86]; k is a scale factor.

The width $\Delta\theta_0$ of the discriminator curve is determined by the equation $F^2(\theta_0-\theta) - F^2(\theta_0+\theta) = 0$, and the angle $0.5\Delta\theta_e$ is determined on the assumption that the maximum voltage u_{df} of the main lobe of the difference pattern for $\theta > 0$ is determined by the expression $du_{df}/d\theta = 0$.

Analysis of the family of difference patterns shows that for a given direction diagram $F(\theta)$ the beamwidths $\Delta\theta_0$ and $\Delta\theta_e$ depend on the ratio $\theta_0/\theta_{0.5}$. The parameters $\Delta\theta_0$ and $\Delta\theta_e$ are depicted in Figure 7.18 as functions of $\theta_0/\theta_{0.5}$ and show that the maximum $\Delta\theta_e$ is obtained when $\theta_0 = 0.5\theta_{0.5}$. Lobe width is reduced substantially by reducing the parameter θ_0. If $\theta_0 > 0.5\theta_{0.5}$, then $\Delta\theta_e$ decreases because of the influence of the side lobes of the pattern.

The normalized difference pattern of phase monopulse systems is described by the expression [24]

$$u_{df} = \frac{2k\mu^2 F^2(\theta) \sin 2\phi}{[1 + 2\mu F(\theta) \cos \phi]^2}, \tag{7.2.5}$$

where $\phi = (2\pi d/\lambda) \sin \theta$; 2d is the distance between the phase centers of the elementary antennas.

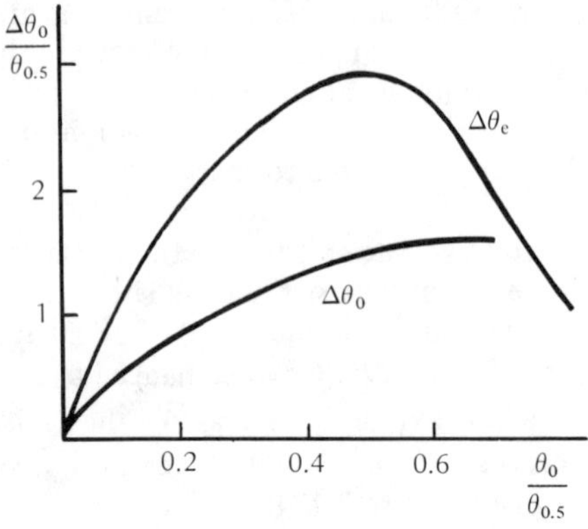

Figure 7.18

For a high amplitude signal (7.2.5) is rewritten as the familiar expression

$$u_{df} = ktg \left(\frac{2\pi d}{\lambda} \sin \theta \right) . \tag{7.2.6}$$

From (7.2.6) we obtain

$$\Delta\theta_e = 2\Delta\theta_0 \approx 2\theta_{0.5} , \tag{7.2.7}$$

where $\theta_{0.5} = \lambda/L$ (L = 4d is the antenna aperture).

Consequently $\Delta\theta_0$ and $\Delta\theta_e$ in phase direction finding methods can be reduced by reducing the parameter λ/L.

Analysis of the fine structure of difference patterns (7.2.4) and (7.2.5) indicates the feasibility of reducing $\Delta\theta_e$ by adjusting the equivalent gain of the AGC system. If μ changes from 1 to 100, then $\Delta\theta_e$ is a linear function of μ and the value of $\Delta\theta_0$ remains constant.

Angle Gating with Side Lobe Cancellers. The capabilities and features of the application of side lobe cancellers in angle tracking systems are examined below by way of example of monopulse systems that incorporate sum-difference signal processing.

All known canceller designs change the radiation pattern $F_0(\theta)$ of the main antenna by multiplying it by some weight factor $h(\theta)$. The resulting radiation pattern is described by the formula

$$F(\theta) = F_0(\theta) \, h(\theta). \tag{7.2.8}$$

The signal of the difference channel of the most common design of monopulse systems of the sum-difference type is written as

$$u_\Delta = u_1 - u_2 = U\,[F_0(\theta_0 - \theta)\,h_1(\theta)\cos(\omega t + \phi)$$

$$- F_0(\theta_0 + \theta)\,h_2(\theta)\cos(\omega t - \phi)]\,, \tag{7.2.9}$$

where $h_1(\theta)$, $h_2(\theta)$ are the weight functions of the elementary antennas (these functions in the general case reflect the amplitude and phase patterns of cancelling antennas); $\phi = (2\pi d/\lambda)\sin\theta$ is the phase shift in the elementary antennas, caused by the displacement of the target by angle θ relative to the tracking axis.

The channel voltage u_Σ is used in AGC systems to normalize the output signal. This voltage should differ from zero in the cancellation interval (in the region of the side lobes). In the sum channel, therefore, weight functions $h_3(\theta)$ and $h_4(\theta)$ are formed for the radiation patterns of the elementary antennas, and consequently we obtain the voltage of the sum channel

$$u_\Sigma = u_1 + u_2 = U[F(\theta_0 - \theta)\,h_3(\theta)\cos(\omega t + \phi)$$

$$+ F(\theta_0 + \theta)\,h_4(\theta)\cos(\omega t - \phi)]\,. \tag{7.2.10}$$

The gains of the sum and difference channels in the steady state mode are

$$k_\Delta = k_\Sigma = k_0/(1 + \mu' U_\Sigma), \tag{7.2.11}$$

where

$$U_\Sigma = U[F^2(\theta_0 - \theta)\,h_3^2(\theta) + F^2(\theta_0 + \theta)\,h_1^2(\theta) + 2F(\theta_0 - \theta)$$

$$\times F(\theta_0 + \theta)\,h_3(\theta)\,h_4(\theta)\cos 2\phi]^{\,1/2} \tag{7.2.12}$$

is the amplitude of voltage u_Σ; k_0 is the value of k_Σ for $U = 0$; $\mu' = \mu/U$.

Assuming that the phase detector multiplies its input voltages and averages the output signals, we obtain, in consideration of relations (7.2.9)–(7.2.12), the following formula for the normalized difference pattern:

$$u_{df} = k\,\{u_\Sigma u_\Delta\}_{av} \tag{7.2.13}$$

$$= k\,\frac{\mu^2\,\{[F^2(\theta_0-\theta)h_1(\theta)h_3(\theta) - F^2(\theta_0+\theta)h_2(\theta)h_4(\theta)]\cos\psi_0 + L_1(\theta)\}}{1 + u\,\sqrt{F^2(\theta_0-\theta)h_3^2(\theta) + F^2(\theta_0+\theta)h_4^2(\theta) + L_2(\theta)}}$$

where

$$L_1(\theta) = F(\theta_0 - \theta)\,F(\theta_0 + \theta)\,[h_1(\theta)h_4(\theta)\cos(2\phi - \psi_0)$$

$$- h_2(\theta)h_3(\theta)\cos(2\phi + \psi_0)]\,;$$

$$L_2(\theta) = 2\,F(\theta_0 - \theta)\,F(\theta_0 + \theta)\,h_3(\theta)\,h_4(\theta)\cos 2\phi;$$

ψ_0 is the phase shift in the phase inverter of the difference channel.

In the special case when the weight functions for the elementary antennas are identical, i.e., $h_1(\theta) = h_2(\theta) = h_I$ and $h_3(\theta) = h_4(\theta) = h_{II}$, expression (7.2.13) is simplified to the form

$$u_{df} = k \frac{\mu^2 \left\{ [F^2(\theta_0-\theta) - F^2(\theta_0+\theta)] h_\Delta^2 \cos \psi_0 + L_1(\theta) \right\}}{\left\{ 1 + \mu \sqrt{F^2(\theta_0-\theta) + F^2(\theta_0+\theta)]} \; h_\Sigma^2 + L_2(\theta) \right\}^2} \tag{7.2.14}$$

where

$$L_1(\theta) = F(\theta_0-\theta) \, F(\theta_0+\theta) \, h_I \, h_{II} \sin 2\phi \sin \psi_0;$$

$$L_2(\theta) = 2F(\theta_0-\theta) \, F(\theta_0+\theta) \, h_2^2 \cos 2\phi;$$

$$h_\Delta^2 = h_1 h_3 = h_2 h_4 = h_I \, h_{II};$$

$$h_\Sigma^2 = h_3 h_4 = h_3^2 = h_4^2 = h_{II}.$$

Assuming that $\phi = 0$ and $\psi_0 = 0$ in (7.2.14), we derive for the normalized difference pattern of an amplitude monopulse system the formula

$$u_{df} = k \frac{\mu^2 \left\{ [F^2(\theta_0-\theta) - F^2(\theta_0+\theta)] \, h_\Delta^2(\theta) \right\}}{\left\{ 1 + \mu [F(\theta_0-\theta) + F(\theta_0+\theta)] \, h_\Sigma(\theta) \right\}^2} \tag{7.2.15}$$

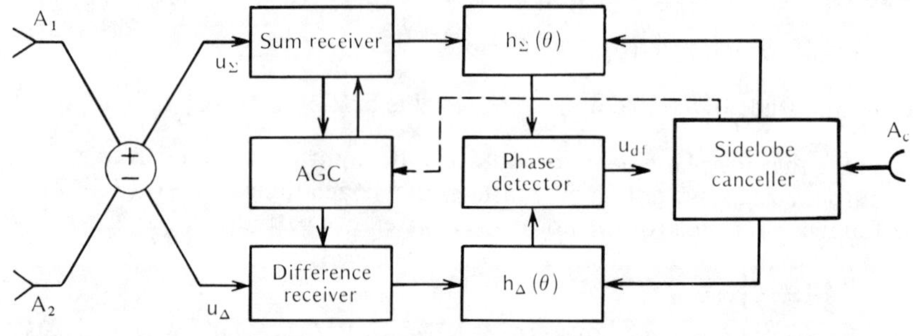

Figure 7.19

Expression (7.2.15) corresponds to the case when the side lobes are cancelled by changing the difference and sum radiation patterns (Figure 7.19).

For high-amplitude signals ($u \to \infty$), when the AGC best normalizes the signal, we obtain

$$u_{df}(\theta) = k\, \frac{[F(\theta_0-\theta) - F(\theta_0+\theta)]\, h_\Delta^2(\theta)}{[F(\theta_0-\theta) + F(\theta_0+\theta)]\, h_\Sigma^2(\theta)} \qquad (7.2.16)$$

Weight function $h_\Delta(\theta)$ should have a cutoff capability (Figure 7.20), and the function $h_\Sigma^2(\theta)$ must provide stable normalization of the error signal for all values of θ. The reason for this is that for certain angles θ, when the sum pattern is close to zero, the denominator of expressions (7.2.15) and (7.2.16) has a very small value, which may result in overemphasizing of the side lobes.

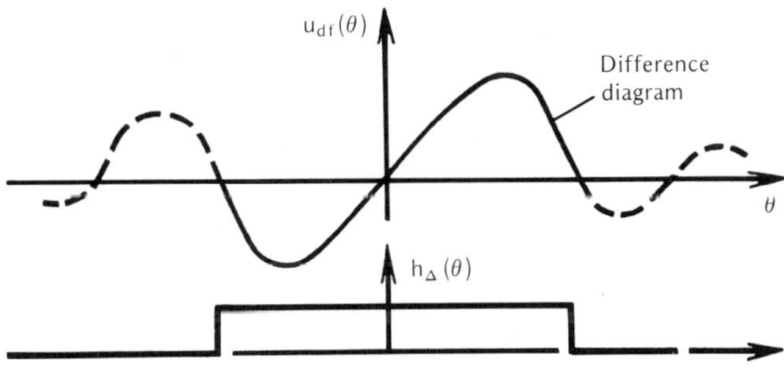

Figure 7.20

To avoid this it is necessary to satisfy the equation

$$[F(\theta_0-\theta) + F(\theta_0+\theta)]\, h_\Sigma(\theta) = c = \text{const.}$$

Thus the weight function of the sum channel should be selected from the condition

$$\frac{\mu^2}{\left\{1 + \mu\,[F(\theta_0-\theta) + F(\theta_0+\theta)]\, h_\Sigma(\theta)\right\}^2} = c.$$

Hence, for $c = 1$

$$h_\Sigma(\theta) = \frac{\mu-1}{\mu[F(\theta_0-\theta) + F(\theta_0+\theta)]}.$$

and for large μ

$$h_\Sigma(\theta) - \frac{1}{F(\theta_0-\theta) + F(\theta_0+\theta)}.$$

The graph of the weight function of the sum channel is illustrated in Figure 7.21. It is noteworthy that this function can be obtained as a result of the combined operation of the AGC system and the side lobe canceller.

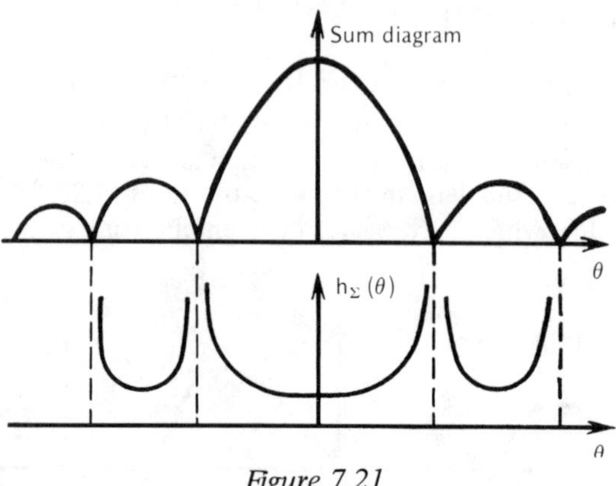

Figure 7.21

Angle Gating with Nonlinear Systems. An angle gating system performs various functions, depending on the purpose of the angle tracking system and type of jamming signal.

In the case of coherent two-point jamming, for example, it prevents the passage of angle error signals into the tracking system during segments of the effective time of action of the jamming, when the cross section of a paired target falls below the tolerable limit [15, 124, 187]. In the case at hand the nonlinear components of the system should perform logic switch-on and switch-off operations, i.e., should be relay elements. In other cases nonlinearities of the limiting and dead zone types may be used.

The amplitude characteristics of nonlinear systems $u = f(u_c)$; used in systems that perform angle gating of angle discriminator $u_{df} = u_{df}(u_c)$, are shown in Figure 7.22. Here u_c is the control voltage, generated by the phase detector.

Nonlinearity of the limiting type (Figure 7.22a) does not alter the width of the angle discriminator curve, but only reduces $\Delta\theta_e$. Such nonlinearity is used to prevent overloads in angle tracking circuits. Nonlinearity of the relay type (Figure 7.22b) has the greatest effect on the angle discriminator curve. This kind of nonlinearity is equivalent in terms of the effect that it produces to side lobe cancellers (Figure 7.20). However, the operating principle of an angle gating system with nonlinearity of the relay type is completely different. It is based on the gating (shutting off) of the angle channel during the time when the control signal u_c exceeds the prescribed clipping threshold U_{cl}.

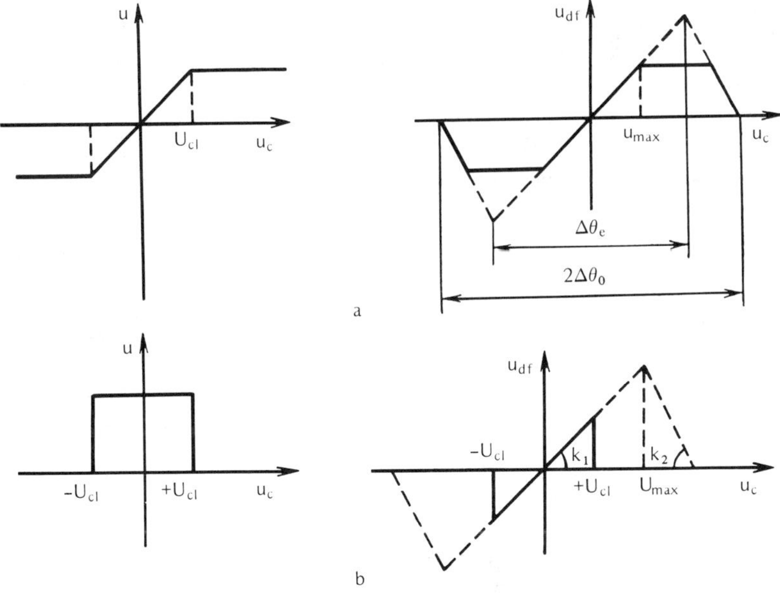

Figure 7.22

Relay nonlinearity reduces both parameters of the angle discriminator curve, both the width $\Delta\theta_0$ and the value $\Delta\theta_e$ (Figure 7.22b). The resolving power of a system increases. Furthermore, the accuracy with which a selected target is tracked is improved, since information does not pass into the angle channel (it is gated) during the time of jamming.

In addition to beneficial effects gating also has the undesired effect of detracting somewhat from the dynamic properties of an angle tracking system. This is related to the inclusion of nonlinear systems and to the digital nature of the transmission of information due to relay gating.

To quantitatively analyze the influence of nonlinear components on the quality of an angle tracking system we shall use the harmonic linearization method [24], in accordance with which nonlinear components, if angle errors are small, can be replaced by linear components with the equivalent gain k_{eq}. In the general case k_{eq} depends not only on the parameters of the system itself, but also on the amplitude A and frequency Ω of an input. The latter, if it contains components with frequencies not exceeding the frequency of the self-excited oscillations of the tracking system, may be written as

$$\theta_{in}(t) = \theta_{in}^0 + A \sin \Omega t. \tag{7.2.17}$$

Here θ_{in}^0 is the constant component of the input signal A and Ω are the amplitude and average frequency of the variable components of the input.

An angle tracking system exhibits basic nonlinearity, caused by the angle discriminator curve. For convenience we shall approximate it with straight lines (Figure 7.22b). By linearizing this curve it is possible to determine the equivalent gain of a direction finding system [24]

$$k_{edf} = 2 \frac{k_1 + k_2}{\pi} \arcsin \frac{\theta_{max}}{A} - k_2, \qquad (7.2.18)$$

where k_1 and k_2 are the angular coefficients of the slope of the straight lines that approximate the discriminator curve (Figure 7.22b); θ_{max} is the value of the input for which the discriminator output voltage is maximum: $A > \theta_{max}$.

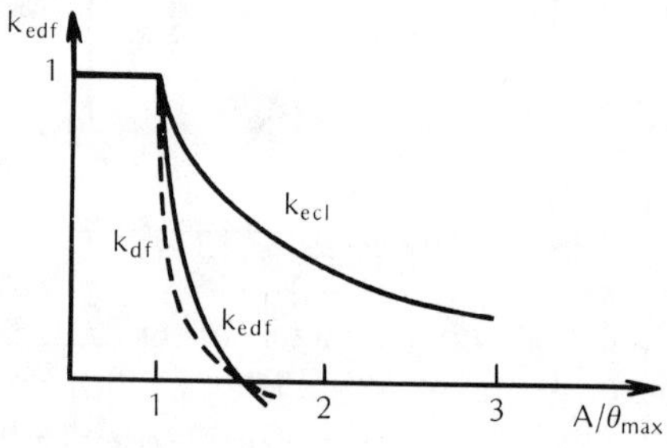

Figure 7.23

Function (7.2.18) is shown graphically in Figure 7.23. Also shown there is a curve, which characterizes the equivalent gain k_{ecl} of a nonlinear component with clipping

$$k_{ecl} = \frac{2k_1}{\pi} \arcsin \frac{\theta_{max}}{A} . \qquad (7.2.19)$$

The broken line in Figure 7.23 shows the change of gain k_{df} of the system when it contains a nonlinear components of the limiter type ($k_{df} = k_{e\,df} k_{e\,cl}$).

As can be seen in Figure 7.23, the presence of nonlinearities in an angle tracking system causes its gain to decrease as the amplitude A of the external input increases. If $A = \theta_{max}$, then the gain vanishes and becomes negative when $A > \theta_{max}$, which indicates that the system has gone into an unstable state.

Analyses show that as A increases self-excited oscillations begin to occur in the system first, and then, when $A > A_{max}$, the tracking system may be knocked out of the automatic tracking mode. If blinking jamming is used, then destabilization at $k_{df} \approx 0$ means that the geometric center of a paired blinking target becomes the point of unstable equilibrium and the system begins to track one of the sources, i.e., the blinking targets are resolved.

The critical amplitude of the external action may be found from the equation

$$k_{df} = 2 \, \frac{k_1 + k_2}{\pi} \, \text{arc sin} \, \frac{\theta_{max}}{A_{max}} - k_2 = 0,$$

whence

$$A_{max} = \theta_{max} \, \text{csc} \, \frac{\pi k_2}{2(k_1 + k_2)} \, . \tag{7.2.20}$$

Analysis of (7.2.20) shows that amplitude A_{max} has the minimum value $(A_{max})_{min}$ when $k_2 = \infty$. The coefficient $k_2 = \infty$ determines the shape of the shortened direction finding characteristic, obtained by relay angle gating (Figure 7.22b).

Consequently, relay gating influences the quality of an angle tracking system in two ways. On the one hand, the width of the angle discriminator curve is reduced, which improves the resolving capability of the system in relation to jamming that causes periodic or random perturbations with a high amplitude (blinking jamming, for example). On the other hand, relay nonlinearity detracts from the dynamic properties of the system and reduces the stability margin due to a reduction of the discriminator. This deficiency of gating systems must be taken into account in the design of angle tracking systems.

The equation of the dynamics of a linearized angle tracking system is written in the form [24]

$$Q(D)\theta^0 + R(D)k_{eq}\theta^0 = S(D)\theta^0_{in} \, , \tag{7.2.21}$$

where $Q(D)$, $R(D)$, $S(D)$ are polynomials with constant coefficients (polynomial $R(D)$ has lower power than polynomial $Q(D)$).

Because the equivalent gain depends on parameters A and Ω of jamming, these parameters affect the stability of the system; stability is usually defined as the attenuation of transient processes in terms of the slowly changing components. The stability margin within the space of the parameters of a system and of active jamming, is determined by familiar methods [155].

There are various ways of generating the control signal of a gating system. A simplified diagram of a monopulse automatic tracking radar with an angle gating system [214] is illustrated in Figure 7.24. The technique of creating an energy contrast between two targets T_1 and T_2, located within the space of resolution, is used in this system for controlling the gating system.

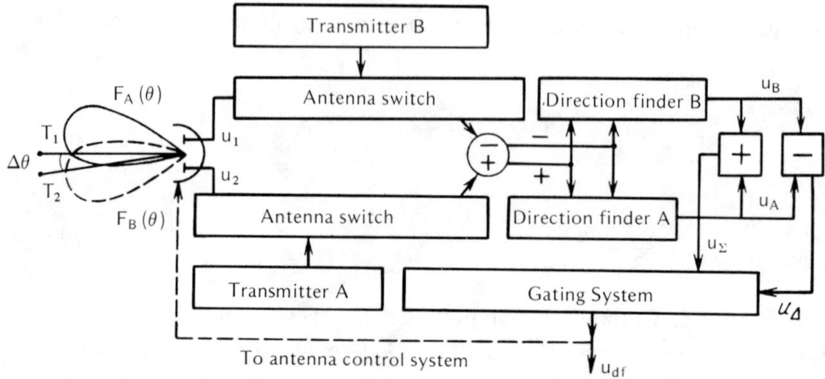

Figure 7.24

A target energy contrast is created by illuminating the targets with the transmitting antenna, with unequal gain in the directions to the targets being resolved. The easiest way to do this is to shift the maximum of the radiation pattern of the transmitting radar antenna relative to the equal signal line. Then the signals received, even from identical targets, will have different strengths.

The system has two direction finders. Direction finder A operates with a transmitting antenna, which has radiation pattern $F_A(\theta)$, shifted upward relative to the tracking axis, and direction finder B has a transmitting antenna with raditation pattern $F_B(\theta)$, shifted downward. Signals A and B can be resolved by means of frequency or time selection.

The control signal u_Σ is generated by adding voltages u_A and u_B from direction finders A and B, respectively, i.e., $u_\Sigma = u_A + u_B$. Difference signal $u_\Delta = u_A - u_B$ is used to control the gating system.

When a single target is being located the sum signal u_Σ is the same as the output voltage of an ordinary direction finder, and signal $u_\Delta = 0$. The situation is completely different when another jamming target appears in the resolution cell. The $u_\Delta \neq 0$, and voltage u_Σ, as before, characterizes the angular coordinates of the first target.

We shall examine in greater detail the operation of a system when resolving a paired target.

Suppose transmitter A is operating. Then the elementary antenna output signals are written as

$$u_1 = U_1 F^2 (\theta_0 - \theta) \cos \omega_1 t + U_2 F^2 (\theta_0 - \theta + \Delta\theta) \cos \omega_2 t, \tag{7.2.22}$$

$$u_2 = U_1 F(\theta_0 + \theta) F(\theta_0 - \theta) \cos \omega_1 t + U_2 F (\theta_0 + \theta$$

$$- \Delta\theta) \times F(\theta_0 - \theta + \Delta\theta) \cos \omega_2 t, \tag{7.2.23}$$

where the definitions presented in §2.2 are used.

At the output of the phase detector of direction finder A we have

$$u_A = k [(u_1 + u_2)(u_1 - u_2)]_{av} = k \left\{ B_A^2 \left[F^2 (\theta_0 - \theta) - F^2 (\theta_0 + \theta) \right] \right. \tag{7.2.24}$$

$$\left. + F^2 (\theta_0 - \theta + \Delta\theta) - F^2 (\theta_0 + \theta - \Delta\theta) \right\} ,$$

where

$$B_A = \frac{U_1 F(\theta_0 - \theta)}{U_2 F(\theta_0 - \theta + \Delta\theta)}$$

is a coefficient, which characterizes the energy ratio of the targets.

Replacing the function $F(\theta)$ by the first two terms of the power series in the vicinity of the equal signal line, we obtain

$$u_A = k [(1 + B_A^2) \theta - \Delta\theta]. \tag{7.2.25}$$

We derive the analogous expression for the voltage generated by direction finder B:

$$u_B = k [(1 + B_B^2) \theta - \Delta\theta], \tag{7.2.26}$$

where $B_B = \dfrac{U_1 F(\theta_0 + \theta)}{U_2 F(\theta_0 + \theta - \Delta\theta)} .$

For sum voltage u_Σ we have

$$u_\Sigma = u_A + u_B = k [(2 + B_A^2 + B_B^2) (\theta - 2\Delta\theta)]. \tag{7.2.27}$$

The difference control signal is

$$u_\Delta = k (B_A^2 - B_B^2) \theta \neq 0, \tag{7.2.28}$$

where $B_A \neq B_B$.

Formula (7.2.28) shows that when selected target T_1 is tracked without error and $\theta = 0$ control voltage $u_\Delta = 0$ and the gating system does not receive a signal. When error θ appears voltage $u_\Delta \neq 0$ and the angle channel is gated for as long as the control voltage exceeds some threshold. Here the control voltage is generated in the general case in the gating system by means of logic comparison of u_Σ and u_Δ. This is used to open the angle tracking system for the period of time necessary for performing automatic tracking with the required accuracy. During the time of gating, the control signal is obtained in the angle tracking system from the storage system. Sum voltage u_Σ carries information about both the targets and is used for generating an error signal in the system. Here the fact that there are two targets is established with the aid of voltage u_Δ.

Chapter 8
Functional, Structural and Combined Selection

8.1. Functional Selection

Functional signal processing techniques provide an opportunity to perform separate measurement of the angular coordinates of several radio emission sources and of the parameters of radio signals (amplitude, frequency, etc.) Consequently, useful signal selection can be performed in the video signal processing stage.

The problem of the simultaneous analysis of several parameters requires the development of a multichannel measuring system, representing a multi-channel radar with a receiving antenna array [8, 167, 209].

The measurement of several parameters of radio signals, in the absence of noise, amounts to the derivation and solution of a certain system of equations with unknowns as the desired signal parameters. Such an equation system usually is of the form

$$f_1 (x_1, \ldots, x_n) = U_{u1},$$

$$f_2 (x_1, \ldots, x_n) = U_{u2},$$

$$\cdots \cdots \cdots \cdots \cdots$$

$$f_n (x_1, \ldots, x_n) = U_{un},$$

where $f_i (x_1, \ldots, x_n)$ is a function of the desired parameters $x_1, \ldots, x_n$; U_{ui} ($i = 1, 2, \ldots, n$) are the measured combined field amplitudes at the antenna aperture or the combined auto- and cross-correlation functions of signals in independent measurement channels.

In practice, functions of desired parameters $f_i (x_1, \ldots, x_n)$ are the input signals u_{Ai} of the elementary channels of a multichannel radar estimator:

$$u_{Ai} = \sum_{\varrho=1}^{n} U_\varrho \exp [- i(\omega_\varrho t + \phi_{u\varrho} + \phi_\varrho)],$$

where ω_ℓ is the frequency of the ℓ-th signal; U_ℓ is the amplitude of the ℓ-th input signal of the i-th antenna, $\phi_{0\ell}$ is the initial phase of the ℓ-th signal. Phase ϕ_ℓ is determined by the relative position of the ℓ-th source in relation to the i-th antenna. For a linear antenna array

$$\phi_\ell = (2\,\pi/\lambda)\,x_i\,\sin\theta_{t\ell},$$

where x_i is the displacement of the i-th antenna relative to the center of the radar antenna array, and $\theta_{t\ell}$ is the angle of arrival of signals from the ℓ-th target. It is assumed here that all n targets are located at the same range.

If frequency ω_ℓ is assumed to be known and identical for all n signals, then the unknown values are the amplitudes U_ℓ, initial phases $\phi_{0\ell}$ and angular coordinates $\theta_{t\ell}$ of sources ($\ell = 1, 2, \ldots, n$). To determine all unknowns we must solve a system of M = 3n independent equations.

In the case of incoherent signals the initial phases need not be taken into account (due to averaging). Then, to determine angular coordinates $\theta_{t\ell}$ and amplitudes there must be M = 2n independent equations of the form

$$\sum_{\ell=1}^{n} U_\ell \exp\left(-j\phi_\ell\right) = U_{t\ell}\ . \tag{8.1.1}$$

   The problem of determining the unknowns amounts essentially to the reproduction of the field distribution at the aperture of the array on the basis of M = 2n measurements.

Each spatial distribution of targets has its own field distribution at the array aperture and its own radiation pattern, called radar response.

The problem of resolving two sources T_1 and T_2, the signals of which have amplitudes U_1 and U_2, is examined below. The angular position of the sources is characterized by angles θ_{t1} and θ_{t2}. Each source produces response $F(\theta)$ in the meter. To measure four unknowns an antenna array should have four elements.

Let the standard radar response to a single signal be

$$F(\theta) = \text{sinc}\,\theta\ = \sin\pi\theta/\pi\theta, \tag{8.1.2}$$

where $\theta = (D/\lambda)\sin\theta$ is the generalized angular coordinate, D is the antenna aperture and θ is the current angle between the line to the target and the axis of the antenna.

The function sinc θ is used here in connection with the fact that it describes the diffraction pattern in a uniformly distributed field at a radar antenna aperture (the response of the estimator) and is used in V. A. Kotel'nikov's theorem of measurements.

The responses of a meter to single signals are depicted in Figure 8.1 by the broken lines. According to Kotel'nikov's theorem response $F(\theta)$ can be perfectly reproduced if the measurements are taken with shift $\Delta\theta = 1$.

In the case of the action of two signals with amplitudes U_1, U_2 and with angular coordinates θ_{t1}, θ_{t2} the sum response (Figure 8.1) is the super-position of two single responses of the form (8.1.2):

$$F_\Sigma(\theta) = U_1 \, \frac{\sin \pi(\theta - \theta_{t1})}{\pi\theta} + U_2 \, \frac{\sin \pi(\theta - \theta_{t2})}{\pi(\theta - \theta_{t2})} \, , \qquad (8.1.3)$$

where U_1, U_2 are dimensionless amplitudes. Therefore it may be reproduced by measurements taken at angular intervals $\Delta\theta = 1$.

If the measured amplitudes are U_{ui} (i = 1, 2, 3, 4), then a system of four independent transcendental equations of the type

$$U_1 \, \text{sinc} \, (\theta_i - \theta_{t1}) + U_2 \, \text{sinc} \, (\theta_i - \theta_{t2}) = U_{ui}, \qquad (8.1.4)$$

may be derived. Here θ_i is the generalized coordinate of the i-th measurement point (Figure 8.1).

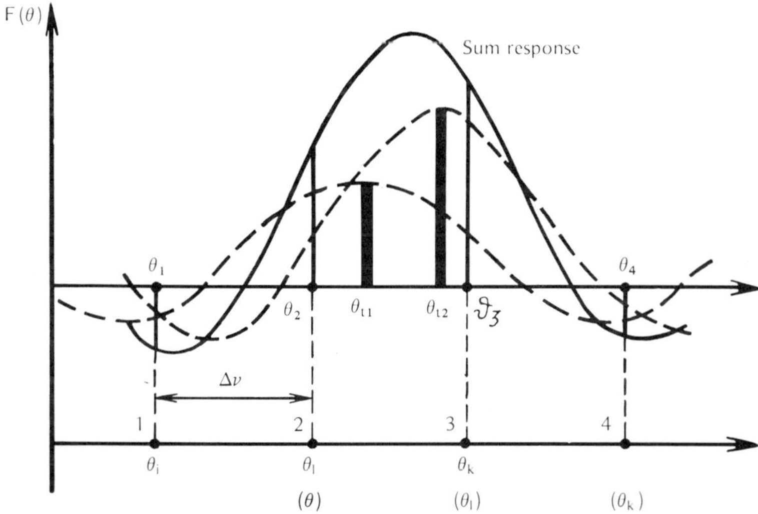

Figure 8.1

With a system of equations of the type (8.1.4) it is possible to find the desired angles θ_{t1} and θ_{t2}. Dividing the i-th equation by the ℓ-th, we obtain

$$\frac{\text{sinc} \, (\theta_i - \theta_{t1}) + (U_2/U_1) \, \text{sinc} \, (\theta_i - \theta_{t2})}{\text{sinc} \, (\theta_\ell - \theta_{t1}) + (U_2/U_1) \, \text{sinc} \, (\theta_\ell - \theta_{t2})} = U_{ui}/U_{u\ell} = U_{ui\ell}. \qquad (8.1.5)$$

Hence

$$\frac{U_{ui\ell}\,\text{sinc}\,(\theta_\ell-\theta_{t1}) - \text{sinc}\,(\theta_i-\theta_{t1})}{U_{ui\ell}\,\text{sinc}\,(\theta_\ell-\theta_{t2}) - \text{sinc}\,(\theta_i-\theta_{t2})} = -\frac{U_2}{U_1}. \tag{8.1.6}$$

By analogy, we obtain from the ℓ-th and k-th equations $(i, \ell, k = 1, 2, 3, 4;\ i \neq \ell \neq k)$

$$\frac{U_{uk\ell}\,\text{sinc}\,(\theta_\ell-\theta_{t1}) - \text{sinc}\,(\theta_k-\theta_{t1})}{U_{uk\ell}\,\text{sinc}\,(\theta_\ell-\theta_{t2}) - \text{sinc}\,(\theta_k-\theta_{t2})} = -\frac{U_2}{U_1}. \tag{8.1.7}$$

From (8.1.6) and (8.1.7) we obtain

$$\frac{U_{ui\ell}\,\text{sinc}\,(\theta_\ell-\theta_{t1}) - \text{sinc}\,(\theta_i-\theta_{t1})}{U_{ui\ell}\,\text{sinc}\,(\theta_\ell-\theta_{t2}) - \text{sinc}\,(\theta_i-\theta_{t2})} =$$

$$= \frac{U_{uk\ell}\,\text{sinc}\,(\theta_\ell-\theta_{t1}) - \text{sinc}\,(\theta_k-\theta_{t1})}{U_{uk\ell}\,\text{sinc}\,(\theta_\ell-\theta_{t2}) - \text{sinc}\,(\theta_k-\theta_{t2})}. \tag{8.1.8}$$

Let $\theta_i = \theta_\ell-1$, $\theta_k = \theta_\ell+1$ (Figure 8.1). Then, recalling that the function $\sin\theta$ is odd, we find

$$\sin\pi(\theta_\ell-\theta_k) = -\sin\pi(\theta_i-\theta_k). \tag{8.1.9}$$

From (8.1.8), we obtain, in consideration of (8.1.9),

$$\frac{\dfrac{U_{ui\ell}}{\theta_\ell-\theta_{t1}} + \dfrac{1}{\theta_i-\theta_{t1}}}{\dfrac{U_{uk\ell}}{\theta_\ell-\theta_{t1}} + \dfrac{1}{\theta_k-\theta_{t1}}} = \frac{\dfrac{U_{ui\ell}}{\theta_\ell-\theta_{t2}} + \dfrac{1}{\theta_i-\theta_{t2}}}{\dfrac{U_{ui\ell}}{\theta_\ell-\theta_{t2}} + \dfrac{1}{\theta_k-\theta_{t2}}}. \tag{8.1.10}$$

After transformations we obtain a system of linear equations of the type

$$U_{ui}(\theta_i-\theta_{t1})(\theta_i-\theta_{t2}) + 2U_{u\ell}(\theta_\ell-\theta_{t1})(\theta_\ell-\theta_{t2}) \tag{8.1.11}$$

$$+ U_{uk}(\theta_k-\theta_{t1})(\theta_k-\theta_{t2}) = 0,$$

where the subscripts i, ℓ, k may acquire any values of the series 1, 2, 3, 4 (for example 1, 2, 3 or 2, 3, 4).

A more convenient form of notation of equation (8.1.11) is

$$S_\ell^2 - S_\ell^1 (\theta_{t1} + \theta_{t2}) + S_\ell^0 \theta_{t1} \theta_{t2} = 0, \tag{8.1.12}$$

where $S_\ell^m = \theta_\ell^m U_{ui} + \theta_\ell^m U_{u\ell} + \theta_k^m U_{uk}$;

m is the exponent (m = 0, 1, 2).

After jointly solving two equations of the type (8.1.11) for specific angular positions of the elementary antennas, for example for $\theta_i = 1, \theta_\ell = 2, \theta_k = 3$ and for $\theta_i = 2, \theta_\ell = 3, \theta_k = 4$ (Figure 8.1), we obtain

$$\theta_{t1}\theta_{t2} = \frac{S_2^0 S_3^2 \quad S_3^1 S_2^2}{S_2^0 S_3^2 - S_3^0 S_2^2} = q_\theta, \tag{8.1.13}$$

$$\theta_{t1} + \theta_{t2} = \frac{S_2^0 S_3^2 - S_3^0 S_2^2}{S_2^2 S_3^1 - S_3^0 S_2^1} = p_\theta. \tag{8.1.14}$$

Since $\theta_{t1} = q_\theta/\theta_{t2}, \theta_{t2} + q/\theta_{t2} = p_\theta$, then

$$\theta_{t1,t2}^2 - p_\theta \theta_{t1,t2} + q_\theta = 0. \tag{8.1.15}$$

Hence the desired coordinates of sources T_1 and T_2 are

$$\theta_{t1}, \theta_{t2} = -\frac{p_\theta}{2} \pm \sqrt{\frac{p_\theta^2}{4} - q_\theta}. \tag{8.1.16}$$

The above example shows that computational operations are quite unwieldy, even in the simplest case of the measurement of the parameters of two signals.

For example, the three-antenna radar, described in [209-211], in addition to phase and amplitude meters, also has a special computer, from whose output are taken two angular coordinates θ_{t1} and θ_{t2}. Therefore a functional processing radar includes computers, intended for the solution of complex non-linear equation systems.

One of the advantages of a functional processing radar is the fact that signal processing is basically transferred to the video channel. This eliminates a number of rigorous requirements on the high-frequency components of the receiving antenna array, which may have a "fixed" aperture field distribution. In this respect functional radars have the same advantages as holographic processing radars. The deficiency of functional processing radars is the complexity of the computer system and the need to know the number of targets a priori for the "exact" choice of the required computation algorithm.

It is important to note in conclusion that the application of functional processing principles in classical operational radars affords an opportunity to enhance their jamming immunity. For example, a typical monopulse radar with four receiving antennas can successfully solve the problem of the simultaneous determination of the coordinates of two targets and measure the strength of their signals [209-211]. However, this is achieved at the cost of substantial complication of the equipment.

8.2. Structural Selection

1. Structural Selection without Feedback

It was mentioned in Chapter 4 that structural secondary selection without feedback is often based on the use of correcting binary codes with the detection or simultaneous detection and correction of errors. Also used extensively is a time code, which is a group of pulses with a priori known pauses.

The essence of any error detection correcting code consists in the fact that one part of its code combinations is used for transmission (for example of guidance commands) and comprises what are called permitted code combinations. The other part comprises prohibited code combinations.

Error detection codes are sometimes called safety codes. Safety can be achieved on the basis of the following principles:

permitted code combinations contain an even number of elementary symbols, in which case the parity of the processed symbols is checked in the receiver;

a message is represented by two mirror-symmetrical code combinations: the zeros in one combination are replaced in the other by ones;

each permitted code combination has the identical number of ones.

Distortions of individual or of many elementary symbols can be detected, depending on the safety principle employed. For example, by using a constant number of ones in permitted code combinations it is possible to detect all individual errors, but not when the number of symbols suppressed by jamming is equal to the number of false symbols formed. When an even number of ones is used in permitted code combinations it is possible to protect against any odd number of distorted symbols.

In a correcting code with simultaneous error detection and correction the necessary number of so-called attendant code combinations is added to each permitted (basic) code combination. When a basic or any attendant code combination is received a decision is made as to whether a basic combination was transmitted.

In addition to basic and attendant code combinations it is also possible to identify prohibited code combinations, on the basis of which errors are only detected.

The above-mentioned groups of code combinations, borrowed from [126, 194], are presented in Table 8.1 as an illustration of what we have said above. For example, if the received code combination is 00010, then the code combination 00110 should be regarded as having been transmitted.

Table 8.1

Permitted code combinations	11000	00110	10011	01101
Attendant code combinations	11001 11010 11100 10000 01000	00111 00100 00010 01110 10110	10010 10001 10111 11011 00001	01100 01111 01001 00101 11101
Prohibited code combinations	11110 01010	00000 10100	01011 11111	10101 00001

Many monographs [126, 177, 194] have been published on the subject of correcting codes, and therefore this subject will not be discussed here. We shall simply point out that when one code combination differs from another by d elements it is possible to correct errors of multiplicity $0.5(d - 1)$ when the number d is odd and of multiplicity $0.5d - 1$ when d is even.

Error detection and detection-correction codes are used for combatting noise jamming and random pulse jamming, which in the general case suppress transmitted code symbols and give rise to the formation of false code symbols.

The noise immunity of an error detection correcting code is estimated through the undetected error probability p_{iun} and detected error probability p_{ide} in the case of the transmission of a specific i-th code combination. Also used are the average probabilities p_{aun} and p_{ade}, defined as the average values of p_{iun} and p_{ide} of all code combinations.

Probabilities p_{iun}, p_{ide}, p_{aun} and p_{ade} are computed on the assumption that a synchronizing signal, sent from the transmitting end before the beginning of a code combination, is not distorted by jamming and are therefore conditional probabilities.

An error will be detected in cases when the transmitted i-th combination is transformed under the influence of jamming into a forbidden combination. Denoting permitted and forbidden combinations with the common number N as the numbers $1, 2, \ldots, M$, and $M + 1, M + 2, \ldots, N$, respectively, we find

$$p_{ide} = \sum_{j=M+i}^{N} p_{ij}.$$

Here p_{ij} is the probability of the transformation of the i-th code combination to the j-th.

The appearance of an undetected error is related to the transformation of one permitted code combination to another. Therefore

$$p_{iun} = \sum_{j=1 (j \neq i)}^{M} p_{ij}.$$

Probabilities p_{ij} are comparatively easy to compute for each specific code when the probabilities p_{10} and p_{01} of the transformation of a one to zero and of a zero to one are known; here p_{10} and p_{01} are calculated as a result of the solution of the problem of the receiver response to a useful signal and jamming. There are many books in which one can find necessary information about computations of p_{ide}, p_{iun}, p_{ade} and p_{aun} [183, 194]. Therefore this subject will not be discussed at length here.

The probability p_{dei} of the erroneous reception of the i-th transmitted combination, or the average error probability p_{dea} of all code combinations is used to evaluate the noise immunity of an error detection-correction code.

Since an i-th code combination includes not only the main code combination, but its companion combinations as well,

$$p_{0i} = 1 - \sum_{\ell=i_0}^{i_{Ni}} p_{i\ell}.$$

Here $p_{i\ell}$ is the probability that an ℓ-th code combination, where $\ell = i_0$, $i_1, \ldots, i_{Ni}$ denotes the number of the main code combination (for $\ell = i_0$) and of its companions (for $\ell = i_1, \ldots, i_{Ni}$), will be formed during the transmission of the i-th code combination as a result of its interaction with jamming.

The average p_{dei} for all M code combinations, by means of which useful information is transmitted, is found as

$$p_{dea} = \sum_{i=1}^{M} p_i p_{dei},$$

where p_i is the probability of the transmission of the i-th code combination.

If the probabilities p_{10} and p_{01} of the suppression of the transmitted and of the formation of false elementary symbols are identical and equal to p_e, then p_{dea} can be found by a simpler method. Actually, in view of the fact that a code permits the correction of errors when v elementary symbols are distorted by interference, we have

$$p_{dea} = \sum_{i=v+1}^{M} C_n^i \, p_e^i \, (1 - p_e)^{n-1}, \tag{8.2.1}$$

where n is the number of elementary symbols in the utilized code, and the combination C_n^i characterizes the number of possible ways of distortion of i symbols in an n-value code.

By calculating the dependence of p_{dea} on p_e using formula (8.2.1) for n = 5, v = 0; n = 9, v = 1 and n = 12, v = 2, it is possible to plot the graphs shown in Figure 8.2 [66]. As can be seen in Figure 8.2, an error detection-correction code has high noise immunity. The probability p_{dea} decreases as n increases and as p_e decreases. When jamming is strong and p_e cannot be assumed much smaller than one, error correction-detection codes lose their advantages over nonredundant codes.

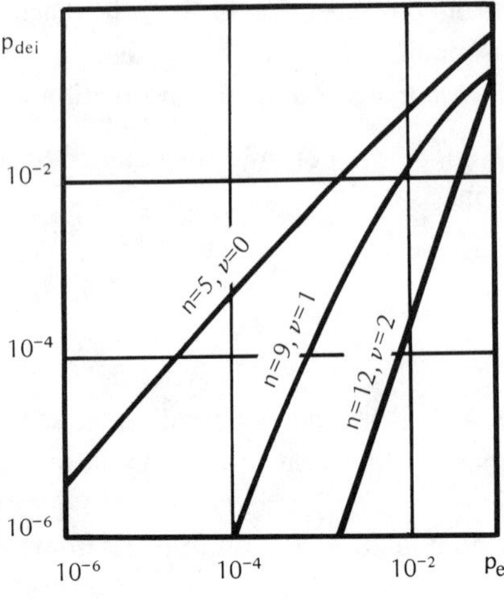

Figure 8.2

The results are the same when the fact that message transmission time t_{cc} for an error detection-correction code is longer than that of a nonredundant code with the same symbol duration is ignored. If time t_{cc}, required for the formation of a code combination of an n_r-value code is spent on the transmission of one code combination of an n_{nr}-value nonredundant code, then the energy of an elementary symbol in the nonredundant code can be increased.

Codes are usually constructed in accordance with the active pause principle, in which the symbols that determine zeros and one in a code combination are represented by electrical signals with the identical energy. Under these conditions the energy ratio of the symbols of a nonredundant n_{nr}-value and redundant n_r-value codes is $q_e = n_r/n_{nr}$. Since $q_e > 1$ the probability p_{e1} of the distortion of an elementary symbol in a nonredundant code by jamming will be less than p_e; it is assumed here that the suppression of transmitted symbols and the formation of false symbols have the identical probabilities.

In order to make a correcting code more noise-immune than a nonredundant code for $q_e > 1$, it is necessary to satisfy the inequality

$$p_{dea} < p_{denr} \tag{8.2.2}$$

where p_{denr} is the average error probability when a nonredundant code is used.

When the transmission of all code combinations is equally probable

$$P_{denr} = 1 - (1 - p_{e1})^{n_{nr}} \tag{8.2.3}$$

It may be written that

$$p_{e1} = p_{e1}(\alpha^2) \text{ and } p_e = p_e(\alpha^2/q_e).$$

where α^2 is a value related to the signal energy and which depends on the method used to transmit and process the elementary symbols.

In view of what has been said above we obtain, on the basis of relations (8.2.1)–(8.2.3),

$$\sum_{i=v+1}^{n_{nr}q_e} C_{q_e n_{nr}}^{\ell} p_e^i \left(\frac{\alpha^2}{q_e}\right) \left[1 - p_e\left(\frac{\alpha^2}{q_e}\right)\right]^{q_e n_{nr}-1} < 1 - [1 - p_{e1}(\alpha^2)]. \tag{8.2.4}$$

With this formula we find the values of q_e that satisfy inequality (8.2.2). For example, if the receiver is acted upon by additive white noise and coherent signal processing is employed, then for weak jamming we have [66]

$$p_{e1}(\alpha^2) \cong \frac{1}{\sqrt{2\pi}\,\alpha} e^{-0.5\alpha^2}.$$

Here α^2 is the ratio of the specific difference energy of the signals that represent the elementary symbols 0 and 1 to the variance of the jamming.

Similarly, we obtain

$$p_e \frac{\alpha^2}{q_e} = \frac{\sqrt[4]{q_e}}{\sqrt{2\pi}\,\alpha} e^{-0.5\,a^2/qe^2}.$$

Substituting $p_{e1}(\alpha^2)$ and $p_e(\alpha^2/q_e)$ into formula (8.2.4) and assuming that inequality $n_{nr}q_e p_e \ll 1$ to be valid, we find that relation (8.2.4) acquires the following form:

$$\frac{C_{q_e n_{nr}}^{v+1}}{n_{nr}} < \frac{(\sqrt{2\pi}\,\alpha)^v}{q^{0.25(v+1)}} \exp \frac{\alpha^2}{2q_o}(v+1 - q_e). \tag{8.2.5}$$

It follows from (8.2.5) that it is valid beginning at certain values of α, but under the necessary condition

$$q_e > v + 1. \tag{8.2.6}$$

Consequently, for given v greater noise immunity is imparted to a correcting code with simultaneous error detection and correction when the signal energy exceeds the spectral density of jamming by a comparatively wide margin and $n_r = q_e \times n_{nr} > v + 1$. In this case the gain in noise immunity increases with the signal-to-noise ratio. However, it decreases as the redundance factor q_e increases. The reason for this is that the elementary symbol energy increases in a nonredundant code.

Equations (8.2.2) and (8.2.5) determine the so-called correcting code applicability conditions. It turns out in practice that by no means all presently known codes satisfy the applicability condition. Detailed information on this subject can be found in the literature [126].

Time codes are an effective weapon against noise and random pulse jamming. This is because jamming can enter a final control system without interacting with a useful signal only in cases when false codes with a prescribed structure are formed. Jamming can also suppress one or many pulses in the useful time code, with the result that the transmitted message will be suppressed.

The probability p_{un} that a transmitted n-pulse time code, in which the probability p_{10} of the suppression of any pulse is identical, will not be suppressed is

$$p_{un} = (1 - p_{01})^n. \tag{8.2.7}$$

Equation (8.2.7) is valid for systems in which a transmitted code can pass through the decoding system only when none of n pulses is suppressed. It shows that as n increases at a given noise level, to which corresponds constant value p_{10}, the probability p_{un} decreases. However, as n increases the condi-conditions of the formation of false codes deteriorate. In the case, for example, of pulse jamming with random and exponentially distributed pauses between pulses, the probability p_{fc} of the formation of a false n-pulse code is determined by the following approximate formula [102]:

$$p_{fc} = n(M_p \tau_p)^n. \tag{8.2.8}$$

Here M_p is the average number of jamming pulses that appear at the decoder input per second, and τ_p is the jamming pulse duration, which is assumed to be constant. It is important to note that formula (8.2.8) sufficiently accurately determines p_{fc} when $M_p \tau_p \ll 1$.

Considering the varying character of the dependence of p_{un} and p_{fc} on n, and also the fact that as n increases the requirements on the delay lines, used in decoding systems, become increasingly more rigid, we may arrive at the conclusion that it is not advisable to use codes with $n > 4\text{-}5$.

In order to reduce the possibility of the formation of false codes for a given n
due to the interaction of transmitted pulses with jamming, it is necessary to
use different and, as a rule, nonmultiple pulse pauses in a code [212]. Then
an n-pulse false code is formed only as a result of the addition of n − 1
jamming pulses to one of the transmitted elementary signals. If two or more
pauses in a code are identical, then a smaller number of interference pulses
need be added to the transmitted signals to arrive at the desired code.

2. Structural Selection with Feedback

Structural selection with feedback is performed in radio communications
systems with two data transmission channels. The required messages (for
example telegraph symbols and guidance commands) are transmitted through
one (the direct) channel and signals with which the performance of components
of the main channel is monitored are transmitted through the other (feedback)
channel.

Electronic systems with feedback are classified as interrogation systems (with
resolving feedback) and as comparison systems (with information feedback)
[75].

In interrogation systems the receiver, with an output final control system, by
analyzing the incoming signals determines whether or not the received code
combination is permitted or prohibited. If the analyzed code combination is
prohibited, then the receiver, without passing that code combination, acts
upon the transmitter, which sends an interrogation signal (command) through
the feedback channel. After the interrogation command is received the data
source is activated. The latter transmits the entire previous code combination,
or just that part of it which was distorted by jamming, in the direct communications channel through the radio transmitter. This operation is repeated a
certain number of times or until interrogation commands cease to be transmitted.

In comparison systems the receiver of the main channel, through the transmitter of the feedback channel, informs (advises) the transmitter of the main
communications channel regarding incoming messages. Signals transmitted
through the feedback channel are called receipts. The latter may be copies
of the signals of the main receiver or may differ from them in some fashion.
The transmitter of the main channel compares a receipt with the transmitted
message and, if there is a discrepancy, first generates commands that prohibit
the recording of the previous message, and then a signal that describes the
necessary message. If the receipt corresponds to the transmitted message,
signals cease to be transmitted through the direct communications channel and
the receiver records the received message.

Comparison systems, the receipts of which are the retransmitted signal of the
main communications channel, are called retransmission systems.

Feedback systems offer protection from jamming and can be used to improve the noise immunity of radio telegraph and radio telephone communications systems, command guidance radio links and data transmission systems in aircraft and missile guidance complexes, etc., if messages are transmitted as code combinations. Radar measuring systems cannot be designed in this manner, since their transmitters are the targets or the radio responders of missiles, airplanes, etc., which are not capable of supplying information about measured coordinates.

The noise immunity of a feedback system is evaluated through the residual probability p_{res} that a message will be distorted after final transmission. Another important characteristic of such systems is the average number of transmissions of a given message, which determines the additional time required on the average for transmitting one message.

We shall first examine an entire interrogation system. When the number of repetitions for the transmission of one message is unlimited and the feedback channel operates free of errors p_{res} can be calculated in the following manner. During the first transmission of a given message the alternatives in the receiver are the correct reception of a code combination, the detection of an error or nondetection. Denoting the probabilities of these events through p_{rec}, p_{ide} and p_{iun}, we have

$$p_{rec} + p_{ide} + p_{iun} = 1.$$

If an error is detected during the first transmission of a message an interrogation signal is sent through the feedback circuit and the feedback channel is reactivated. The result of the second transmission of the message may be the same as that of the first.

A distorted message is recorded when the following incompatible messages are received:

an error was not detected during the first transmission;

an error was detected during the first transmission, but not during the second;

errors were detected in the first and second transmissions, and an error was not detected in the third transmission, etc.

Therefore, in interrogation communications systems with p_{iun} = const, p_{ide} = const and there is no limitation on the time of transmission of a given message,

$$p_{res} = p_{iun} \sum_{r=0}^{\infty} p_{ide}^{r} = \frac{p_{iun}}{1 - p_{ide}}. \tag{8.2.9}$$

As can be seen in (8.2.9) p_{res} is directly proportional to p_{iun} and increases with p_{ide}. The probabilities p_{iun} and p_{ide} for jamming of a given type and intensity depend on the structure of the code employed. If jamming is not very strong $p_{ide} \ll 1$ and $p_{res} \approx p_{iun}$. Under these conditions p_{res} is approximately equal to the probability p_{iun} that an error is not detected. The value of p_{iun} is always smaller than the probability of erroneous reception in the case when a correcting code is used. Therefore an interrogation system is more noise-immune than a system in which a correcting code with simultaneous error detection and correction is used. The presence of errors in the feedback channel increases p_{res}, and its computation under these conditions and its analysis are presented in [75].

The average number N_{av} of transmissions of messages through the direct circuit without regard for errors in the feedback channel is

$$N_{av} = \sum_{N=1}^{\infty} N \, p(N). \tag{8.2.10}$$

Here $p(N)$ is the probability that a message will be recorded after N transmissions. However $p(N)$ can be found as the product of the probability p_{erde}^{N-1} that an error will be detected during the transmission of the $(N-1)$-st code combination, multiplied by the probability of message recording $1 - p_{erde}$, i.e.,

$$p(N) = p_{erde}^{N-1} \, (1 - p_{erde}). \tag{8.2.11}$$

In consideration of (8.2.10) and (8.2.11) we obtain

$$N_{av} = \sum_{N=1}^{\infty} N \, p_{erde}^{N-1} \, (1 - p_{erde}). \tag{8.2.12}$$

The formula derived above gives the sum of the arithmetic-geometric progression

$$N_{av} = 1/(1 - p_{ide}). \tag{8.2.13}$$

Hence, even when $p_{ide} = 0.5\text{-}0.6$ it is necessary that $N_{av} = 2\text{-}2.5$. Consequently the increase of transmission time and of the associated signal energy in an interrogation system is negligible.

A similar examination of a retransmission system discloses the following [66, 75]. In the absence of jamming in the feedback channel and when the transmission time of a given message is unlimited, probability $p_{res} = 0$. This is because all distortions are detected. In cases when the jamming in the direct and feedback channels is identical in terms of structure and intensity the probability p_{res} and the average number of transmissions for retransmission systems may be calculated through formulas (8.2.9) and (8.2.13), respectively. It should be recalled, however, that these formulas are valid only for the case of the independent distortion of the elements of code combinations by jamming.

Consequently retransmission systems have better noise immunity than re-solving feedback systems, but the difference in their noise immunities is insignificant.

The general feature of feedback systems is that in the presence of very strong jamming their noise immunity may be worse than that of systems without a feedback channel. The conditions under which feedback systems lose their advantages depend on the energy characteristics of the signal and jamming, the kind of jamming, etc. These conditions are examined in [75] in relation to fluctuation and random pulse jamming.

8.3. Amplitude-Frequency Selection

In amplitude-frequency selection the detection of signals in the presence of jamming is based on the simultaneous utilization of their amplitude and frequency differences. The WLN system described below is typical of amplitude-frequency selection. Such a system is used extensively in radio communications systems for combatting high-amplitude, short-duration pulsed jamming in receiving channels [7, 41, 42, 58, 84].

A receiver designed for receiving amplitude-modulated signals including a WLN system, is shown in Figure 8.3. The WLN system is connected to the receiver input. It includes three components: wide-band amplifier, bilateral symmetrical amplitude limiter and narrow-band amplifier. Such a system is known by the initials of these components: WLN.

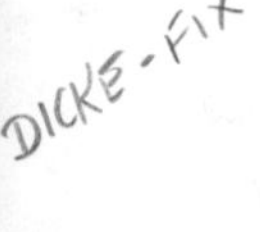

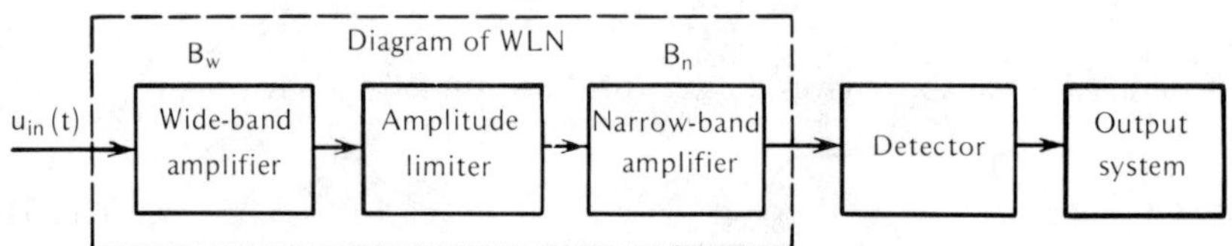

Figure 8.3

The wide and narrow passbands B_w and B_n of the amplifiers are symmetrical with respect to the central frequency f_0 of the input voltage. Let us assume that the frequency responses $K(f)$ of the amplifiers are rectangular (Figure 8.4). Passband B_n is matched with the spectral width of the signal B_s ($B_n \approx B_s$), but band B_w is selected in consideration of jamming pulse duration τ_j ($B_w \leqslant 1/\tau_j$) and is many times larger than band B_n. The average jamming spectral width is substantially greater than the spectral width of the useful signal.

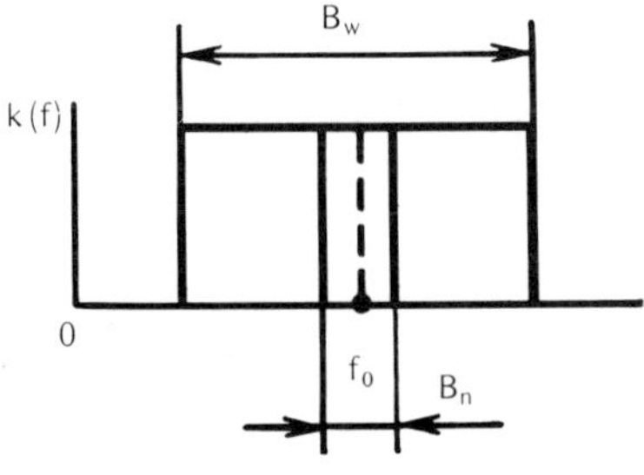

Figure 8.4

The limit level is selected in accordance with the amplitude of the output signal $u_s(t)$ of the wide-band amplifier. In the case, for example, of the reception of an amplitude- or frequency-keyed telegraph signal the cutoff level u_1 is equal to the amplitude u_s of signal $u_s(t)$, and in the case of the reception of amplitude-modulated broadcast signals level u_1 should be double the amplitude u_0 of the unmodulated carrier. If indicated conditions are satisfied the useful signal at the WLN output will have maximum amplitude and the transmitted messages will not be distorted. In practice the amplitude of received signals may vary in a wide range, depending on the distance between the terminals, transmitter power, etc. The gain of the wide-band amplifier can be controlled for optimizing the ratio between the cutoff level and the amplitude of the signal.

If jamming consists of short, nonoverlapping radio pulses the WLN system substantially increases the signal-to-noise ratio. The principle whereby a WLN system suppresses pulsed jamming is explained below on the assumption that the limiting threshold is regulated in accordance with the amplitude of the signal. The receiver input is acted upon by high-frequency jamming pulse $u_j(t)$ with a rectangular envelope, duration τ_j and amplitude U_j (Figure 8.5a). Pulse $u_{jw}(t)$ with an exponential envelope (Figure 8.5b) appears at the output of the wide-band amplifier. The duration of the leading edge of this pulse is determined by τ_j (we recall that band B_w is selected in consideration of τ_j and is equal to $B_w \leqslant 1/\tau_j$) and the duration of the trailing edge is determined by passband B_w. The limiter sharply reduces the amplitude, and consequently the energy of pulsed jamming. The jamming at the output of the bilateral symmetrical limiter (see 57.2) will consist of a pulse with a trapezoidal envelope amplitude U_{c1} and duration τ_{jw} (Figure 8.5b). If we assume that $B_w \approx 1/\tau_j$ and consequently, the jamming pulse is not substantially distorted by the wideband amplifier, then τ_{jw} may be estimated through the equation

$$U_{c1} = U_j k_w \exp\left(-\frac{0.5\tau_{jw}}{\tau_w}\right).$$

(8.3.1)

Here k_w is the gain of the wide-band amplifier; $\tau_w = 1/3B_w$ is the time constant of the filter with passband B_w. It is assumed in (8.3.1) that the time of exponential buildup of the amplitude of the jamming from U_{c1} to its maximum and the time of its falloff from maximum to U_{c1} are equal. From (8.3.1) we obtain

$$\tau_{jw} = \frac{2}{3B_w} \ln \frac{U_j k_w}{U_{c1}} . \tag{8.3.2}$$

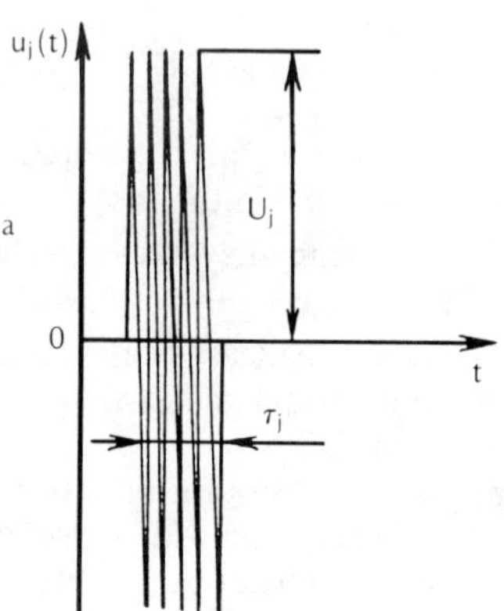

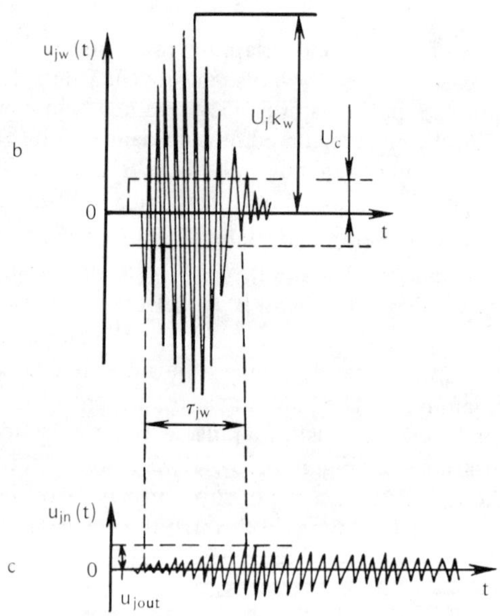

Figure 8.5

This relatively short jamming pulse is the input to the narrow-band amplifier. The time constant of the main filter system is

$$\tau_n \approx 1/3B_n, \tag{8.3.3}$$

At the output of the second amplifier the oscillation will build up during time τ_{jw}. At moment $t = \tau_{jw}$ the amplitude of the output reaches the maximum

$$U_{jout} = k_n U_{cl} \left[1 - \exp\left(-\frac{\tau_{jw}}{\tau_n}\right) \right],$$

where k_n is the gain of the narrow-band amplifier.

Recalling (8.3.2) and (8.3.3), we find

$$U_{jout} = k_n U_{cl} \left[1 - \exp\left(-2 \frac{B_n}{B_w} \ln \frac{k_w U_j}{U_{cl}}\right) \right]. \tag{8.3.4}$$

Interference $u_{jn}(t)$ at the output of the WLN system is illustrated in Figure 8.5c.

In the case of the reception of an AM telegraph signal voltage $u_{sout}(t)$, illustrated in Figure 8.6, appears at the output of the narrow-band amplifier. If the input signal amplitude is U_s, the wide-band amplifier output amplitude will be $U_s k_w$, and consequently it is advisable to assume that

$$k_w U_s = U_{cl}, \tag{8.3.5}$$

i.e., that the cutoff level is controlled in accordance with the amplitude of the signal. The narrow-band amplifier output amplitude reaches the steady state value

$$U_{sout} = k_w k_n U_s, \tag{8.3.6}$$

since the band of that amplifier is matched with a single signal ($B_n \geqslant 1/\tau_s$). Pulsed jamming, in application to this kind of signal, is especially dangerous when it fills the pause and distorts the code. Figure 8.6 shows a jamming pulse which could be generated at the output of a narrow-band amplifier in the absence of a WLN system.

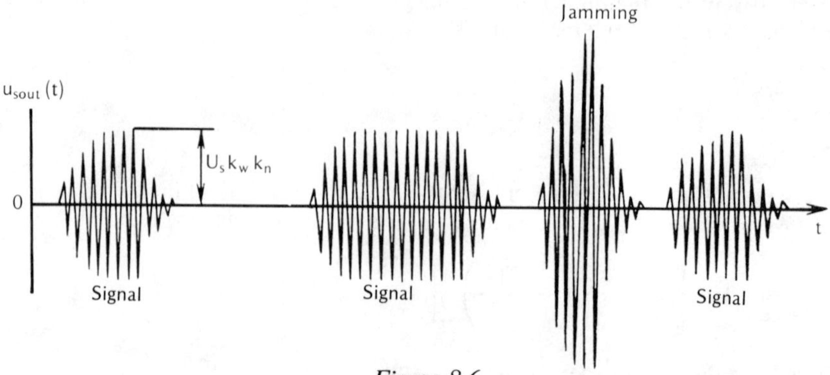

Figure 8.6

The damaging effect of jamming is reduced sharply when the amplitude of
the signal is made to exceed that of the jamming by a factor of 2-3. It is
easy to show that the WLN output signal is stronger than jamming while
the receiver input signal may be many times weaker. It follows from
formulas (8.3.4)–(8.3.6) that

$$\frac{U_s}{U_{jout}} = \left\{ 1 - \exp\left[-\frac{2B_n}{B_w} \ln\left(\frac{U_j}{U_s} \right)_{in} \right] \right\}^{-1}. \tag{8.3.7}$$

As can be seen by formula (8.3.7), when $U_j/U_s)_{in} > 1$ a WLN system improves
the signal-to-noise ratio and satisfies the condition $(U_s/U_j)_{out} > 1$. Formula
(8.3.7) gives the ratio of the receiver passbands B_n and B_w for which the
required excess of the signal over the jamming at the WLN output is guaranteed
for given receiver input levels. For example, the requirement to obtain at the
output $(U_s/U_j)_{out} \geqslant 2$ for $(U_j/U_s)_{in} = 10$ necessitates that the inequality
$B_w/B_n \geqslant 7$ be satisfied.

We have examined the case of the arrival of signals and jamming at different
moments of time. We shall now examine the action of pulsed jamming on a
continuous signal. We shall also assume that the wide passband, which is many
times larger than the narrow band, is at the same time considerably narrower
than the effective spectral width of the jamming, i.e.,

$$B_n \ll B_w \ll 1/\tau_j \tag{8.3.8}$$

It will be shown below that when a WLN system contains a nonlinear element
any signal that enters the wide-band then passes into the narrow-band channel.
Therefore, in order to improve the jamming immunity of a receiver the wide
band should be narrowed as much as possible and need not necessarily depend
on the jamming pulse duration.

Let us estimate the increase of the signal-to-noise ratio of a WLN system under these more general assumptions. We assume that a signal is a constant-amplitude wave $u_s(t) = U_s \cos \omega_0 t$ for $-\infty < t < \infty$, and the jamming $u_j(t)$ is a short radio pulse with a rectangular envelope on the same frequency ω_0, but with a phase shift π, i.e., $u_j(t) = U_j \cos (\omega_0 t + \pi)$ for $0 \leqslant t \leqslant \tau_j$. We also assume that $(U_s/U_{int})_{in} \ll 1$.

The assumption that the frequencies of the signal and jamming are equal is valid because free oscillations on natural frequency ω_0 occur and exist for a long time in the filter under the influence of a short stimulating jamming pulse. The assumption that the phases of the jamming and signal are opposite enables us to examine the dangerous case of the action of input of which is acted upon by a mixture of jamming and signal, differs strongly from the case when the input receives only a signal, and consequently receiver error is probable. In view of what we have said above the output voltage of the wide-band amplifier during the response time of an jamming pulse may be written as the sum of the steady state signal and noise with an exponentially increasing amplitude:

$$u_w(t) = k_w U_s \cos \omega_0 t + k_w U_j(1 - e^{-t/\tau_w}) \cos (\omega_0 t + \pi).$$

The maximum $U_{w\,max}$ of the envelope of voltage $u_w(t)$, generated at moment $t = \tau_j$, is

$$U_{w\,max} = k_w [U_j(1 - e^{-\tau_j/\tau_w}) - U_c].$$

For $\tau_j \ll \tau_w = 1/3B_w$, after expanding $e^{-\tau_j/\tau_w}$ into a series and using the first two terms of the expansion, we obtain

$$U_{w\,max} \approx k_w (3 U_j \tau_j B_w - U_s). \tag{8.3.9}$$

After the end of the jamming pulse the amplitude of the output voltage will decrease exponentially from the value determined by formula (8.3.9). The high-frequency duty factor will be determined by the parameters of the jamming, which is substantially stronger than the signal. Thus, after the end of the jamming pulse

$$u_w(t) = k_w(3U_j\tau_j B_w - U_s)e^{-t/\tau_w} \cos (\omega_0 t + \pi).$$

It may be assumed for approximation that the jamming is active for period of time τ_{jw}, corresponding to a reduction of the envelope of the output voltage to the clipping level. Consequently the equation (for $t = \tau_{jw}$)

$$k_w (3U_j\tau_j B_w - U_s) e^{-\tau_{jw}/\tau_w} = U_{cl}$$

is valid, from which we may determine τ_{jw}:

$$\tau_{jw} = \tau_w \ln [a(U_j/U_s)_{in} - 1]. \tag{8.3.10}$$

Here a $= 3\tau_j B_w$.

Only the signal appears at the output of the amplitude clipper in the absence of jamming:

$$u_{os}(t) = U_{cl} \cos \omega_0 t; \qquad\qquad (8.3.11)$$

however, during the time of action τ_{jw} of the jamming, when the amplitude difference of the jamming and signal exceeds the threshold level, the output voltage acquires the phase of the jamming and is

$$u_{oj}(t) = U_{cl} \cos (\omega_0 t + \pi). \qquad\qquad (8.3.12)$$

The effect would be the same if we were to assume that at the clipper output there exists a continuous signal, described by formula (8.3.11), and the equivalent jamming pulse $2U_{cl} \cos (\omega_0 t + \pi)$ exists for time τ_{jw} under the influence of jamming. Then the amplitude of the narrow-band output signal would be $U_{sout} = k_n U_{cl}$. The envelope of the jamming would have the maximum value at the end of the jamming pulse, i.e., when $t = \tau_{jw}$. The amplitude of the jamming at the narrow-band filter output will increase exponentially under the influence of the input voltage, with amplitude $2k_n U_{cl}$. Consequently

$$U_{jout} = 2k_n U_{cl} (1 - e^{-\tau_{jw}/\tau_n}). \qquad\qquad (8.3.13)$$

When $\tau_{jw} \ll \tau_n = 1/3B_n$ formula (8.3.13) acquires the form

$$U_{jout} \approx 2U_{cl} k_n \tau_{jw}/\tau_n. \qquad\qquad (8.3.14)$$

In consideration of (8.3.10), and recalling that $\tau_n/\tau_w = B_w/B_n$, we obtain

$$\left(\frac{U_s}{U_j}\right)_{out} = \frac{B_w}{2B_n \ln [a(U_j/U_s)_{in} - 1]}. \qquad\qquad (8.3.15)$$

Graphs for determining the WLN output signal-to-noise ratio as a function of the parameters of the system, jamming and input signal-to-noise ratio, are shown in Figure 8.7; the parameter of the family of curves $b = 3\tau_j B_w (U_j/U_s)_{in}$.

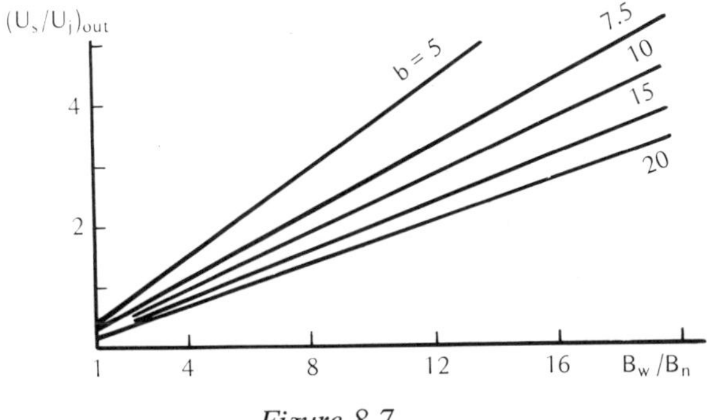

Figure 8.7

In some cases, for example, when different types of radio equipment are installed aboard a ship or aircraft, strong pulsed jamming may occur as a result of a considerable frequency difference between the jamming and resonance frequency f_0 of the protected receiving channel. In such cases a more complicated WLN system may be used, whose effect consists of the fact that the amplitude of the jamming at the output of the system is virtually independent of the input amplitude. This version of WLN system is illustrated in Figure 8.8. The input voltage is supplied to two wide-band amplifiers with frequency response $k_{w1}(f)$ and $k_{w2}(f)$ (Figure 8.9). The frequency f_0 to which the narrow-band amplifier is tuned is determined by the formula

$$f_0 = (f_1 + f_2)/2. \tag{8.3.16}$$

Since the passbands of the wide-band filters overlap the useful signal and frequency f_0 passes through both filters. The frequency difference Δf_0 between the filters is somewhat smaller than their passband B_w ($\Delta f_0 < B_w$). The frequency difference between the jamming and the receiver is assumed to substantially exceed the passband of the wide-band amplifiers.

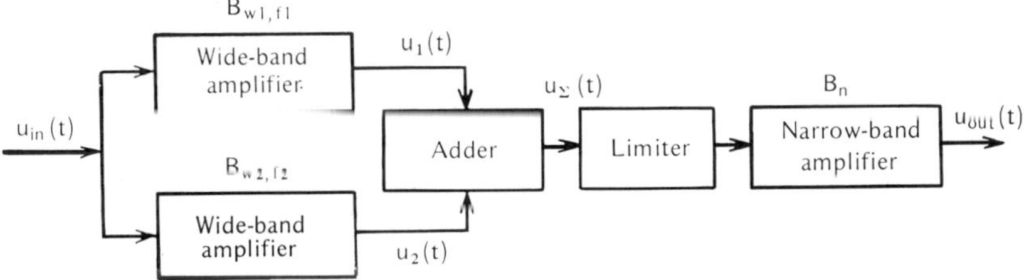

Figure 8.8

Under the influence of short jamming pulses (in comparison with $\tau_w = 1/3B_w$)

$$u_1(t) = U(t) \cos 2\pi f_1 t,$$

$$u_2(t) = U(t) \cos (2\pi f_2 t + \Delta\phi), \qquad (8.3.17)$$

will appear at the output of the wide-band amplifiers. Here $U(t) = U_{max} e^{-t/\tau_w}$; $\Delta\phi$ is the initial phase shift.

These voltages are supplied to the adder, which generates the voltage

$$u_\Sigma(t) = 2u(t) \cos \left(\pi\Delta f_0 t - \frac{\Delta\phi}{2}\right) \cos \left(2\pi f_0 t + \frac{\Delta\phi}{2}\right), \qquad (8.3.18)$$

where $\Delta f_0 = f_2 - f_1$.

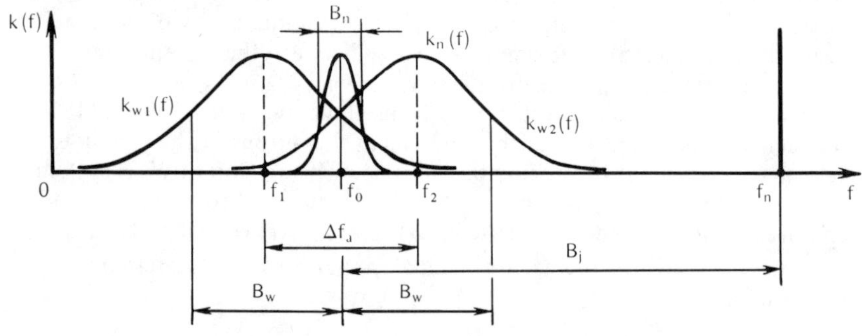

Figure 8.9

The sum voltage is an AM signal with envelope period $T_m = 1/\Delta f_0$ and carrier frequency f_0. The sum voltage, no matter how high the maximum value of the output jamming envelope of wide-band amplifiers, vanishes at moments of time determined by the relation

$$\pi\Delta f_0 t - \frac{\Delta\phi}{2} = (2n + 1) \frac{\pi}{2}.$$

Here n is an integer, which may have the values $0, 1, 2, \ldots$. At those moments of time when the envelope vanishes the phase of the high-frequency signal shifts by π. If by selecting special parameters (number of amplifier stages, ratio $\Delta f_0/\Delta f_w$, etc.) it is possible to satisfy the condition $\Delta\phi = \pi$, then, at the times of change of phase, the high-frequency component will pass through zero, i.e., $\cos (2\pi f_0 t + \Delta\phi/2) = 0$, and there will be no voltage jumps related to a phase shift of the waves by π.

The maximum output jamming envelope of the narrow-band amplifier is determined by the trivial relation

$$(U_{jout})_{max} = 3k_n U_{cl}(B_n/\Delta f_0) \qquad (8.3.19)$$

This value depends only on the parameters of the circuit and is not related to the input jamming amplitude. Experimental analysis of such a version of WLN disclosed that when the amplitude of jamming pulses at the receiver input changes by 60 dB the corresponding change of the maximum output jamming envelope is approximately 10% [41].

A modification of WLN known as WLON, which has an auxilliary local oscillator in front of the receiver amplitude detector (Figure 8.10), is used to further weaken pulsed jamming in radio telegraphic reception. The oscillator generates constant-amplitude waves, the frequency of which is precisely the same as the frequency of the signal, and the phase ϕ_h is either equal to the phase ϕ_s of the signal or differs from it by π. During the reception of useful signals the constant component of the detector current i_{ded} will either increase ($\phi_h = \phi_s$), or will decrease ($\phi_h = \phi_s - \pi$) by a completely defined value kU_s, where k is the gain of the system.

When the receiver input is acted upon by pulsed jamming the phase of the induced waves at the output of the narrow-band amplifier is random with uniform density distribution in the 0 to 2π range. Because the phase of the jamming voltage is random the average increment of the detector current decreases significantly in the presence of an auxiliary oscillator in comparison with the case when an ordinary WLN system is used.

Let the oscillator and jamming voltages be

$$u_h(t) = U_h \cos \omega_0 t \quad \text{and} \quad u_j(t) = U_j \cos (\omega_0 t + \phi_j)$$

respectively. Then the sum voltage will be

$$u_\Sigma(t) = U_h \left[\left(\frac{U_j}{U_h} \right)^2 + 2 \frac{U_j}{U_h} \cos \psi_j + 1 \right]^{1/2} \cos (\omega_0 t + \alpha), \qquad (8.3.20)$$

where $\alpha = \mathrm{arctg} \left(- \dfrac{U_h + U_j \cos \psi_j}{U_j \sin \psi_j} \right) + \dfrac{\pi}{2}$.

The relative amplitude increment is determined by the relation

$$\frac{\Delta U}{U_j} = \frac{U_h}{U_j} \left[\left(\frac{U_j^2}{U_h^2} + 2 \frac{U_j}{U_h} \cos \psi_j + 1 \right)^{1/2} - 1 \right].$$

Here $\Delta U = U_\Sigma - U_h$ is the amplitude increment of the detector input waves, caused by the appearance of jamming, and

$$U_\Sigma = U_h \left(\frac{U_j^2}{U_h^2} + 2 \frac{U_j}{U_h} \cos \psi_1 + 1 \right)^{1/2}$$

It should be added that in the absence of oscillator voltage the amplitude increment of the waves and, accordingly, the increment of the constant component of detector current, are determined by the jamming amplitude U_j. In the special case when $U_j = U_h$ we have

$$\Delta U/U_j = \sqrt{2(1 + \cos \psi_j)} - 1, \tag{8.3.21}$$

and the average relative increment is

$$\left(\frac{\Delta U}{U_j}\right)_{av} = \frac{1}{\pi} \int_0^{j} [\sqrt{2(1 + \cos \psi_j)} - 1] \, d\psi_j = 0.27.$$

The graph of $\Delta U/U_j$ as a function of ψ_j is illustrated in Figure 8.11.

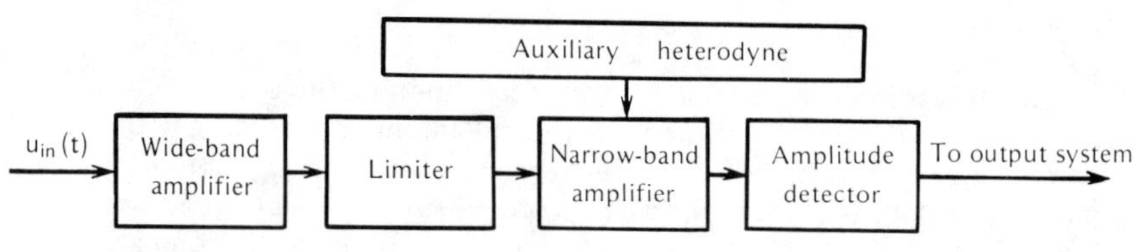

Figure 8.10

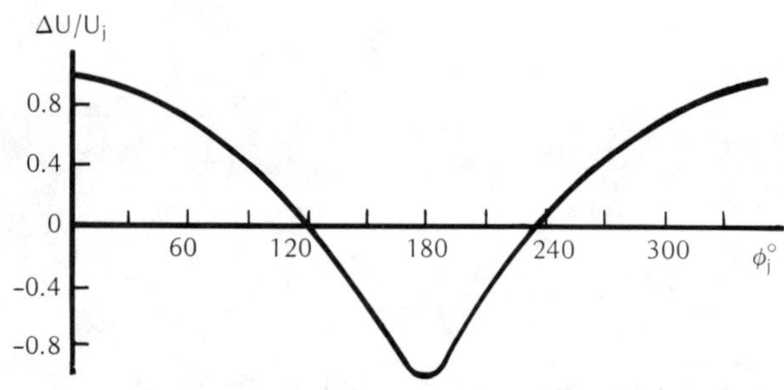

Figure 8.11

By using an auxiliary oscillator it is possible to substantially reduce the effectiveness of jamming. If the law of diminution of the ratio $\Delta U/U_j$ as a function of ψ_j is known it is possible to determine the probability with which this ratio will fall within prescribed limits:

$$p\left(\frac{\Delta U}{U_j} \leqslant \gamma\right) = \int_{\psi_{j1}}^{\psi_{j2}} \omega(\psi_j) \, d\psi_j \tag{8.3.22}$$

where $w(\psi_j)$ is the distribution density of the phase of the jamming; ψ_{j1} and ψ_{j2} are the boundary values of the range of change of ψ_j, within which the relation $\Delta U/U_j \leqslant \gamma$ is valid. When the probabilities ψ_j have a uniform distribution density

$$p\left(\frac{\Delta U}{U_j} \leqslant \gamma\right) = \frac{\psi_{j2} - \psi_{j1}}{2\pi}. \tag{8.3.23}$$

Thus the probability that the amplitude increment of the sum signal will not exceed one-half the amplitude of the jamming is 0.55. However, the probability that its diminution will not exceed one-half the amplitude of the jamming is 0.83 for $U_j = U_h$.

A WLN system that effectively controls pulsed jamming not only does not improve, but even detracts somewhat from the signal-to-noise ratio in the presence of white noise. As is known [39], if an ideal band limiter with the characteristic $U_{out} = 0$ for $U_{in} < 0$ and $U_{out} = U_{cl}$ for $U_{in} > U_{cl}$ is used and the input of this limiter receives a sinusoidal signal and white noise, the amplitude of the limiter output signal is determined by the formula

$$U_{sout} = \frac{U_{cl}}{\sqrt{\pi}} q_{cl} e^{-0.5q_{cl}^2} [I_0(0.5q_{cl}^2) + I_1(0.5q_{cl}^2)], \tag{8.3.24}$$

where U_{cl} is the limiting threshold; $q_{cl}^2 = U_{s\,in}^2/2\sigma_{w\,in}^2$ is the limitier input signal-to-noise ratio, i.e., after the wide-band filter; $I_0(x)$ and $I_1(x)$ are Bessel functions of imaginary argument x. For very large values of q_{cl}

$$I_0(0.5q_{cl}^2) \approx I_1(0.5q_{cl}^2) \approx \frac{\exp(0.5q_{cl}^2)}{q_{cl}\sqrt{\pi}}.$$

Under these conditions the amplitude U_{sout} of the limiter output signal is maximum:

$$(U_{sout})_{max} \sim 2\,U_{cl}/\pi. \tag{8.3.25}$$

Consequently the maximum limiter output signal level (for $R = 1$ ohm) is

$$S_{cl} = (P_{sout})_{max} = \frac{1}{2}(U_{sout})_{max}^2 - 2U_{cl}^2/\pi^2. \tag{8.3.26}$$

Two circumstances must be emphasized here. First, this power (with a sufficiently strong signal) does not depend on the input signal power and is determined only by the clipping level U_{cl}. Second, this power is redistributed between the signal and noise; when $q_{cl}^2 \ll 1$ it may be assumed equal to the clipper output noise power within the main spectral band (near frequency f_0), and the square of the ratio of the amplitude of the limiter output signal to the maximum amplitude may be assumed equal to the signal-to-noise ratio at the limiter input. For $q_{cl}^2 \ll 1$ the following equations are approximately satisfied:

$$I_0(0.5q_{cl}^2) \approx 1, \quad I_1(0.5q_{cl}^2) \approx 0.25q_{cl}^2, \quad e^{-0.5q_{cl}^2} \approx 1.$$

Therefore

$$U_{sout} = \frac{U_{cl}}{\sqrt{\pi}} q_{cl} \tag{8.3.27}$$

which leads, finally, to the expression

$$\left(\frac{S}{N}\right)_{cl} = \frac{\pi}{4} q_{cl}^2. \tag{8.3.28}$$

Consequently, in the case of a weak signal the presence of a limiter in the receiver results in a reduction of the signal-to-noise ratio by a factor of $\pi/4$. After filtering in the narrow-band filter the signal-to-noise ratio increases by a factor of B_w/B_n and becomes equal to

$$\left(\frac{S}{N}\right)_{out} = \frac{\pi}{4} q_{cl}^2 \frac{B_w}{B_n}. \tag{8.3.29}$$

The same signal-to-noise ratio can be achieved by replacing the WLN system with a narrow-band filter, matched to the signal. The only difference in the case when a linear filter is used is the factor 0.25π.

Analysis of the change of the signal-to-noise ratio in a WLN system for $q_{cl}^2 \approx 1$ poses serious problems. Such an analysis was carried out by the mathematical modeling method [58]. Two systems were compared in terms of the output signal-to-noise ratio at a given input ratio. A diagram of one of the systems is shown in Figure 8.3, and in the second system the WLN system is replaced by a filter, matched to individual pulsed signals.

A signal, representing a sequence of N radio pulses with a rectangular envelope, was modeled. The signal storage element of the output system was matched to the duration of the pulse train and the narrow-pass filter of the WLN system to individual pulsed signals. The passband of the wide-band filter could be changed in a wide range. The effective input signal-to-noise ratio was

$$\left(\frac{S}{N}\right)_{in} = \frac{U_{s\ in}^2}{2\sigma_{w\ in}^2},$$

where $\sigma_{w\ in}^2$ determines the noise level in the band B_w, and varied from 0 to 18. The results of modeling show that a reduction of the signal-to-noise ratio as a result of the use of a WLN system depends on the ratio of bands B_w and B_n (Figure 8.12). When $B_w/B_n < 10$ losses increase sharply, reaching 3-4 dB.

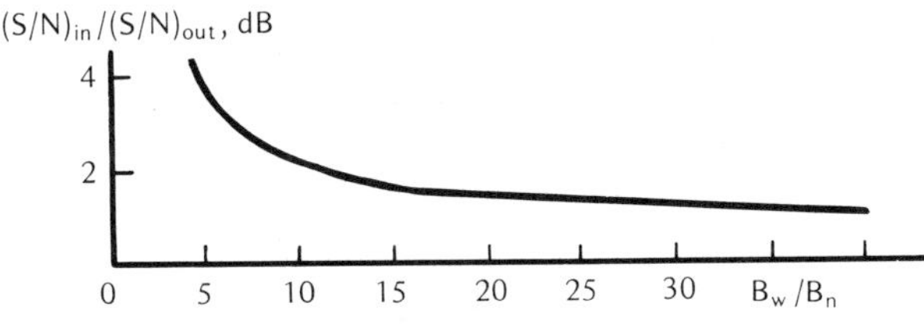

Figure 8.12

The appearance of crosstalk distortion as a result of the simultaneous entry of a useful signal and the signal of a strong jammer, the frequency f_j of which differs substantially from frequency f_0, into the wide band of the input amplifier is a serious deficiency of a WLN system. If a narrow-band filter were used as the input element of the receiver, then for $|f_j - f_0| > 0.5B_n$ the jamming would be filtered out. The fact that a wide-band amplifier is connected to the receiver input and is followed by an amplitude limiter leads to the penetration of jamming into the narrow-band channel of the receiver. To demonstrate this we will assume that the receiver input picks up the sum of the voltages of the signal and jamming

$$u_{in}(t) = U_s \cos \omega_0 t + U_j \cos \omega_j t,$$

where $U_s \ll U_j$ and $|\omega_0 - \omega_j| > 2B_n$.

Denoting $\omega_0 \quad \omega_j - \Omega$ and substituting

$$\cos \omega_0 t = \cos (\omega_j + \Omega)t = \cos \omega_j t \cos \Omega t - \sin \omega_j t \sin \Omega t,$$

we obtain

$$u_{in}(t) - (U_s \cos \Omega t + U_j) \cos \omega_j t - U_s \sin \Omega t \sin \omega_j t.$$

The factors of $\cos \omega_j t$ and $\sin \omega_j t$ may be viewed as slowly changing functions of time (since $\Omega \ll \omega_j$), and $u_{in}(t)$ may be written as

$$u_{in}(t) = U(t)\cos\left[\omega_j t + \phi(t)\right], \tag{8.3.30}$$

where

$$U(t) = \sqrt{(U_j + U_s \cos \Omega t)^2 + U_s^2 \sin^2 \Omega t} \ ;$$

$$\phi(t) = \arctan \frac{U_s \sin \Omega t}{U_j + U_s \cos \Omega t} \approx \frac{U_s}{U_j} \sin \Omega t.$$

Assuming that the clipping threshold is below the lowest envelope of the resulting wave, we obtain at the limiter output

$$u_{out}(t) = U_{cl} \cos\left(\omega_j t + \frac{U_s}{U_j} \sin \Omega t\right). \tag{8.3.31}$$

It follows from formula (8.3.31) that a phase-modulated wave, the spectrum of which for $(U_s/U_j) \ll 1$ includes three components with the frequencies ω_j, $\omega_j + \Omega = \omega_0$ and $\omega_j - \Omega = 2\omega_j - \omega_0$, is generated at the limiter output. Only one component with the frequency ω_0 and with amplitude $U_{cl} = U_s/U_j$ passes through the receiver; this fact is responsible for the dependence of the stated amplitude on the amplitude of the jamming signal.

If, for example, the jamming is a signal from an AM broadcasting station, the program of that station will be heard on frequency ω_0. Since the passband of the input amplifier with a WLN system must be selected many times larger than the passband of the main channel of the receiver, the probability that jamming stations will enter this wide band is quite high.

A WGN system [84, 87], in which the amplitude limiter is replaced by a controlled gate (Figure 8.13), is used in radio receivers to correct this deficiency. The received signal is analyzed with an jamming selection system. If the input voltage has the characteristics of jamming the indicated system generates control voltage U_c, which, by acting on the gate, blocks the receiver during the response time of the jamming. But in the presence of a useful pulse the gate is not activated and the receiver is not blocked. The probability of the extraction of jamming and not of the signal should be close to one.

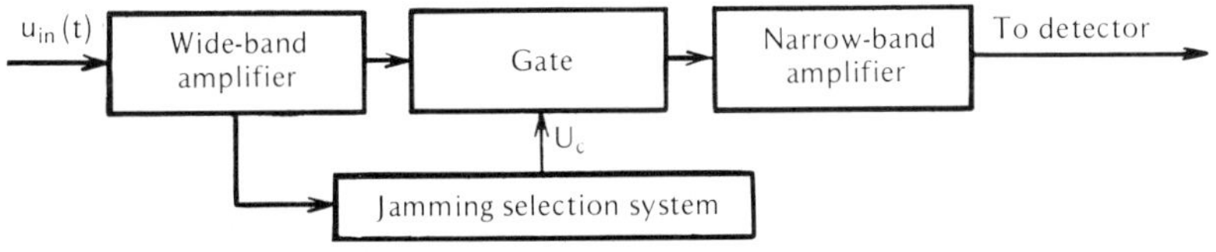

Figure 8.13

A modified WLN system, called WLT [196], which includes a wide-band
amplifier, maximum and minimum amplitude limiters, trap filter, detector
and trigger system (Figure 8.14), is used as the jamming selector. Signals and
jamming pass undistorted through the wide-band amplifier. The limiter
equalizes the amplitudes of the output voltages, irrespective of fluctuations
of the signals and jamming at the receiver input. The trap filter, with
amplitude-frequency response $\phi_{tr}(f)$, cuts off the region of frequencies
corresponding to the spectrum of the useful signal, the average frequency of
which is f_0. After the limiter and trap filter the amplitude of the jamming
voltage substantially exceeds the amplitude of the signal, irrespective of the
situation at the input. High-frequency waves are detected. Then comes the
minimum limiter, the purpose of which is to prevent the relatively low voltage
of the useful signal from passing to the output. The output element of the
jamming selection system is a trigger system of the blocking oscillator type,
the pulses of which block the receiver.

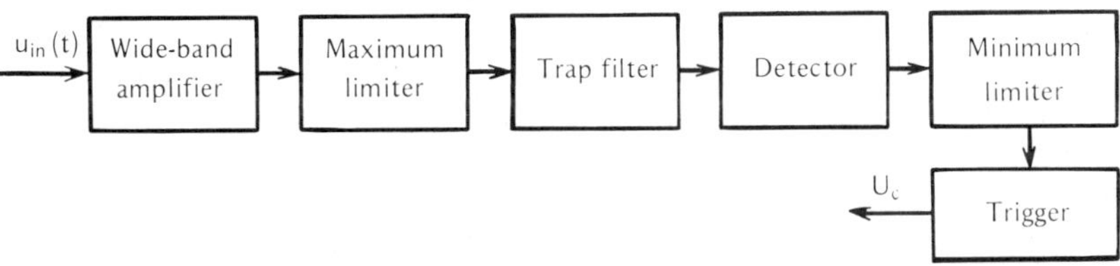

Figure 8.14

Experimental analysis of the WGN + WLT system shows that it extracts useful
signals with a probability of more than 0.95 in the presence of 50-60 dB changes
of the signal-to-noise ratio.

The WLT system operates effectively at relatively low jamming pulse repetition
frequencies (tens-hundreds of hertz). When the jamming pulse repetition frequency is high (tens of thousands of hertz) a nonlinear negative feedback

circuit may be used in the high-frequency amplifier of the receiver to suppress the jamming. A diagram of such a receiver is shown in Figure 8.15.

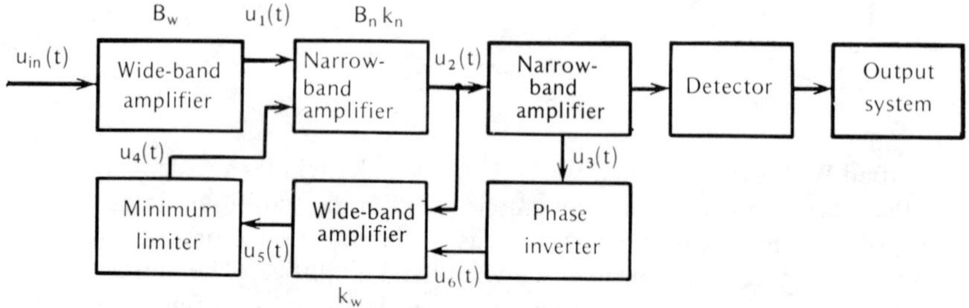

Figure 8.15

The narrow-band amplifier with gain k_n is subject to nonlinear negative feeback. When a jamming pulse reaches the receiver input the amplitude of the narrow-band amplifier output signals $u_2(t) = U_2(t) \cos \omega t$ increases rapidly. Voltage $u_2(t)$ is amplified by a wide-band amplifier with gain k_w. The resulting voltage $u_3(t) = k_w u_2(t)$ passes through the minimum limiter, where a small part near the zero value is cut from this voltage. The output voltage $u_4(t)$ of the feedback loop is in antiphase in relation to voltage $u_1(t)$, which is the input voltage of the narrow-band amplifier. When negative feedback is used the amplitude of the narrow-band amplifier output signals decreases approximately by a factor of k_n. And, under the influence of feedback, the passband of the amplifier expands by the same amount and the signals stimulated by the jamming pulses attenuate very rapidly.

When only a useful signal passes through the system the feedback circuit should be open. This is done by applying voltage $u_6(t)$ to the input of the wide-band amplifier of the feedback loop from the second narrow-band amplifier. Voltage $u_6(t)$ is set equal in amplitude to $u_2(t)$, and the phases of these voltages are set opposite to each other by the phase inverter. If the useful signal were continuous waves it would be possible to completely compensate one voltage with the other, i.e., to satisfy the condition $u_2(t) = u_6(t) = 0$. But when, for example, an AM signal is used the sum $u_2(t)$ and $u_6(t)$ will not be equal to zero because of a lag of $u_6(t)$ in relation to $u_2(t)$. However, in the case of relatively slow changes of the signal envelope the modulus of the sum $u_\Sigma(t) = u_2(t) + u_6(t)$ is sufficiently small and the amplitude of voltage $u_\Sigma(t)$ after being amplified in the wide-band amplifier does not reach the minimum cutoff level. This is precisely the way the cutoff level is set during adjustment of the system.

Consequently the signal voltage does not reach the output of the feedback loop and this loop is open in relation to the signal. In application to jamming this effect does not occur because of a substantial (in relation to the jamming pulse duration) lag of voltage $u_5(t)$ in comparison with $u_2(t)$. The amplitudes of voltages $u_2(t)$ and $u_5(t)$ of the jamming differ greatly at any given moment of time and it may be assumed approximately that only voltage $u_2(t)$ acts upon the input of the wide-band amplifier.

Experimental analysis of this receiver system disclosed that for jamming pulse repetition frequencies of 8-10 kHz the jamming-to-signal ratio in terms of the high-frequency amplifier output voltage is 0.1-0.2 for an input ratio of the order of 10.

8.4. Space-Time Process of Signals

1. General Information

A complete description of signals and jamming requires a knowledge of their time and spatial characteristics, which permit a mixture of a signal and jamming to be viewed as a space-time signal. Only that part of the mixture which enters the antenna aperture can take part in processing.

Antennas are classified as linear, two-dimensional and three-dimensional, depending on the number of spatial coordinates for which signals are processed. The field at each point of the aperture may be represented as a scalar value, which depends on the field polarization of the signal and antenna and is a function of the linear coordinates in the aperture and of time.

The utilization of additional field characteristics complicates the receiver and necessitates the use of a multichannel receiver or additional signal processing.

The problem of space-time processing of signals arises in the detection, analysis of the parameters, filtering and resolution of signals. Space-time selectivity is one of the important characteristics of radars. It determines the capacity of radars to select received signals in accordance with space and time parameters.

Space-time selectivity depends on the characteristics of the antenna system, the parameters of the received signals and jamming and on the method used to process them in the receiver. Space-time selectivity can be most rigorously analyzed on the basis of analysis of the generalized space-time ambiguity function of a signal [25, 167a, 213, 180]. The generalized ambiguity function may be written in most cases as the product of the space and time ambiguity functions.

The space-time representation of signals may be useful for the solution of many radar problems. By way of illustration of its capabilities we shall limit the examination to the problem of the angular resolution of two signals.

As the space-time selectivity criterion of a radar it is helpful to use the integral mean square difference [134]

$$\epsilon = \int_T \int_V |s_1(t, r) - s_2(t, r)|^2 \, dtdr, \qquad (8.4.1)$$

where $s_1(t, r)$, $s_2(t, r)$ are the signals to be resolved in the antenna aperture; r is a three-dimensional coordinate; $T = T_1 - T_2$ is the time processing interval; V is the range of space processing (antenna aperture). The decision as to whether there is one or two signals is made by comparing ϵ with some threshold.

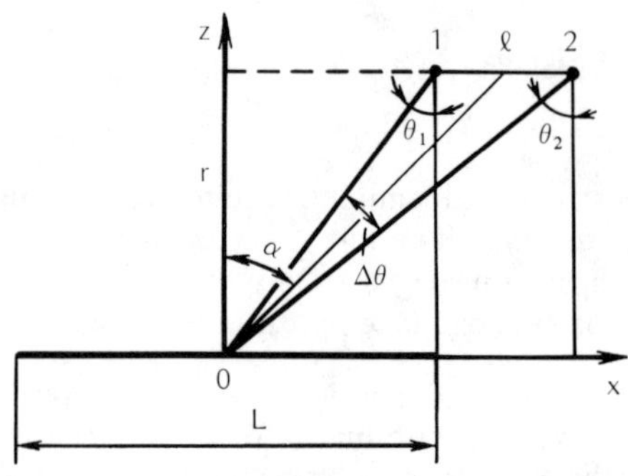

Figure 8.16

We will determine the ways of improving the space-time selectivity of radar when receiving two signals, generated by sources 1 and 2, separated in space (Figure 8.16). To explain the principle we shall examine a linear antenna with length L, situated on the $0x$ axis, as shown in Figure 8.16. In the general case these signals in the aperture of a linear antenna may be written as

$$u_1(t, x, \theta_1) = U_1 a_1(t - u_\theta x) \exp j\left[\omega_0(t - u_\theta x) - \phi_1\right], \qquad (8.4.2)$$

$$u_2(t, x, \theta_2) = U_2 a_2(t - u_\theta x - u_\theta x) \exp j\left[\omega_0(t - u_\theta x - v_\theta x) - \phi_2\right], \qquad (8.4.3)$$

where $u_\theta = \sin \theta_1 / c$; $v_\theta = (\sin \theta_2 - \sin \theta_1)/c$, θ_1, θ_2 are the angles that characterize the position of the sources in the plane of the antenna; c is the speed of light; $a_i(t)$ is the amplitude modulation function ($i = 1, 2$); x is the antenna coordinate; ϕ_1, ϕ_2 are phase shifts, determined by the distances to the targets and by the initial phases.

The phase difference of the signals due to the difference of the propagation distance is determined by the formula

$$\phi_1 - \phi_2 = \Delta\phi = (\ell\omega/c) \sin\alpha, \qquad (8.4.4)$$

where ℓ is the distance between the sources; α is the angle at which the sources to be resolved are viewed.

The values $\omega_0 u_\theta$, $\omega_0 v_\theta$, measured in units of $1/m$, are called space frequencies, since they determine the rate of change of phase on the x axis.

The signals received by element dx of the linear aperture of the antenna are

$$u_{a1} = \dot{I}(x)\, u_1(t, x, \theta_1)\, dx,$$

$$u_{a2} = \dot{I}(x)\, u_2(t, x, \theta_2)\, dx,$$

(8.4.5)

where $\dot{I}(x)$ is a weight function, determined by the distribution of the current through the linear aperture.

We write the integral mean square difference between these signals:

(8.4.6)

$$\epsilon = \int_{-L/2}^{L/2} \int_{-\infty}^{\infty} |u_{a1}(t, x) - u_{a2}(t, x)|^2 \, dt, dx.$$

Recalling (8.4.2) and (8.4.3), we obtain

$$\epsilon = U_1^2 \int_{-L/2}^{L/2} \int_{-\infty}^{\infty} a_1^2(t - u_\theta x)\, |\dot{I}(x)|^2 \, dxdt$$

$$+ U_2^2 \int_{-L/2}^{L/2} \int_{-\infty}^{\infty} a_2^2(t - \omega x)\, |\dot{I}(x)|^2 \, dxdt$$

$$- 2\,\mathrm{Re}\, U_1 U_2 e^{j\Delta\phi} \int_{-L/2}^{L/2} \int_{-\infty}^{\infty} \Big\{ |\dot{I}(x)|^2\, a_1(t - u_\theta x)$$

(8.4.7)

$$a_2(t - u_\theta x - v_\theta x)e^{-j\omega_0 v_\theta x} \Big\} \, dxdt,$$

where $w = \sin\theta_2 / c$.

We examine the first term in (8.4.7):

$$B_1 = U_1^2 \int_{-L/2}^{L/2} |\dot{I}(x)|^2 \, dx \int_{-\infty}^{\infty} a_1^2(t - u_\theta x)\, dt.$$

(8.4.8)

If the signal is assumed to have finite duration the internal integral in (8.4.8) is its effective value

$$T_{ef} = \int_{-\infty}^{\infty} a_1^2(t-u_\theta x)\, dt. \tag{8.4.9}$$

By analogy the effective antenna length is

$$L_{ef} = \int_{-L/2}^{L/2} |\dot{I}(x)|^2\, dx. \tag{8.4.10}$$

Then (8.4.8) is simplified and acquires the form

$$B_1 = U_1^2 T_{ef} L_{ef} = E_1 L_{ef}, \tag{8.4.11}$$

where E_1 is the energy of the first signal.

Likewise, for the second term in expression (8.4.7) we obtain

$$B_2 = U_2^2 T_{ef} L_{ef} = E_2 L_{ef}. \tag{8.4.12}$$

Here E_2 is the energy of the second signal.

The third term in formula (8.4.7) may be rewritten as[1]

$$B_3 = 2\,\mathrm{Re}\, U_1 U_2 e^{j\Delta\phi} \int_{-\infty}^{\infty} |\dot{I}(x)|^2\, C_{1,2}(v_0 x)\, e^{j\omega_0 \theta x}\, dx, \tag{8.4.13}$$

where $C_{1,2}(v_\theta x) = \displaystyle\int_{-\infty}^{\infty} a_1(t-u_\theta x)\, a_2(t-u_\theta x-v_\theta x)\, dt$

is a function of the cross correlation of the envelopes of the received signals, which may be represented through the Fourier transform [39, 155],

$$C_{1,2} = \frac{1}{2\pi} \int_{-\infty}^{\infty} S_c(\omega) e^{j\omega v}\theta\, x d\omega, \tag{8.4.14}$$

---

[1] The limits of integration in (8.4.13) are assumed to be infinite, since function $I(x)$ exhibits a limiting property.

where $S_c(\omega) = S_{a1}(\omega)S'_{a2}(\omega)$ is the convolution spectrum; $S_{a1}(\omega)$ is the spectral density of the modulation function of the first signal $a_1(t)$; $S'_{a2}(\omega)$ is the spectral density of the function that is a mirror image of the modulating function of the second signal $a_2(t)$.

Transforming (8.4.13) in consideration of (8.4.14), we obtain

$$B_3 = \mathrm{Re} U_1 U_2 e^{j\Delta\phi} \frac{1}{2\pi} \int\!\!\int_{-\infty}^{\infty} |\dot{I}(x)|^2 S_c(\omega)e^{j(\omega_0+\omega)vx}\,d\omega dx. \tag{8.4.15}$$

Using the formula for the spatial autocorrelation function of the radiation pattern [25, 134]

$$\psi(\omega_0 v_\theta) = \omega_0 \int_{-\infty}^{\infty} F^*(\omega_0 u_\theta)F[\omega_0(u_\theta+v_\theta)]\,du_\theta \tag{8.4.16}$$

$$= \frac{1}{2\pi} \int_{-\infty}^{\infty} |\dot{I}(x)|^2 e^{j\omega_0 v_\theta x}\,dx,$$

where $F(\omega_0 u_\theta)$ is the radiation pattern of the antenna, we may write (8.4.15) in the form

$$B_3 = \mathrm{Re} U_1 U_2 e^{j\Delta\phi} \int_{-\infty}^{\infty} S_s(\omega)\,\psi\,[(\omega_0+\omega)\,v_\theta]\,d\omega \tag{8.4.17}$$

$$= \mathrm{Re} U_1 U_2 e^{j\Delta\phi}\,\psi(\omega_0 v_\theta) \int_{-\infty}^{\infty} S_s(\omega)\,d\omega.$$

It is assumed here that for small variations of frequency ω (narrow-band signals) the radiation pattern changes only slightly, i.e.,

$$\psi[(\omega_0+\omega)\,v_\theta] \approx \psi(\omega_0 v_\theta).$$

We assume that the amplitude modulation of both signals is identical, i.e., $a_1(t) = a_2(t)$, and to it corresponds a real spectrum. In the case of symmetry in time $S_c(\omega) = |S_1(\omega)|^2$, where $S_1(\omega) = S'_{a2}(\omega) = S_{a1}(\omega)$. In accordance with Parseval's equation

$$\int_{-\infty}^{\infty} |S_c(\omega)|^2\,d\omega = 2\pi \int_{-\infty}^{\infty} |a_1(t)|^2\,dt = 2\pi T_{ef}. \tag{8.4.18}$$

In consideration of (8.4.18) we obtain for (8.4.17)

$$B_3 = 2\pi U_1 U_2 T_{ef} \cos \Delta\phi\, \psi\,(\omega_0 v_\theta).$$ (8.4.19)

Recalling (8.4.11), (8.4.12) and (8.4.19), we find the integral estimate

$$\epsilon = k[1 + \beta^2 - 2\beta \cos \Delta\phi\psi_i\,(\omega_0 v_\theta)],$$ (8.4.20)

where

$$\psi_i(\omega_0 v_\theta) = \frac{1}{2\pi L_{ef}}\, \psi\,(\omega_0 v_\theta)$$ (8.4.21)

is the standard ambiguity function in terms of angular coordinates; $\beta = U_2/U_1$; $k = U^2 T_{ef} L_{ef}$.

Using once again Parseval's equation

$$\int_{-\infty}^{\infty} |\dot{I}(x)|^2\, dx = 2\pi \int_{-\infty}^{\infty} |F(\omega_0 v_\theta)|^2\, d(\omega_0 v_\theta),$$

we may rewrite formula (8.4.21) as

$$\psi_i(\omega_0 v_\theta) = \frac{\omega_0 \displaystyle\int_{-\infty}^{\infty} F^*(\omega_0 u_\theta)\, F[\omega_0(u_\theta + v_\theta)]\, du_\theta}{\displaystyle\int_{-\infty}^{\infty} |F(\omega_0 v_\theta)|^2\, d\,(\omega_0 v_\theta)}.$$ (8.4.22)

For small v_θ $(v_\theta \to 0)$ $\psi_i\,(\omega_0 v_\theta) = 1$.

Analysis of expression (8.4.20) shows that the resolving power of a radar depends not only on the characteristics of the antenna, which determine its effective length L_{ef} and ambiguity function, but also on the parameters of the resolved signals: amplitude ratio β and phase difference $\Delta\phi$. If the signals are coherent, then analysis of (8.4.20) to the extremum yields the conditions of the worst and best resolution, which are written as follows:

the best resolution for

$$\Delta\phi = (2m + 1)\pi, \quad m = 0, 1, 2, \ldots,$$ (8.4.23)

the worst resolution for

$$\Delta\phi = 2m\pi, \quad \beta = \psi_i(\omega_0 v_\theta).$$ (8.4.24)

The best resolution is achieved with antiphased sources and the worst for cophasal sources.

Substituting (8.4.24) into (8.4.20) we obtain for cophasal sources

$$\epsilon_{min} = k\left\{1 - [\psi_i'(\omega_0 v_\theta)]^2\right\}. \qquad (8.4.25)$$

For antiphased sources we obtain from (8.4.20) and (8.4.23) for $\beta = 1$

$$(\epsilon_{max})_{min} = 2k\left[1 + \psi_i(\omega_0 v_\theta)\right]. \qquad (8.4.26)$$

For small angles $\Delta\theta$ (8.4.20) is simplified and has the form

$$\epsilon = k(1 + \beta^2 - 2\beta \cos \Delta\phi). \qquad (8.4.27)$$

Hence it follows that the conditions of the worst resolution for small v are

$$\Delta\phi = 2m\pi, \quad \beta = 1. \qquad (8.4.28)$$

Averaging (8.4.27) in terms of phase we obtain for incoherent sources

$$\epsilon = k(1 + \beta^2). \qquad (8.4.29)$$

Conditions (8.4.28) and expression (8.4.29) show that resolution is worse for sources of identical strength.

Let us compare the resolving power of radars for the case of cophase and antiphase signals, for which purpose we divide (8.4.26) by (8.4.25). As the result we obtain

$$\eta = \frac{2}{1 - \psi_i(\omega_0 v_\theta)}. \qquad (8.4.30)$$

If we estimate the resolving power of radars through the width of the ambiguity function at the level $\psi_i = 0.5$ we obtain $\eta = 4$.

Consequently, when antiphase sources are being resolved one may expect four times better resolving power than in the case of cophase sources.

The fact that the resolving power of a radar depends on the signal strength ratio and phase difference is of great practical importance. A radar in which the feasibility of controlling the relative strength of received signals is used to improve resolving power is described in §7.2. Resolving power can be improved substantially by controlling the phase difference of the received signals. The frequency of the transmitted signals may be changed for this purpose. It follows from condition (8.4.23) and formula (8.4.4) that

$$\Delta\psi = (\ell\omega/c) \sin \alpha = (2m + 1)\pi.$$

For $m = 0$ we find the required frequency shift that yields the best resolution of paired sources:

$$f = c/2\ell \sin \alpha.$$

2. Holographic Processing of Radio Signals

The invention of holography led to the simplest realization of the concept of optimum space-time signal processing. The very first analyses of holographic systems revealed that they have numerous advantages in relation to new effective jamming control techniques, based on the most complete utilization of phase information contained in the received useful radio signal.

In traditional radar information about the initial phase of the useful signal either is not used or is used only partially to achieve the maximum signal-to-noise ratio in the receiver (for example in optimum receivers using time processing of signals with known initial phase). Usually only the amplitude distribution is recorded during target imaging (signal processing), and phase information is irretrievably lost, which detracts from the jamming control capabilities of systems.

In holographic systems it is possible, by using all the information contained in a signal, which enters the antenna aperture $\Delta s = \Delta x \times \Delta y$, to optimize signal processing not only in time, but also in space through the antenna aperture. What essentially happens is that holographic processing makes the wave front, i.e., the amplitude and phase distributions in the antenna aperture, visible. To do this, obviously, it is necessary and sufficient to transform the wave front by some means or another from the radio frequency range to the visible range. The wave front can be reproduced only when a multichannel and coherent radar is used.

To explain the possible ways of improving the jamming immunity of radars by using holographic processing methods we shall examine the basic design principles of holographic systems.

Information about a target (subject) is contained in the amplitude and phase distributions of the electromagnetic field created by the radiating subject.

In the general case the combined field distribution is written as

$$\dot{e}(x, y, t) = E(x, y, t)e^{j\phi(x, y, t)},$$

where $E(x, y, t) = |\dot{e}(x, y, t)|$ is the field amplitude distribution in image plane $x0y$; $\phi(x, y, t)$ is the phase distribution.

For the sake of simplicity we shall examine below a stationary hologram, which enables us to ignore time. In other words, the signal frequency stability is assumed to be such that the phase difference of the added reflected and reference waves changes negligibly during the hologram recording time. The target is assumed to be stationary.

In holography the imaging process takes place in two stages. First a hologram is recorded, and then the wave front (image) is reconstructed. The purpose of the holographic system is to produce a pattern of a target (subject) by remembering the field scattered by it and then reproducing that field. To make a hologram it is necessary to use a source of coherent waves, i.e., a laser. A hologram is produced by the interaction (interference) of two waves: the wave scattered by the target and the reference wave from the coherent radiation source.

A recording instrument (signal storage system), for example a photographic plate, records the pattern that occurs as a result of the interference of the reference and scattered waves. To ensure that interference takes place the target (subject) is irradiated with the reference wave, which is scattered by the target. Part of the scattered wave strikes the recorder (photographic plate) where it interacts with the reference wave. Let us assume that the reference wave is plane and strikes the photographic plate at angle θ. Phase changes linearly along the x axis of the linear hologram:

$$e_r = E_0 \exp\left[-j\,\omega_2 x\right],$$

where $\omega_x = (2\pi/\lambda)\sin\theta$ is traditionally called the space frequency. When added in space the scattered wave and reference wave create in the plane of the photographic plate a field, the combined amplitude of which is described by the expression

$$e(x, y) = E_0 e^{-j\omega_x x} + E(x, y)\,e^{j\phi(x, y)}.$$

The photographic emulsion is exposed approximately in proportion to the strength of the incident signal. Therefore an interference pattern of the form

$$I(x, y) = |e(x, y)|^2 = E_0^2 + E^2(x, y) + E_0 E(x, y)e^{j[\omega_x x + \phi(x, y)]}$$
$$+ E_0 E(x, y)e^{-j[\omega_x x + \phi(x, y)]}. \qquad (8.4.31)$$

will be reproduced on the photographic plate.

It follows from the expressions derived above that the third and fourth terms of the sum carry complete information about the amplitude and phase distributions of the reflected wave. Here the phase information is coded in the form of phase modulation of a "three-dimensional" signal:

$$E_0 E(x, y)\cos\left[\omega_x x + \phi(x, y)\right].$$

The task of reproducing the wave front consists in phase demodulation of (8.4.31) and elimination of jamming, which is $E_0^2 + E^2(x, y)$. For phase detection it is necessary and sufficient to multiply (8.4.31) by a signal of the form

$$c_r - E_0 \exp\left(-j\omega_x x\right). \qquad (8.4.32)$$

However, reproduction of the amplitude distribution is possible if the transmission $T(x, y)$ of the processed plate is proportional to its exposure, so that

$$T(x, y) = k_0 I(x, y), \qquad (8.4.33)$$

where k_0 is the proportionality factor. Expression (8.4.33) in consideration of (8.4.31) is called the holography equation.

In the second step, called the step of reproduction of subject wave $E(x, y)$, the hologram is irradiated with the same plane coherent wave. The incident wave experiences diffraction on the interference pattern of the hologram. Beyond the hologram, when condition (8.4.32) is satisfied, is formed the electromagnetic field

$$E_{out}(x, y) = e_r(x, y) \, T(x, y) = k_0 e_r(x, y) \, I(x, y)$$

$$= k_0 \left\{ E_0^2 \left[1 + \frac{E^2(x, y)}{E_0^2} \right] e^{-j\omega_x x} + E_0 E(x, y) e^{j\phi(x, y)} \right. \qquad (8.4.34)$$

$$\left. + \left[E_0 E(x, y) e^{-j\phi(x, y)} \right] e^{-j2\omega_x x} \right\}.$$

Three waves, corresponding to the three terms of expression (8.4.34), are generated behind the hologram. The first component retains the direction of the reference wave ($\omega_x = (2\pi/\lambda) \sin \theta$) and is jamming. It may be made small by selecting the desired amplitude E_0. The second component is proportional to the desired distribution $E(x, y) \times \exp[j\phi(x, y)]$ and is the reproduced scattered wave. This wave propagates perpendicular to the plane of the hologram, since $\omega_x = 0$. The third component is also proportional to the reproduced field, but the wave generated by this component propagates at angle $-2\theta(-2\omega_x = (4\pi/\lambda) \sin(-\theta)$. This wave is complex conjugate to the initial wave.

In the second step of the holographic process two images are formed:

    an imaginary image, which perfectly corresponds to the true subject; this image generates the second component of (8.4.34);

    a real image, produced by the third component; this image exhibits the property of pseudoreproducibility, i.e., it has relief that is opposite that of the true image (valleys are seen as hills and vice versa).

Both images can be used in holography. The imaginary image is often used in practice.

There is a large variety of holographic systems, which may also be used in the radio frequency range in various signal processing stages [47, 142, 157]. Familiarization with them does not fall within the scope of this work.

There is nothing fundamentally different about holography in the radio
frequency range, since the nature of electromagnetic waves does not depend
on wavelength. If elements capable of recording the interference pattern of
reference and reflected radio waves are assumed to exist (detectors analogous
to photographic plates used in optics), information about the amplitude and
phase distributions of the reflected field can be reproduced by known holographic techniques. The image of a target can be reproduced by irradiating a
radio hologram with a reference coherent radio wave. However, this image of
the target will be invisible, because it is reproduced in the radio frequency
range. To make a visible image it is necessary to irradiate the radio hologram
with a coherent wave in the optical range. However, this kind of imaging is
frought with major technical difficulties, and therefore a different processing
method is utilized.

To make a visible image in holographic radar a radio hologram is "transformed"
into an optical hologram, which is then developed and is a permanent carrier of
information about the observed target [157, 142].

Holographic signal processing is also broken down into two steps in radar.

First a radio hologram is made, for which purpose the transmitter generates
a reference signal, used for irradiating the target. The radio signal reflected
by the target interacts on the receiving end with the reference signal and
produces a radio hologram, which is converted to an optical hologram. In
the second step the image is reproduced. For this purpose the above-described
operations are carried out. Consequently radio holography and optical holography differ only in the hologram-taking step because of the difference of
the wavelengths of the electromagnetic fields of the radio- and optical-
frequency ranges.

Holographic radars usually employ an antenna system of the array type.
If the antenna array receives a reference plane coherent wave and a wave
scattered by the target, then the intensity of the total signal varies through
the antenna aperture as a result of their interference. A system of interference bands is formed. It should be re-emphasized that the radio hologram
development time should be such that time and space coherence is not
disrupted, i.e., the interference pattern does not travel along the elements
of the array.

Let us examine the making of a radio hologram of a point target. A hologram
of a point target, as is known [163], is a Fresnel zone grid. The dark rings
correspond to the intensity antinodes (maxima) of the radio wave interference

pattern and the light rings correspond to the nodes (minima). Only those elements of the grid which are located within the antinodes are stimulated (Figure 8.17).

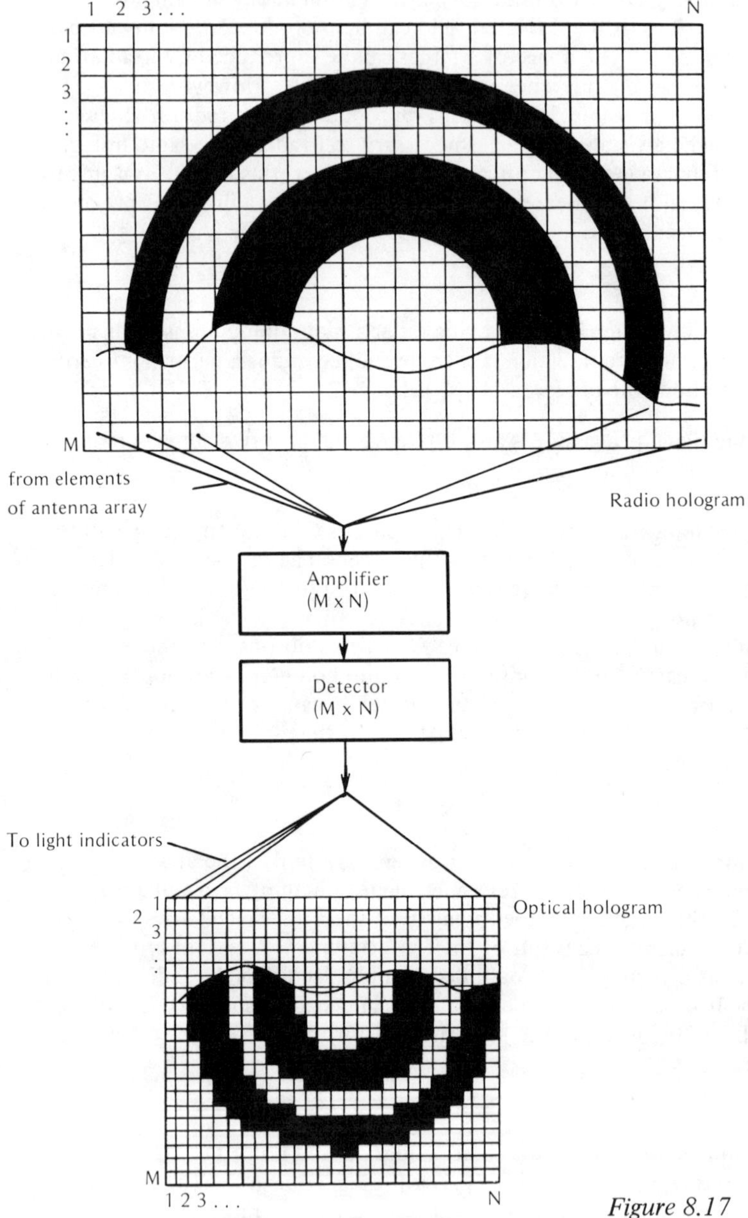

Figure 8.17

To each i-th element (of the total number M X N of elements) corresponds an i-th receiving channel (amplifier and detector), the load of which is a light indicator. Light indicators are arranged in the same order as the elements of the antenna array and comprise a matrix of elements M X N, analogous to the antenna array.

Only the indicators whose position corresponds to the maximum intensity rings of the radio hologram light up under the influence of signals from the receivers. The intensity variation pattern duplicates the interference pattern of the radio hologram in the matrix of the optical elements. The result is an optical hologram. The completeness of information about the interference pattern in the resulting hologram depends on the spatial sampling step. The limiting correspondence is achieved when the conditions of Kotel'nikov's theorem are satisfied in relation to the space frequencies.

The image that appears on the light element matrix may be recorded on photographic film and later reproduced by the above-examined method.

A hologram of complex targets is viewed as the superposition of the Fresnel zone grids made by each point of the target. It has a complex interference pattern, the "transformation" of which to an optical hologram is accomplished in accordance with the procedure above described.

One of the chief advantages of holographic systems is their high resolving power in terms of angular coordinates and range during simultaneous scanning of large regions of space. The observation zone of a holographic radar is determined by the width of the beam of each element or by the width of the beam of the transmitting antenna.

By using holographic pseudoprocessing it is theoretically possible to take full advantage of the potential resolving power of an entire receiving antenna system, consisting of M X N elements. Consequently, the effectiveness of "angle" jamming of various types, radiated from points in space beyond a target, is reduced.

Holographic radio signal processing, using high resolving power in combination with reprocessing, also makes it possible to effectively filter out repeater jamming.

In relation to all kinds of incoherent jamming a holographic radar will behave like an ordinary M X N-channel coherent radar, the observation time of which is equal to the hologram exposure time T_{exp}. Noise jamming produces an incoherent background on a hologram, since the spatial coherence time of such jamming is short and the interference pattern produced by the jamming and reference waves cannot be recorded. If the jamming has spectral width B_j, then the coherence distance is determined by the formula $\ell_{int} = c/B_j$. The coherence distance necessary for the stable recording of a hologram is determined by the exposure time T_{exp} and is $\ell_{exp} = cT_{exp}$. To preclude the possibility of recording a noise hologram it is necessary to have $\ell_{int} \ll \ell_{exp}$, whence the hologram exposure time is $T_{exp} \gg 1/B_j$. If $T_{exp} \ll 1/B_j$, then it is possible in principle to take a sharp picture of the source of jamming.

If the antenna aperture is L and the number of linear aperture elements is N, then the simultaneous scanning zone can be determined in the first approximation by the formula $\theta_{sc} = \lambda/(L/N)$. In the limiting case, when $N = L/(\lambda/2)$, we have $\theta_{sc} = 2$ rad.

Let us estimate the potential resolving power of a holographic radar on the basis of the Abbe criterion [163], which is equivalent to the Rayleigh criterion. For simplicity we shall assume that the radar antenna system is an array in which the size of the elementary antennas is d (Figure 8.18).

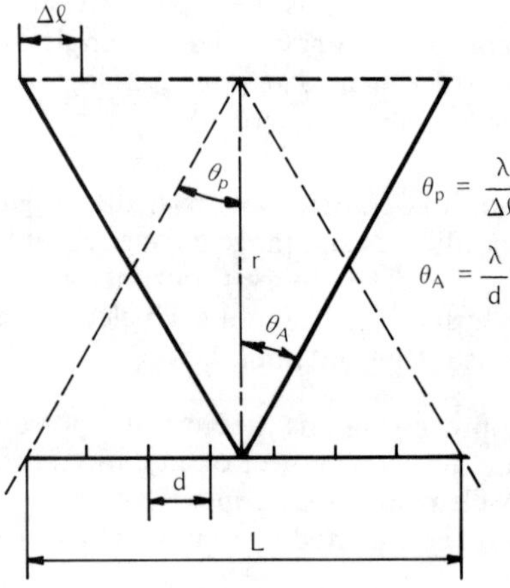

Figure 8.18

If the target as a periodic structure, i.e., consists of alternating reflecting and nonreflecting bands (squares) resolving power is defined as the smallest distance ($\Delta\ell$) between the bands, which can be reproduced by a given radar. Such a radar as a space frequency filter has the boundary space frequency

$$\omega_{max\ A} = \frac{d\phi}{dx} = \frac{2\pi}{\lambda}\sin\theta_A \approx \frac{2\pi}{\lambda}\theta_A, \tag{8.4.35}$$

where $\theta_{0.5} \cong \theta_A = \lambda/d$. Angle θ_A determines the direction from which it is still possible to receive signals.

A target representing a linear three-dimensional array reradiates a coherent radar signal, the space frequency of which is $\omega_p = 2\pi/\lambda \sin \theta_p$. Angle θ_p determines the direction of propagation of radio waves that carry information about the spatial structure of the target. Since $\sin \theta_p = \lambda/\Delta\ell$,

$$\omega_p = 2\pi/\Delta\ell. \tag{8.4.36}$$

The Abbe resolution condition [143, 163] is written as

$$\omega_p \leqslant \omega_{max\ A}. \tag{8.4.37}$$

From (8.4.35)–(8.4.37) we derive an expression for the potential resolving power

$$\Delta\ell \geqslant d. \tag{8.4.38}$$

Consequently a holographic radar has a resolving power in the antenna plane commensurable with the size of the elementary antenna, which is much smaller than the overall antenna size ($d \ll L$). Formula (8.4.36) determines the resolving power of holographic radar, operating in the semiactive mode. For radars with the same transceiving antenna, the potential resolving power is

$$\Delta\ell = d/2. \tag{8.4.39}$$

That does not mean, however, that the resolving power of a holographic radar does not depend on the overall antenna aperture L. For a given L, as is known, the angle-resolving power of radars is determined by the value $\theta_L \approx \lambda/L$.

Expressions (8.4.38) and (8.4.39) indicate only the feasibility of achieving linear resolving power $\Delta\ell = d$ or $\Delta\ell = d/2$ at target ranges r, determined by the focusing condition

$$\lambda/\Delta\ell \leqslant L/2r. \tag{8.4.40}$$

Hence, for given linear resolution $\Delta\ell$ it is possible to determine range r, at which the limiting resolution is achieved:

$$r \leqslant L\Delta\ell/2\lambda. \tag{8.4.41}$$

It follows from (8.4.38) and (8.4.39) that a reduction of elementary antenna size d leads to an improvement of resolving power and at the limit $\Delta\ell = \lambda/2$. What we have just said above must be understood in the following sense: for any (small) size $d \geqslant \lambda/2$ it is always possible to find processing techniques that are capable of providing the limiting resolving power $\Delta\ell = d$ and $\Delta\ell = d/2$. An example is radar antenna aperture synthesis [136]. In this case, in order to achieve the prescribed resolution $\Delta\ell = d$, aperture L must be synthesized in accordance with condition (8.4.40), i.e., $L \geqslant (2\lambda/\Delta\ell)r$, which agrees perfectly with the Rayleigh criterion [160, 143, 163]. Holographic signal processing techniques are used to reproduce the three-dimensional pattern of a target, and consequently holographic radars with a monochromatic transmission exhibit range resolving power. In order to determine its quantitative value it is necessary to find the dependence of the space frequency of the hologram on range.

A radar transmission may be written as

$$u(t) = U_0 e^{-j\omega_0 t}. \tag{8.4.42}$$

The signal received by an elementary antenna is

$$u_a(t) = U_a e^{-j\omega_0 (t - \tau_r)}, \tag{8.4.43}$$

where $\tau_r = 2r_s/c = (2/c)\sqrt{x^2 + r^2}$ is the lag of the signal received by the antenna element with current coordinate x (Figure 8.16). For $r \gg x$

$$\tau_r = \frac{2}{c}\left(r + \frac{x^2}{2r}\right). \tag{8.4.44}$$

In consideration of (8.4.44) we may write signal (8.4.43) as

$$u_a = U_a e^{-j\omega_0 t} e^{j\phi(x, r)},$$

where

$$\phi(x,r) = \frac{4\pi}{\lambda} r + \frac{2\pi}{\lambda r} x^2. \tag{8.4.45}$$

To make a hologram in a receiving system signals $u(t)$ and $u_a(t)$ are added together and detected by a quadratic detector, the result of which is an output signal, characterizing a one-dimensional hologram (one-dimensional Fresnel zone pattern):

$$u_{out}(x) = k_d [u(t) + u_a(t)]^2 = k_d [U_a^2 + U_0^2 + 2U_0 U_a \cos \phi(x, r)]. \tag{8.4.46}$$

The dependence of the space frequency on range coordinate r in the antenna aperture will be determined by the relation

$$\omega_r = \frac{d\phi}{dr} = \frac{4\pi}{\lambda} - \frac{2\pi}{\lambda r^2} x^2. \tag{8.4.47}$$

If the aperture is L, then the received signal will occupy in terms of coordinate r the space frequency band

$$\Delta\omega_r = \omega_{r\,max} - \omega_{r\,min}$$

where $\omega_{r\,min} = \dfrac{4\pi}{\lambda} - \dfrac{2\pi}{\lambda r^2} L^2$; $\omega_{r\,max} = \dfrac{4\pi}{\lambda}$. Hence

$\Delta\omega_r = (2\pi/\lambda)(L^2/r^2)$.

Consequently the radar has the linear range resolving power

$$\Delta r = 2\pi/\Delta\omega_r = (\lambda/L^2)r^2 . \qquad (8.4.48)$$

The quadratic dependence of resolving power Δr on range r of the targets being resolved has a simple physical explanation. Since the carrier of information about all three coordinates of a target is the field interference pattern, fixed on a hologram, the number of separately reproducible details of the target is fixed. As r increases the number of details that fall within the field of view of a holographic radar increases in proportion to r^2. Consequently only those details whose linear dimensions also increase by a factor of r^2 can be distinguished.

At great ranges resolving power deteriorates substantially. However, the very fact of the existence of range resolution during operation with continuous signals indicates the possibility of improving radar characteristics in the presence of jamming. For example, holographic systems have better passive jamming filtering capabilities than ordinary radars with continuous signals.

Let us determine the signal-to-noise ratio at the output of a processing system in the presence of passive jamming.

In optimum holographic radars, which focus the receiving antenna array on each target, the processing system changes the phase of the signal in accordance with the change of target range. This operation, called antenna focusing, is used to compensate the quadratic phase shift of the signals, which occurs in elements of the receiving array due to a change of phase of the received signal in the aperture.

In holographic processing the signals are combined in the image plane in consideration of their phases, and therefore we obtain at the output

$$u_{out} = \sum_{i=1}^{Q} (u_{si} + u_{ji}),$$

where u_{si} and u_{ji} are elementary antenna input signals and jamming; $Q = M \times N$ is the number of elements of the array. In consideration of focusing

$$u_{s1} = u_{s2} = \ldots = u_s NM = u_s,$$

$$u_{out} = Qu_s + \sum_{i=1}^{Q} u_{ji}. \qquad (8.4.49)$$

The strength of a useful signal at a load impedance of 1 ohm is

$$P_s = (Qu_s)^2 = Q^2 P_{si}, \qquad (8.4.50)$$

where P_{si} is the useful input signal strength of the i-th channel.

Let us assume that passive jamming is white space-time noise with spectral density G_0 [$V^2/Hz \cdot m^2$]. The cloud produced by this jamming overlaps the elementary antenna radiation pattern. Then, representing holographic radar as a space frequency filter, the boundary frequency of which is determined by formula (8.4.35), $\omega_{max\ A} = 2\pi/d$, we find that the interval of the space correlation range in the plane of the radio hologram (antenna) is

$$\delta_c = 2\pi/\omega_{max\ A} = d.$$

The above equation shows that the correlation interval is equal to the elementary antenna size. This enables us to assume that the jamming signals in individual channels are uncorrelated. Therefore the jamming power at a load impedance of 1 ohm is

$$J = \left[\sum_{i=1}^{Q} u_{ji} \right]^2 = QJ_i , \qquad (8.4.51)$$

where J_i is the input jamming power of an elementary channel.

From (8.4.50) and (8.4.51) we find the signal to noise ratio

$$J/S = J_i/QS_i \qquad (8.4.52)$$

Consequently, for an antenna array consisting of Q elements it is possible by focusing to increase the output signal-to-noise ratio of the system by a factor of Q, since the signals in the adder are combined coherently and the jamming incoherently.

If n_{av} is the average concentration of elementary reflectors per unit volume, then the reflectivity of the cloud is

$$\sigma_c = \sigma_1 n_{av} = 0.17\lambda^2 n_{av},$$

where σ_1 is the cross section of a single dipole. The radar resolution space is

$$\Delta V = \Delta r_\phi \Delta r_\theta \Delta r. \qquad (8.4.53)$$

Here $\Delta r_\phi = \lambda r_i/X$ is the linear resolving power in terms of azimuth; $\Delta r_\theta = \lambda r_i/Y$ is the linear resolving power in terms of elevation; $\Delta r = \lambda r_i^2\ L^2$ is resolving power in terms of range; $L = \sqrt{X^2 + Y^2}$, X, Y are the dimensions of the aperture of the receiving antenna array; r_i is the initial target range.

In consideration of (8.4.53) we obtain for the total cross section σ_Σ of the resolution space

$$\sigma_\Sigma = 0.17 n_{av} \frac{r_i^4 \lambda^3}{XY(X^2+Y^2)} \, . \tag{8.4.54}$$

The elementary antenna output jamming power of the holographic radar is

$$J = \frac{P_t G_t}{4\pi r_i^2} \, \frac{\sigma_\Sigma}{4\pi r_i^2} \, A, \tag{8.4.55}$$

where P_t is the radar transmitter power; G_t is the gain of the transmitting antenna; A is the effective antenna area.

The useful signal power is

$$S = \frac{P_t G_t}{(4\pi r_i^2)^2} \, \sigma_t \, A. \tag{8.4.56}$$

In consideration of (8.4.55) and (8.4.56), we obtain

$$(J/S)_{out} = \sigma_\Sigma / Q\sigma_t \, . \tag{8.4.57}$$

For filled antenna arrays, in which the distance between individual elements is $\lambda/2$, the number of elementary antennas is

$$Q = M \times N = 4XY/\lambda^2 \, . \tag{8.4.58}$$

From (8.4.54), (8.4.57) and (8.4.58) we obtain

$$\frac{J}{S}\bigg|_{out} = 0.17 n_{av} \frac{r_i^4 \lambda^7}{4X^2 Y^2 (X^2+Y^2)\sigma_t} \tag{8.4.59}$$

Let $X = Y = L$. Then, recalling that for optimum holographic radar with focused antennas,

$$L = \lambda r_i / d, \tag{8.4.60}$$

we obtain

$$(J/S)_{out} - d^6 \sigma_c / 8\lambda r_i^2 \, \sigma_t. \tag{8.4.61}$$

Expression (8.4.61) shows that the output signal-to-noise ratio may be increased substantially by increasing the antenna aperture. The increase of the signal-to-noise ratio as the target range decreases is attributed to the fact that the aperture L of focused holographic radar is selected in accordance with expression (8.4.60).

For radars in which the aperture of the antenna array is constant, we obtain from (8.4.59)

$$(J/S)_{out} = r_i^4 \lambda^6 \sigma_c / 8L^6 \sigma_t.$$

In this case L = const and the signal-to-noise ratio may be increased substantially at short target ranges.

3. Synthetic Aperture Radar [47, 136]

The techniques examined above using holographic signal processing provide an opportunity to utilize in an optimum way information contained in a reflected signal only when the phase distribution in the antenna aperture is stationary, i.e., when the brightness distribution on the film does not change appreciably during its exposure.

In the case of relative motion of holographic radar and the observed targets, the phase and brightness distributions will change due to the Doppler effect, distorting the hologram and reducing resolving power. But if the law of change of the phase difference of the transmitted and reflected signals is known it can be taken into account during processing. An example of the use of an a priori known law of change of the current phase difference is the processing of signals in a radar with an antenna aperture synthesized as a result of the motion of a target. Radar of this type appeared and was developed independently of progress in holography. It also was ahead of holography in terms of practical application, but comparison of the concepts of holography with aperture synthesis techniques eventually revealed their commonality. The minor difference between these two methods of improving angle resolving power is that in true holographic radar a hologram is taken and developed in the entire aperture, whereas in side-looking radar it is taken and developed step by step during the entire target illumination time. Furthermore a linear, and not a two-dimensional hologram is produced in radar in which the aperture is formed by moving the antenna. Therefore resolution can be improved only in terms of one coordinate, namely that parallel to the direction of movement.

The operating principle of synthetic aperture radar is examined briefly below.

Suppose an airplane, flying at velocity v in the direction of the 0x axis, carries a coherent radar with an antenna, the axis of whose radiation pattern is perpendicular to vector v (Figure 8.19). We assume that point target T is located in the field of view of the radar.

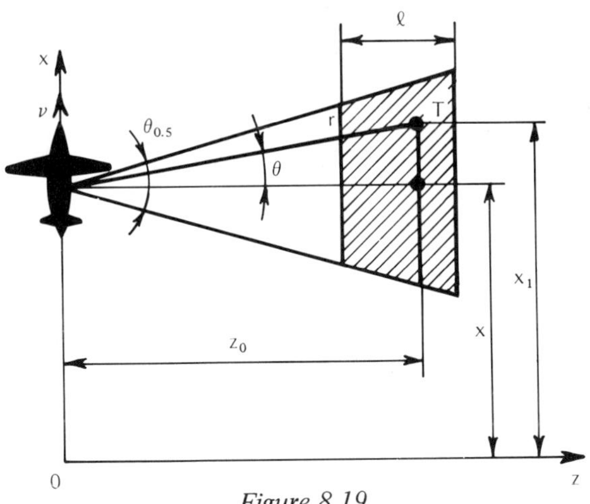

Figure 8.19

If the radar uses the monochromatic signal

$$u = U_0 e^{-j\omega_0 t},\tag{8.4.62}$$

then the signal in the antenna aperture, scattered by the point target, may be written as

$$u_r(t) = U_r(t-\tau_d)\, e^{-j\omega_0(t-\tau_d)}.$$

The envelope $U_r(t-\tau_d)$ is similar in shape to the amplitude pattern of the antenna employed. It may be assumed that the amplitude distribution in the aperture, the length of which is $L \approx z_0\theta_{0.5}$, exerts no significant influence on processing results. The signal lag time is $\tau_d = 2r/c$. When $\theta_{0.5} \leqslant 10°$ we may use for r the approximation

$$r \approx \sqrt{z_0^2 + (x-x_1)^2} \approx z_0 + \frac{1}{2}\frac{(x-x_1)^2}{z_0},$$

where x_1 is the distance between the 0Z axis and the axis of the radiation pattern of the radar antenna. Since $x = vt$ the received signal will be described by the expression

$$u_r(t) = U_r\, \exp\left\{-j\left[\omega_0 t - \frac{2\pi}{\lambda z_0}(vt-x_1)^2 - \phi_0\right]\right\},\tag{8.4.63}$$

where $\phi_0 = (4\pi/\lambda)z_0$ is the constant phase shift (the unknown initial phase). The lack of exact information about target range coordinate z_0 distorts the hologram, and therefore it is necessary to exclude the influence of phase ϕ_0.

Due to the motion of the airplane a pulse with duration $\tau_x = L/v = z_0\theta_{0.5}/v$ is formed at the receiver input. The frequency of the waves that fill the pulse varies within its duration by the law

$$f(t) = \frac{d\phi(t)}{2\pi d_t} = f_0 - \frac{2v}{\lambda z_0}(vt-x_1) = f_0 - f_d(t)$$

for

$$t_0 - \tau_x/2 \leqslant t \leqslant t_0 + \tau_x/2, \qquad\qquad\qquad (8.4.64)$$

where t_0 is the time the target appears on the flight traverse. Targets with different x, z coordinates may be located within the field of view of the radar. For targets separated from x by distance Δx_i the Doppler frequency change will be accompanied by time shift

$$\Delta t_i = \Delta x_i/v.$$

The signals of targets differing in terms of coordinate z by $\pm\Delta z_i$ have a different Doppler frequency shift

$$\Delta f_d(t) = \frac{2v}{\lambda}(vt-x_1)\frac{\Delta z_i}{z_0^2}. \qquad\qquad\qquad (8.4.65)$$

Signals reflected from targets located at different ranges behave as mutual interference. To eliminate this interference the radar is operated in the mode in which short coherent pulses are radiated.

Fast changes of phase $\phi(t) = \omega_0 t$ do not carry any information. Therefore, in cases when it is not necessary to keep the carrier frequency to solve the signal processing problem this noninformative parameter is eliminated by means of phase detection using reference carrier wave (8.4.62). Phase detection is usually preceded by a reduction of the carrier frequency by means of synchronous mixing of the transmittal and received signals. Two quadrature phase detectors, in which, as is known, the reference waves are phase-shifted by $\pi/2$, detector multiplies the signals and then averages the product in the time interval $T_{av} \leqslant 1/F_{d\,max}$.

Assuming that the use of a quadrature phase detector with subsequent combining of the signals eliminates the influence of initial phase ϕ_0, and using the rule of scalar multiplication of vectors (8.4.62) and (8.4.63), we obtain

$$u_{pd}(t) = k_{pd}U_r U_0 \exp j\left[\frac{2\pi}{\lambda z_0}(vt-x_1)^2\right]. \tag{8.4.66}$$

Comparison of (8.4.66) with basic holography equation (8.4.31) shows that their useful components completely coincide. Actually, $U_r = U_r(x)$ contains information about the amplitude distribution of the target signal in aperture L, and the phase factors completely coincide, since by replacing $t = x/v$ it is possible to convert to space frequencies, so that

$$u_{pd} = k_{pd}U_r(x)U_0 \exp\left[-j\frac{2\pi}{\lambda z_0}(x-x_1)^2\right] \tag{8.4.67}$$

$$= k_{pd}U_r(x)U_0 \exp\left[-j\psi(x)\right].$$

Naturally, the real part of signal (8.4.67) is recorded. Since the transparency of the film may vary only from 0 to 1 it is necessary, in addition to signal (8.4.67), to establish some average recording level, whereupon the equation of holography for $z = z_0$ acquires the form

$$I(x, z_0) = 0.5 + s\left(\frac{x}{v}\right)\cos\frac{2\pi}{\lambda z_0}(x-x_1)^2, \tag{8.4.68}$$

where $s(x/v)$ is a function that carries information about the amplitude field distribution in the aperture and varies from 0 to 0.5. Equation (8.4.68) shows that a hologram of a point target for each element of resolution Δz will be a series of dark and light areas, alternating in terms of coordinate x in accordance with the change of the space frequency

$$\omega_x = \frac{d\phi(x)}{dx} = \frac{4\pi}{\lambda z_0}|x-x_1|. \tag{8.4.69}$$

Another type of radar is used, in which the essential concepts of holography are essentially utilized. This type of radar uses pulses with a high-frequency duty factor, modulated by linear frequency modulation (LFM pulses).

Returning to (8.4.64), we notice that the high-frequency carrier of the pulses of a synthetic aperture radar is also linearly frequency-modulated.

The purpose of the processing system in both cases is to determine the center of gravity of a pulse. In view of the above-mentioned analogy it is possible to apply the concepts of holography to LFM radar. The notation of a reflected LFM pulse is completely analogous to (8.4.68), but it will be compressed by a factor of c/v relative to the longitudinal coordinate. We find the equation for a transparency (a sheet of transparent material, on which dark strips are applied), so that by placing the transparency on the recording of hologram (8.4.68) it will be possible to determine coordinate x_1 of a point target, located at range z_0. To do this it is necessary and sufficient to distribute the brightnesses in terms of both coordinates in accordance with (8.4.68).

To the light bands, obviously, will correspond $\cos \phi(x) = 1$, and to the dark $\cos \phi(x) = -1$, i.e., a phase difference π should occur from band to band. The transparency naturally should be prepared in the same scale as the recording, i.e., $x_{re} = q_{re}x$, where q_{re} is the recording scale. Since the transparency is a copy of the hologram for a point source it is necessary and sufficient to shift it relative to the recording and, when the bands on both images coincide, mark target coordinate x_1.

The actual processing procedure is more complicated, since in order to convert a hologram to a panorama it is necessary to process it simultaneously by range intervals. There are many ways of automating this process in radar, in which the aperture is synthesized by moving the antenna [134, 136, 81].

Any of these methods may be used when examining the problem of the noise immunity of this type of radar. They all utilize coherent multichannel storage and are equivalent from the standpoint of noise immunity. Methods of processing LFM pulses using passive "compression" circuits are, naturally, equivalent to optical optimum signal processing techniques in synthetic aperture radar. Therefore, to determine the noise immunity gain of such radar in relation to receiver noise it is helpful to use results published in the literature [95].

Let the envelope of an LFM pulse be rectangular in form, so that it may be described with the expression

$$u(t) = U_0 \cos (\omega_0 t + \mu t^2 /2) \quad \text{for } -\tau_p/2 \leqslant t \leqslant \tau_p/2, \tag{8.4.70}$$

where μ is the frequency increment rate.

A filter matched with (8.4.70) should have a weight function of the form

$$g(t) = k_1 \cos (\omega_0 t - \mu t^2 /2) \quad \text{for } -\tau_p/2 \leqslant t \leqslant \tau_p/2, \tag{8.4.71}$$

where $k_1 = \sqrt{2\mu/\pi}$.

When signal $u(t)$ is supplied to the input of such a filter the envelope of the output pulse will be described by the expression [95]

$$s_{out}(t) = k \sqrt{\frac{2\mu}{\pi}} \, U_0 \, \frac{\sin \left[\dfrac{\mu t}{2} (\tau_p - |t|) \right]}{\mu t}$$

$$\text{for } -\tau_p \leqslant t \leqslant \tau_p, \tag{8.4.72}$$

where k is a proportionality factor, equal to 1 s.

In order to convert to synthetic aperture radar signals it is necessary and sufficient, in accordance with (8.4.63) for $x_1 = 0$, to substitute into (8.4.72) the following values of the parameters μ and τ_p:

$$\mu = \frac{4\pi v^2}{\lambda z_0} \quad \text{and} \quad \tau_p = \tau_x = \frac{z_0 \theta_{0.5}}{v}. \tag{8.4.73}$$

The peak output voltage of the matched filter is found by replacing the values of μ and τ_p in (8.4.72) with their expressions from (8.4.73) and by reducing current time t to zero. As a result we obtain, after simple conversions,

$$s_{out}(O) = kU_0 \theta_{0.5} \sqrt{2z_0/\lambda}.$$

Matched filter (8.4.71) does not alter the energy of standard white noise within its passband. Therefore the signal-to-noise ratio (at maximum signal power) is increased as a result of aperture synthesis by the factor

$$k_s = 2\theta_{0.5}^2 (z_0/\lambda) = 2(L/d)$$

where L is the size of the synthesized aperture.

Let $\theta_{0.5} = 0.2$ rad, $z_0 = 3 \cdot 10^4$ m, $\lambda = 3 \cdot 10^{-2}$ m. Then $k_s = 16 \cdot 10^4$. This gain reflects only potential capabilities. Because of unavoidable and very appreciable losses due to trajectory instability and imperfection of processing systems, the gain will be more modest. It should be borne in mind, however, that when a matched filter is used the ratio of the output signal energy to the spectral noise density does not change. This means that the detection characteristics of synthetic aperture and real aperture radars are identical at the same radiated energy. Also, the pulse compression effect increases the azimuth resolving power of radar.

If a radar receives sinusoidal jamming on frequency ω_j, with frequency difference $\Delta\omega = \omega_j - \omega_0$ relative to carrier frequency ω_0, then the response of the optimum filter will have an envelope of the form

$$s_{out}(t, \Delta\omega) = U_0 \sqrt{\frac{2\mu}{\pi}} \; \frac{\sin \dfrac{\Delta\omega + \mu t}{2}(\tau_p - |t|)}{\Delta\omega + \mu t}. \tag{8.4.74}$$

Comparison of (8.4.72) and (8.4.74) shows that this type of jamming produces a blip on the radar screen, displaced relative to the position of the source by $\Delta x = (\lambda z_0/4\pi v)\Delta\omega_j$. The intensity of the jamming blip diminishes as frequency error $\Delta\omega$ increases in accordance with $(\sin \Delta\omega)/\Delta\omega$. When $\Delta\omega > \Delta\omega_{max}$, where $\Delta\omega_{max}$ is the maximum Doppler frequency of the useful signal, this kind of jamming will have virtually no influence on radar performance.

4. Spatial Filtering in the Video Channel

Holographic radar data processing techniques offer great opportunities for
performing spatial image filtering with optical coherent systems [47, 81, 143,
157, 163, 182]. Spatial filtering in combination with holographic data processing techniques improves the jamming immunity of radar as a result of
signal processing in the image reproduction stage immediately after the field
interference pattern is established. Spatial image filtering may be used to
solve the following basic problems:

detection of signals in a jamming background;

small target pattern recognition;

image defect correction;

reconstruction of the pattern of a target on the basis of a recorded part
of its image.

Spatial filtering systems make use of the remarkable ability of lenses to perform
Fourier transformation. The simplest lens makes in the rear of its focal plane
a Fourier pattern of an image of a subject, placed in the front focal plane. By
making spatial filters with a given transparency distribution in the rear focal
plane it is possible to use familiar electronic frequency filtering techniques in
application to frequency spectrum conversions in space.

Unlike electronic filters, an optical filter exhibits selective properties in the
range of space and time frequencies. Spatial selectivity is achieved by selecting
the corresponding law of change of the coordinate transparency of a filter.
Time frequency filtering can be accomplished by using the right light filters.
The time properties of filters usually are not taken into consideration in
spatial filtering theory. The chief advantage of spatial filters is their capacity
to process signals with many degrees of freedom. For example, it is very
easy to build filters for processing two-dimensional signals (in coordinate
axes Ox and Oy).

Optical filters may be characterized by the pulse transient response
$h(x_1, y_1, x_2, y_2)$, which determines the response of an optical filter to a
point light source. When the input of a system receives signal $s_{in}(x, y)$ the
image is recorded as the two-dimensional convolution

$$s_{out}(x, y) = \int\limits_{-\infty}^{\infty} \int s_{in}(x'\, y')\, h(x-x', y-y')\, dx'\, dy'.$$

Any image can be made in frequency representation as superpositions of
waves of the form $\exp\left[j(\omega_x x + \omega_y y)\right]$ with space frequencies ω_x and ω_y.

A spatial filter changes the amplitudes and phases of the spatial components of a signal and is characterized by the space frequency characteristic

$$H(\omega_x, \omega_y) = \int\limits_{-\infty}^{\infty}\!\!\int h(x, y)\, e^{-j(\omega_x x + \omega_y y)}\, dx\, dy.$$

Input signal $s_{in}(x, y)$, with space frequency spectrum

$$S_{in}(\omega_x, \omega_y) = \int\limits_{-\infty}^{\infty}\!\!\int s_{in}(x, y)\, e^{-j(\omega_x x + \omega_y y)}\, dx\, dy.$$

causes the output reaction

$$s_{out}(x, y) = \frac{1}{4\pi^2} \int\!\!\int S_{in}(\omega_x, \omega_y)\, H(\omega_x, \omega_y)\, e^{j(\omega_x x + \omega_y y)}\, d\omega_x\, d\omega_y.$$

$$(8.4.75)$$

Expression (8.4.75) completely describes the properties of optical filters. Analysis of one-dimensional processes with the mathematical tool of spatial transformations requires the use of one-dimensional Fourier transforms.

The invention of holography greatly improved the capabilities of spatial filtering, since it became possible not only to record bipolar optical signals, but also to preserve all phase information. By virtue of the latter circumstance it is possible to make any complex spatial filter.

A diagram of the simplest optical spatial filter is shown in Figure 8.20 [143]. Image filtering takes place through double Fourier transformation (birefringence). Image $s_{in}(x)$ is illuminated with a coherent light field, which, passing through lens L_1 with focal length F_f, produces in rear plane P_1 the image spectrum $S(\omega_x)$. If a signal with the space frequency $\omega_x = 2\pi/\Delta\ell$ must be extracted the filter, installed in plane P_1, must have transparent regions only near frequency $\omega_x = 2\pi/\Delta\ell$. Then, after the second Fourier transformation (second diffraction), performed by lens L_2, the desired image appears at the output. Various spatial filtering systems, designed for target pattern recognition, are based on this principle.

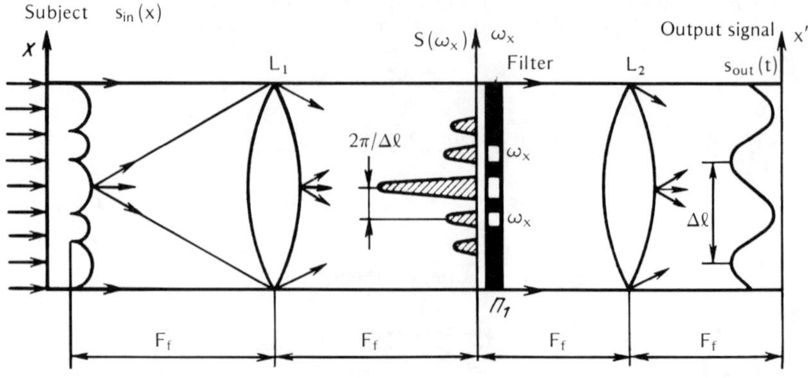

Figure 8.20

A spatial filter (mask), placed in plane P_1 (Figure 8.20), is built on the basis
of the known space frequency spectrum of the subject to be recognized. The
filter passes only "its own" signal. Therefore an image of the subject is
formed in the output plane from a signal-to-noise mixture.

Various spatial filter designs are examined in the literature [47, 143, 157, 163,
182]. Their highly effective recognition of various objects is noted. Such
systems may be used to counteract responsive jamming.

In an examination of the problem of detecting signal $S(x)$ in the background
of spatially distributed noise $n(x)$, signal $\xi(x)$ at the input of a spatial filtering
system is represented as the sum $\xi(x) = s(x) + n(x)$. According to detection
theory the maximum signal-to-noise ratio occurs when a space frequency
filter has an amplitude transmittance, characterized in the space frequency
range by the function

$$T_{opt}(\omega_x) = S^*(\omega_x)/N(\omega_x),$$

where $S^*(\omega_x)$ is the complex conjugate Fourier pattern of useful signal $s(x)$;
$N(\omega_x)$ is the spectral noise density.

For the simplest case, when a jamming signal is represented in the space
frequency range by ideal white noise with constant spectral density N_0, the
optimum filter is described by a function that is complex-conjugate with the
signal spectrum

$$T_{opt}(\omega_x) = S^*(\omega_x). \qquad (8.4.76)$$

Complex conjugation is required for transforming a useful signal into a real
optical signal. The optimum spatial filter compensates the phase lead in the
initial wave $\xi(x)$ with the corresponding phase lag in the filter, so that a plane
wave, which forms a bright point image, appears at the output.

Fourier-holography devices, with which a hologram of the Fourier pattern of useful signal $s(x)$ can be made, are used for building the optimum spatial filter. The hologram is the interference pattern that occurs when plane reference wave $e^{j\omega_0 x}$ interacts with the Fourier pattern of useful signal $S(\omega_x)$ [157].

The amplitude transmittance of such a holographic filter (mask) is

$$T(\omega_x) = |S(\omega_x) + e^{j\omega_0 x}|^2 = 1 + |S(\omega_x)|^2 + S^* e^{j\omega_0 x} + S e^{-j\omega_0 x}.$$

When a signal-noise mixture passes through the filter the signal

$$S_{out}(\omega_x) = S_\xi(\omega_x)T(\omega_x) = S_\xi(\omega_x)[1 + |S(\omega_x)|^2]$$

$$+ S_\xi(\omega_x)S^*(\omega_x)e^{j\omega_0 x} + S_\xi(\omega_x)S(\omega_x)e^{-j\omega_0 x},$$

where $S_\xi(\omega_x) = S(\omega_x) + N(\omega_x)$ is the input signal spectrum, appears at the output.

Three light rays are formed at the output of the optimum filter. The first ray, corresponding to the term

$$S_1(\omega_x) = S_\xi(\omega_x)[1 + |S(\omega_x)|^2],$$

propagates along the optical axis ($\omega_x=0$). The second term $S_\xi(\omega_x)S^*(\omega_x)$ $\exp(j\omega_0 x)$ describes a ray that is deflected upward. The ray has the two components

$$S_\xi(\omega_x)S^*(\omega_x) = [S(\omega_x) + N(\omega_x)]S^*(\omega_x)$$

$$= |S(\omega_x)|^2 + N(\omega_x)S^*(\omega_x).$$

Component $|S(\omega_x)|^2$ is the useful response of the system and, after Fourier transformation by the second lens L_2, is transformed into the autocorrelation function of signal $S(x)$, which produces in the output plane a bright focused spot, at the time of whose appearance a decision is made regarding the presence of a useful signal.

The component of the form $N(\omega_x)S^*(\omega_x)$ is converted at the output lens L_2 to the cross correlation function of noise $n(x)$ and useful signal $s(x)$ due to the random phase distribution along the leading edge of the wave. This component produces in the output plane a dim blurred spot around the bright image of the useful signal.

The third ray, described by the third term of output signal $S_{out}(\omega_x)$, behaves analogously. It also has a useful component, which makes a bright spot on the other side of the screen, and a noise background.

Optimum optical filtering systems can be used to counteract spatially distributed jamming, for example passive jamming. They are also used for pattern recognition.

Chapter 9
Combined Coordinate Sensors

9.1. Design Principles and Applications of Combined Sensors

Problems of guiding moving objects (airplanes, ships and missiles) are solved by using information about their coordinates and motion parameters, obtained from various kinds of sensors, including electronic instruments. It often turns out that the operation of the latter is partially or completely duplicated and can be duplicated with nonelectronic instruments. If the instruments operate independently, then the crew (or an automatic system), at any given moment of time, can use signals from the instrument which provides the best measurement accuracy under specific conditions. Signals from other sensors are virtually useless or at best help to improve the reliability of other data obtained or to eliminate ambiguities.

At the same time, by correctly combining various sensors into a single system with a common indicator or output system in automatic control systems it is possible to improve substantially the accuracy of data received about the co-ordinates of a target or the parameters of its motion. If a combined system includes electronic sensors, then by using information received from other sensors (especially self-contained nonelectronic instruments), it is possible to substantially improve the noise immunity of the system.

A sensor in which electronic tracking systems play the main role and non-electronic systems are auxiliary is called a complex radio automation system. But if nonelectronic sensors are the main instruments and an electronic system is auxiliary, the sensor is called an autonomous radio correction system. Such a classification of combined sensors is purely conventional, since the same combining methods are used in both cases and no distinction is made between them.

Combining sensors for the purpose of improving accuracy has long been used in meteorology in general and in navigation in particular. For example, gyro-magnetic compasses have been installed aboard aircraft since the middle 1930's. These instruments include two sensors: a magnetic system, aimed in the direction of the magnetic field vector, and a three-stage gyroscope, which maintains the position of the measurement axis in inertial space constant. The signals received from the magnetic system and three-stage gyroscope are trans-mitted through a special system, called a correcting filter, to a heading indicator. This kind of combining substantially reduced deficiencies inherent to each sensor and improved airplane heading measurement accuracy.

The extensive incorporation of electronic coordinate sensors of the tracking type in mobile equipment confronted designers with the problem of increasing their sensitivity and noise immunity.

Electronic coordinate sensors usually are tracking systems, which contain sensitive, intermediate and final control elements. The simplest kinds of sensing elements are time, phase, frequency or amplitude detectors. A sensor often consists of a complicated series of radio signal-voltage (current) converters, in which the voltage is proportional to the measured coordinate. For example, the sensitive element of direction finders consists of a directional antenna, frequency converter, a receiver with an automatic gain control or other type of control system, and amplitude and phase detectors. Some of these components are nonlinear. Therefore their conversion properties change as the signal-to-noise ratio changes [8, 18, 106].

All types of electronic sensors without exception have nonlinear discrimination characteristics in terms of the parameter being converted. These features are not important in relation to analysis of the noise immunity of electronic tracking systems if the signal-to-noise ratio is sufficiently high.

The gain of a sensor with high signal-to-noise ratios is close to its gain under noise-free operating conditions. Therefore low-level jamming increases coordinate measurement errors without causing an appreciable increase of the probability of automatic tracking interruption. As the noise level increases the gain decreases, and consequently the dynamic radio signal parameter tracking errors increase. If the dynamic error exceeds the aperture width of the discriminator curve tracking of the useful signal will be interrupted.

The characteristics $U_{fd} = \psi(\Delta f)$ of a frequency detector with detuned circuits, determined experimentally [18], are given in Figure 9.1 as an illustration of

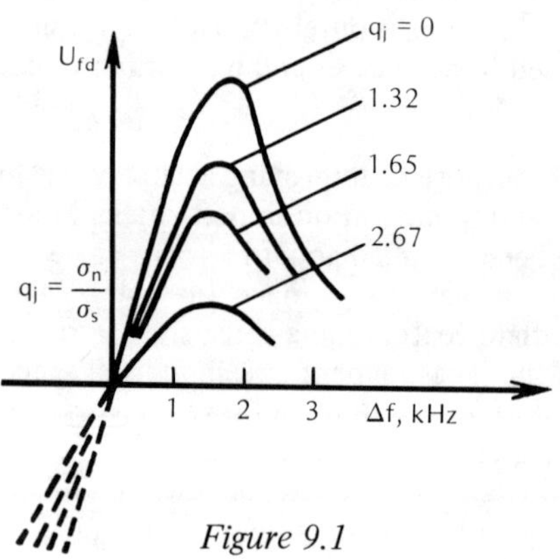

Figure 9.1

what was said above. The operation of a detector in a Doppler velocity vector sensor was analyzed. According to the operating principle of such a sensor the useful signal is wide-band noise with shifted central frequency f_0, which is symmetrical with respect to spectral density. The spectral width of the useful signal is approximately 1 kHz. The jamming is also noise, but with somewhat greater spectral width than the aperture of the discriminator curve of the detector.

Examination of this characteristic shows that as the jamming-to-signal ratio q_j increases the gain of the detector decreases, and consequently the passband of the tracking system in relation to the tracked parameter narrows and the dynamic error increases.

By combining some kind of electronic tracking instrument with autonomous electronic instruments it is possible to operate at extremely low signal-to-noise ratios, at which the gain is substantially lower than nominal. This is related primarily to the fact that the mismatch caused by the dynamic error may be substantially reduced or even eliminated as a result of combining. Under these conditions a reduction of the gain of the electronic tracking system not only does not increase the probability of tracking interruption but, to the contrary, reduces fluctuation error and, consequently, reduces the indicated probability to some extent as a result of narrowing of the passband of the tracking system. It must be stipulated from the outset, however, that these properties are manifested only in methods of complete utilization of information that are rational from the standpoint of the solution of the problem of improving noise immunity.

By using efficient sensor combining methods it is possible to solve the following two most important problems for navigation and radio guidance systems: improvement of precision and sensitivity and improvement of jamming immunity. These two problems should be examined separately in relation to combining methods, even though they are similar to each other.

As is well known, improvement of the sensitivity of electronic coordinate sensors with internal (less frequently external) receiver noise is limited. Narrowing the passband, by means of combining, relaxes requirements on the input signal-to-noise ratio of the systems and thus increases their sensitivity. Since internal noise and external wide-band jamming exert the identical influence on tracking system performance, combining substantially improves the noise immunity of electronic sensors in relation to this kind of jamming.

Combining is a way to combat other kinds of special manmade jamming, aimed at interrupting the tracking of a radio signal in relation to the parameter being measured. These kinds of jamming include range and velocity (Doppler frequency) deception jamming, jamming intended for obliterating the useful signal by overloading the receiver, etc.

It will be shown below how the operating mode in which an electronic system is activated briefly can be used in combined systems for correcting autonomous systems. This mode in electronic systems enhances their security, which impedes jamming. Consequently the combining of electronic and autonomous non-electronic sensors facilitated the solution of the problem of increasing the jamming immunity of coordinate and parameter sensors. However, not just any combination of sensors provides any appreciable gain in jamming immunity. Therefore it is worthwhile to examine in detail techniques used in the design of combined systems and to evaluate them from the standpoint of the problem of improving jamming immunity.

Combined sensors usually consist of complex multichannel systems, which, in turn, are made up of electronic systems, autonomous coordinate sensors, systems for converting signals from one coordinate system to another, matching and interface systems, output systems, etc.

However, the problem of improving the jamming immunity of a complex is solved separately for each electronic sensor that is a part of the complex. Combined systems usually are designed to measure two and more coordinates, i.e., they are multicoordinate systems. Their component autonomous non-electronic and electronic sensors operate in different coordinate systems. In order to combine them it is necessary to convert the signals from the various sensors to a common coordinate system.

To gain an understanding of the basic principles of the combining of sensors it is helpful to start the examination with the simplest single-coordinate complex. We shall proceed from the assumption that measured coordinate $s(t)$ or its derivatives $D^N s(t)$, where $D = d/dt$, are determined simultaneously by tracking electronic and nonelectronic sensors, and measurement errors are independent stationary functions of time.

It becomes necessary to develop a complex in the case when an autonomous sensor is not capable of solving the problem due to limited precision and the operation of an electronic system can be interrupted by jamming. Also encountered are problems that can be solved only by using combined sensors, even in the absence of radio interference.

Let us first explain the statement of the problem of combining sensors into a single complex by way of a specific example. For this purpose let us examine Figure 9.2. In an airplane or any other moving object, the center of mass of which is located at point O_s, which is the origin of nonrotating coordinate system $x_{zs}O_s y_{zs}$, let an onboard radar or angle-range measuring system of some type be used to solve the problem of determining the coordinates of a target located at point O_t, or of an airplane relative to a beacon, located at the same point O_t. In a polar coordinate system there will be two angles and range. For simplicity we shall examine a plane problem. To solve it is sufficient to measure angle ϵ between the lines of the target and nonrotating

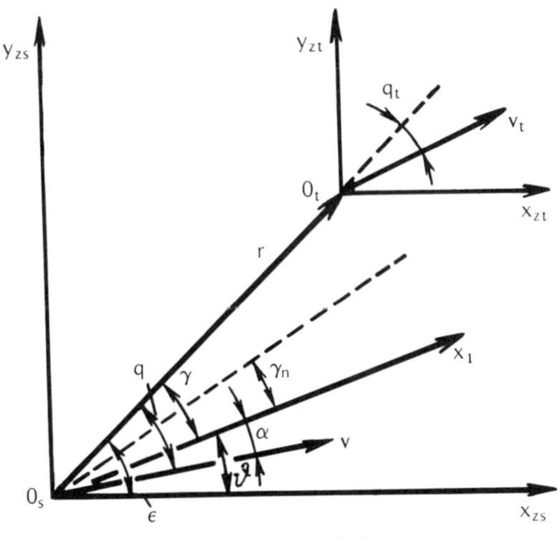

Figure 9.2

axis $O_s x_{zs}$, and range r. This problem corresponds to missile homing in one plane, to the guidance of an airplane to a beacon with an angle-range measuring navigation system, etc.

To solve this problem it is necessary to determine at least two coordinates. However, each coordinate is measured with independent errors. Therefore each measurement must be examined separately. Moreover, the different measurement channels in each sensor are assumed to be independent.

The homing problem is solved variously. In the simplest case it is necessary and sufficient to keep longitudinal axis $O_s x_1$ of the guided missile aimed at the target, so that angle γ will be equal to zero for the entire time of flight (direct guidance). This method is unsuitable for a moving target ($v_t \neq 0$) because of the great curvature of the trajectory of the homing missile.

Methods in which derivative $\dot\epsilon$ of angle ϵ is measured, for example the proportional navigation method, in which $\dot\epsilon = D\epsilon$ must be equal to zero throughout the entire flight time, are more suitable for guidance to moving targets.

In homing it is not always necessary to measure target range r or its derivative $\dot r = Dr$. However, in order to successfully solve the problem of measuring angular coordinates it is necessary to select the useful signal in a background of different kinds of noise. Both range selection (pulse position on the time scale) and Doppler frequency shift selection may be used for this purpose.

Thus we have broken homing down into two independent problems, which are solved by single-coordinate sensors.

Changes of angle γ in time are caused by two factors: the relative motion of points O_s and O_t and fluctuations of longitudinal axis $O_s x_1$ of the target, so that

$$\gamma = \epsilon - \vartheta. \tag{9.1.1}$$

It follows from Figure 9.2 that

$$\dot{\epsilon} = (1/r)(v_s \sin q - v_t \sin q_t).$$

Therefore

$$\epsilon = \epsilon_0 + \int_0^t \frac{v_s}{r} \sin q\, dt - \int_0^t \frac{v_t}{r} \sin q_t\, dt \ . \tag{9.1.2}$$

where $\epsilon_0 = \epsilon \mid_{t=0}$.

Measurement of $\dot{\epsilon}$ is possible only by tracking the target in terms of angle γ. However

$$\dot{\gamma} = \dot{\epsilon} - \dot{\vartheta}, \tag{9.1.3}$$

and consequently the results of measurements of $\dot{\epsilon}$ will be distorted by fluctuations of the longitudinal axis of the target. In this case angle ϑ changes considerably faster than angle ϵ. Angle ϑ or its derivative can be measured with position or rate gyroscopes, respectively.

Therefore, when formula (9.1.1) is used an electronic tracking system should track changes of angle ϵ in time, and angle ϑ may be measured with a gyroscopic sensor, in which case there is no need for a wide-band electronic tracking system. Furthermore, when the target is equipped with the corresponding sensors it is also possible to consider changes of angle ϵ caused by the translatory motion of the sensor. In fact, the component ϵ_s of angle ϵ

$$\epsilon_s = \frac{1}{r} \int_0^t v_s \sin q\, dt + \epsilon_{so}\ ,$$

where $\epsilon_{so} = \epsilon_s \mid_{t=0}$, which is caused by this motion, can be determined with the aid of autonomous nonelectronic systems and is used as a priori data for the angle-measuring channel. Then the angle-measuring channel need track only changes of the angle, caused by the motion of the target:

$$\epsilon_t = \frac{1}{r} \int_0^t v_t \sin q_t\, dt + \epsilon_{to}\ ,$$

where $\epsilon_{t0} = \epsilon_t \mid_{t=0}$.

It is perfectly obvious that when the right sensor is selected the passband of a tracking electronic direction finder can be narrowed sharply when angles ϑ and ϵ_s are known, and consequently its noise immunity can be increased. Since changes of angle ϑ in time act as jamming they may be excluded from the measurement results by using signals from a gyroscopic sensor.

The problems of measuring γ and ϵ in accordance with formulas (9.1.1) and (9.1.3) seem diametrically opposing at first glance. Such, indeed, is not the case. The fact is that an electronic tracking system in both cases should provide reliable target tracking, i.e., track changes of angle γ. In the former case, however, the goniometer output signal must contain both ϵ and ϑ, and in the latter case only $\dot{\epsilon}$.

We shall examine the technical solutions of the stated problems below. For the time being, however, we restrict the examination to the problem of measuring (tracking) range changes. It follows from Figure 9.2 that

$$r = r_0 + \int_0^t v_t \cos q_t dt - \int_0^t v_s \cos q dt, \qquad (9.1.4)$$

where r_0 is the initial distance between points O_s and O_t. Equation (9.1.4) shows that range changes are caused both by the motion of the sensor itself and by the motion of the target. The component

$$r_0 = - \int_0^t v_s \cos q dt, \qquad (9.1.5)$$

where $r_{s0} = r_s \mid_{t=0}$, or its time derivatives, can be measured with navigation systems. The results of these measurements for an electronic system will be used as a priori data, easing requirements on the dynamics of the tracking range finder or velocity sensor.

It follows from the above examination that the problem of combining consists in analyzing a priori information about the motion of an examined electronic sensor, received from other sensors, which herein henceforth are called proper motion sensors (PMS).

With a priori information it is possible to ease requirements on the dynamics of an electronic tracking system and to improve the precision and noise immunity of a complex, especially of electronic sensors. A priori information may be complete or incomplete. In the examples discussed here it may be assumed to be complete on the condition $v_t = 0$. The task of electronic meters here is to reduce PMS errors. Combined systems of this type are called complete information systems. If $v_t \neq 0$ only a system with incomplete information can be built. The more complete the information received from PMS, the more noise-immune an electronic system will be.

It can be shown that the development of combined sensors for $v_t = 0$ is absurd, since autonomous estimators theoretically are capable of yielding all information required for the solution of a guidance problem. This, however, is by no means always the case. The fact that autonomous sensors have limited accuracy and homing is not feasible without electronic sensors. Electronic sensors successfully solve the homing problem, but it is always desirable to increase their noise immunity and precision.

Combined sensors are not confined to analysis of proper motion. They are also useful for stationary sensors. This can be illustrated by way of an example. Let a ground complex measure the motion parameters of a satellite. By virtue of the high stability of the indicated motion it is possible to calculate in advance the coordinates of a satellite, which are used as *a priori* data for an electronic tracking system, i.e., as PMS signals. Consequently the sensors need track not the completely measured coordinates, but only the errors of their forecasting. Thus, whenever it is possible to obtain a priori information about a tracked parameter, the concepts of combining can be used. A coordinate forecasting computer performs as a PMS in the case of satellite coordinate measurement.

The gain in noise immunity acquired by combining estimators depends not only on the completeness of a priori information about the measured parameter of a radio signal, but also on combining techniques, i.e., on the structure of the combined sensor.

There are quite a few ways of making combined utilization of information received from different sensors.

They can all be divided into two groups: systems with independent sensors and systems with sensors connected by correction circuits.

9.2. Combined Systems with Independent Sensors

There are at least two types of combined systems with independent autonomous and electronic sensors, the output signal of which depicts a coordinate with less error than that of each separate sensor. The main problem solved in this case is that of increasing measurement accuracy.

The simplest system is one with parallel sensors and matching filters. A simplified functional diagram of such a system is shown in Figure 9.3. The parameter $S = s(t)$ of a radio signal, measured by an electronic tracking system (ETS) reaches its input along with jamming. The action of the latter is taken into consideration by adding some equivalent real jamming $J' = J'(t)$ to the useful signal. This equivalence stems from the fact that signal J', applied to the input of the tracking system, causes errors with the same statistical characteristics

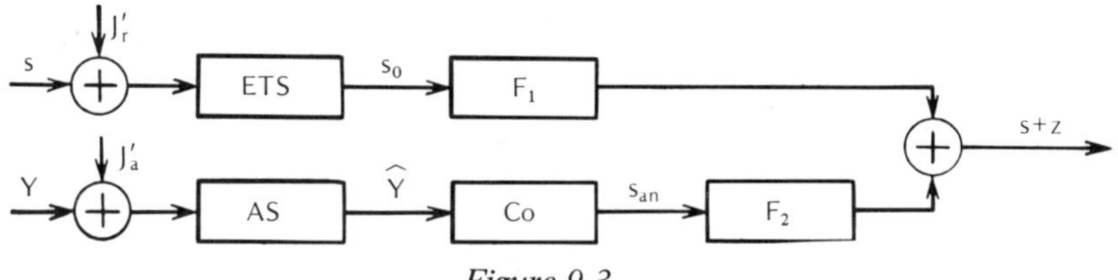

Figure 9.3

as the real jamming from the receiver input to appear on the output. This
substitution is valid for linear or linearized systems and greatly simplifies qualitative analysis of tracking instruments. From the output of the electronic
sensor the coordinate (it is assumed to be proportional to s(t)) is supplied to
the input of filter ϕ_1 in the form of the electrical signal $s_r = s_m + J_r$, where s_m
is the measured value of the coordinate and J_r is the jamming that distorts s.

The autonomous sensor (AS) measures the coordinate Y. Its input also receives
disturbance J'_a, which is equivalent to the real jamming. The computer (Co)
converts the output signal of the autonomous meter Y to a coordinate system,
in which s is determined in such a way that at the computer output we obtain

$$s_a = P (D) W_a (D) s + W_a(D) J'_a ,$$

where $P(D) = D^N$ is the statement that describes the relation between the
measured coordinates and $W_a(D)$ is the transfer function of the AE and computer
for jamming J'_a. Here N may acquire the values $0, \pm 1, \pm 2$. If coordinate Y must
be determined the computer must be connected to the output of the electronic
system. To determine some third coordinate X both channels must have computers. Signal s_{an} goes to the input of filter ϕ_2.

Filters ϕ_1 and ϕ_2 are designed to match the scales and dimensionalities of s_r
and s_{an} and to minimize measurement error z by filtering out errors J and
$J_a = W_a(D)J'_a$. These filters may also perform differentiation or integration.

Through $F_r(D)$ and $F_a(D)$ we denote the transfer functions of matching filters
ϕ_1 and ϕ_2, respectively. The output signal error z = z(t) in the general case
consists of dynamic error, caused by the time constants of the sensors, and the
error caused by disturbances J'_r and J'_a. The second component is conditionally
called the fluctuation component.

Through $W_r(D)$ and $W_a(D)$ we denote the transfer functions of ETS and AS,
and to simplify the calculations we shall assume that the nonelectronic sensor
determines the coordinate $Y(t) = P(D)s(t)$.

In view of all the preceding comments the dynamic structural diagram of the
combined meter may be represented in the form illustrated in Figure 9.4.

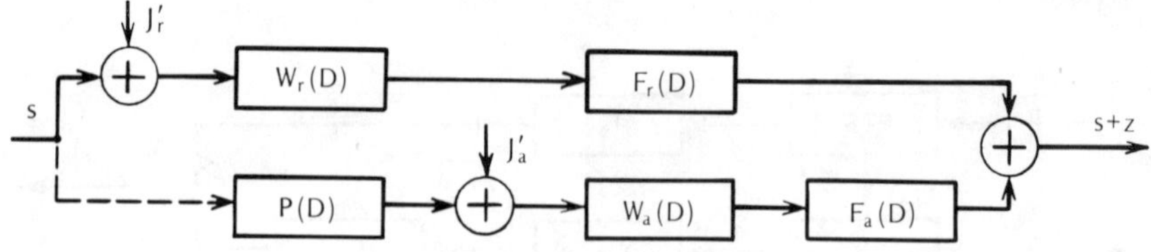

Figure 9.4

The output signal $X(t) = s(t) + z(t)$ of the complex, in accordance with the diagram, may be written as

$$s + z = (W_r F_r + P W_a F_a) s + W_r F_r J'_r + W_a F_a J'_a. \tag{9.2.1}$$

Here and in the ensuing analysis argument D in the transfer functions is omitted for the sake of brevity and s, J, etc., are written in place of $s(t)$, $J(t)$, etc.

If the correlation functions of disturbances J'_r, J'_a and of signal s are known it is then possible to state and solve the problem of finding the transfer functions $F_r(D)$ and $F_a(D)$ which minimize the dispersion σ_z^2 of measurement error z. The problem as stated in this way was solved in [18]. But if the statistics of the useful signal are not known, which is usually the case in practice, the problem is solved with certain constraints. It is necessary, first of all, that the transfer function of the useful signal for s

$$W_s = W_r F_r + P W_a F_a$$

ensure its undistorted reproduction. Function W_s will be nondistorting only when the modulus of the combined system gain is $|W_s(j\omega)| = 1$ for all frequencies of spectrum s. Consequently the condition

$$W_r F_r = 1 - P W_a F_a . \tag{9.2.2}$$

must be satisfied. In complex systems theory equation (9.2.2) is called the condition of invariance (independence) of the measurement error relative to the characteristics of the input signal. This actually corresponds to the selection of the parameters of the system for which the dynamic error vanishes. The invariance condition is conveniently written as

$$F_r = (1 - P W_a F_a)/W_r \tag{9.2.3}$$

The physical feasibility of designing the components of a system requires that the orders of the polynomials of the numerators in transfer functions $W_r(D)$, $F_r(D)$, $W_a(D)$ and $F_a(D)$ be equal to or less than the orders of the polynomials of the denominators.

If transfer functions W_a and W_r are known and condition (9.2.2) is satisfied the structure of filters ϕ_1 and ϕ_2 may be determined on the basis of minimization of the fluctuation error;

$$z_f(t) = W_r F_r J'_r + W_a F_a J'_a \ . \tag{9.2.4}$$

The problem in this statement is considerably easier to solve than in [18], in spite of the fact that condition (9.2.3) places a constraint on the choice of F_r and F_a. Under normal conditions (in the absence of jamming) the precision of ETS is limited by internal receiver noise. The equivalent J'_r of this noise, expressed in terms of the input of the tracking system, may be equated to white noise in terms of parameter s.

Autonomous coordinate sensors [18, 146], in addition to Doppler velocity sensors, have errors, the spectrum of which lies next to the zero frequencies and occupies a band of tenths or hundredths of a hertz. As a rule, the errors of autonomous sensors are nonstationary functions of time. When invariance condition (9.2.2) is satisfied it is possible to select the transfer functions for equivalent jamming J'_r and disturbances J'_a without regard to the spectrum of the useful signal. Then, in order to minimize error $z(t)$, it is necessary that the transfer function for J'_r be the transfer function of a low-pass filter. It should be borne in mind that it is sometimes not necessary to include auxiliary filters ϕ_1 and ϕ_2 (Figure 9.3) in measurement channels. In such cases $F_a = 1$, $F_r = 1$ and the error filtering problem is solved by the estimators themselves. In the general case

$$\phi_r(D) = W_r(D)F_r(D) = \frac{a_0 D^m + \ldots + a_{m-1}D + a_m}{b_0 D^n + \ldots + b_{n-1}D + b_n}, \tag{9.2.5}$$

for $n \geqslant m$.

It is possible to combine coordinate estimators with velocity or acceleration estimators in any of the possible combinations, such that $P(D) = D^N$ for $N = 0$, $\pm 1, \pm 2$, and for the transfer function of the channel of a nonelectronic sensor we obtain, in consideration of (9.2.2),

$$\phi_a = W_a F_a = \frac{1}{p} [1 - W_r F_r]$$

$$= \frac{D^{-N}[b_0 D'' - a_0 D''' + \ldots + (b_{n-1} - a_{m-1})D + (b_n - a_m)]}{b_0 D^n + \ldots + b_{n-1}D + b_n} . \tag{9.2.6}$$

It follows from expression (9.2.6) that the structure of the channel through which the signal from nonelectronic sensors passes through the adder is determined not only by the structure of the signal conversion channel, but also by the type of relationship between s and s_a. Circuits in which the order of polynomial of D of the numerator of the transfer function is lower than the order of the polynomial of the denominator are physically feasible. Therefore it follows from (9.2.6) that the design of a complex in accordance with the diagram in Figure 9.4 for $P(D) = D^{-1}$ or $P(D) = D^{-2}$ requires that the signal of the electronic sensor be integrated once or twice. In this case, however, the component of measurement error $z_r(t)$, attributed to noise J_r', is determined by the expression

$$z_r(t) = \frac{a_0 D^m + \ldots + a_{m-1} D + a_m}{D^N (b_0 D^n + \ldots + b_{n-1} D + b_n)} J_r'$$

and for $m \leqslant n$, $a_m \neq 0$, $a_{m-1} \neq 0$, $N = 1, 2$ it is a nonstationary function with a variance that increases in time, which makes such a combination of sensors undesirable.

A need arises in practice to combine electronic velocity estimators (of the Doppler type, for example) with autonomous coordinate estimators (different types of celestial instruments). However, they are combined in a manner that differs from that shown in Figure 9.3. To achieve the best filtering of measurement errors under the above assumption concerning the character of their spectra, and to make transfer functions $\phi_r(D)$ and $\phi_a(D)$ correspond to the stationary components, it is necessary to satisfy the following conditions:

$\phi_r(D)$ and $\phi_a(D)$ must be the transfer functions of the low- and high-pass filters, respectively;

the output adder must receive signals with the same dimensionalities.

The above conditions can be satisfied with filters of different complexity. If $Y = s$ ($N = 0$), then $\phi_a = W_a F_a$ will be the transfer function of a high-pass filter only when $b_n = a_m$ in (9.2.6). When $N = 1$, i.e., when an electronic coordinate sensor and an autonomous velocity estimator are combined it is necessary to use the additional equation $b_{n-1} = a_{m-1}$. Doppler velocity sensors and longitudinal acceleration sensors (accelerometers) can be combined in the same manner. When $N = 2$ the condition that requires the addition of values with the same dimensionality will be satisfied when $b_n = a_m$; $b_{n-1} = a_{m-1}$; $b_{n-2} = a_{m-2}$.

By satisfying the condition of the invariance of reproduction error in relation to the characteristics of the input signal it is possible to state and solve the problem of the synthesis of the optimum structure of channels. An example of such solutions can be found in [16, 18, 133]. But since we are faced with the problem of substantiating efficient methods, and not of optimum systems for the specific statistics of jamming and disturbances J_r' and J_a', we shall restrict the analysis to the simplest circuitry decisions which satisfy measurement error filtering in consideration of the specific features of the spectra of the errors. Hence it follows directly that when $N = 0$ the simplest filters will be those with the transfer functions

$$\phi_r(D) = \frac{1}{T_r D + 1} \qquad (9.2.7)$$

and in accordance with (9.2.2),

$$\phi_a(D) = 1 - \phi_r(D) = \frac{T_r D}{T_r D + 1} . \qquad (9.2.8)$$

If $N = 1$ it is necessary that

$$\phi_r(D) = \frac{b_1 D + b_2}{b_0 D^2 + b_1 D + b_2} , \qquad (9.2.9)$$

$$\phi_a(D) = \frac{b_0 D^2}{b_0 D^2 + b_1 D + b_2} ; \qquad (9.2.10)$$

Finally, when $N = 2$

$$\phi_r(D) = \frac{b_1 D^2 + b_2 D + b_3}{b_0 D^3 + b_1 D^2 + b_2 D + b_3} \qquad (9.2.11)$$

and, consequently,

$$\phi_a(D) = \frac{b_0 D^3}{b_0 D^3 + b_1 D^2 + b_2 D + b_3} . \qquad (9.2.12)$$

Finally, higher-order dynamic components may also be used as filters for jamming and disturbances, but the character of the qualitative analysis presented below, and the conclusions drawn on its basis remain unchanged, in view of which we shall limit the discussion to the design of the simplest types of channels.

Now, after deriving specific expressions for the transfer functions of the channels, we can give an extremely simple explanation of the possibilities of improving the precision and noise immunity of sensors by combining them. For simplicity we shall assume that $s_a = s$, and the transfer functions of the channels are determined by expressions (9.2.7) and (9.2.8).

We convert from time to frequency and examine the steady state mode. For this purpose we replace D by $j\omega$ in (9.2.7) and (9.2.8) and write the frequency characteristics of the channels in exponential form:

$$\phi_r(j\omega) = |\phi_r(j\omega)| e^{j\psi_r(\omega)}, \qquad (9.2.13)$$

$$\phi_a(j\omega) = |\phi_a(j\omega)| \, e^{j\phi_a(\omega)}, \tag{9.2.14}$$

where

$$|\phi_r(j\omega)| = \frac{1}{\sqrt{T^2\omega^2 + 1}}, \qquad \phi_r(\omega) = -\operatorname{arctg} T\omega;$$

$$|\phi_a(j\omega)| = \frac{T\omega}{\sqrt{T^2\omega^2 + 1}}, \qquad \phi_a(\omega) = -\operatorname{arctg} \frac{1}{T\omega}.$$

The functions $|\phi_r(j\omega)|^2$ and $|\phi_a(j\omega)|^2$ are illustrated in Figure 9.5a and b, respectively.

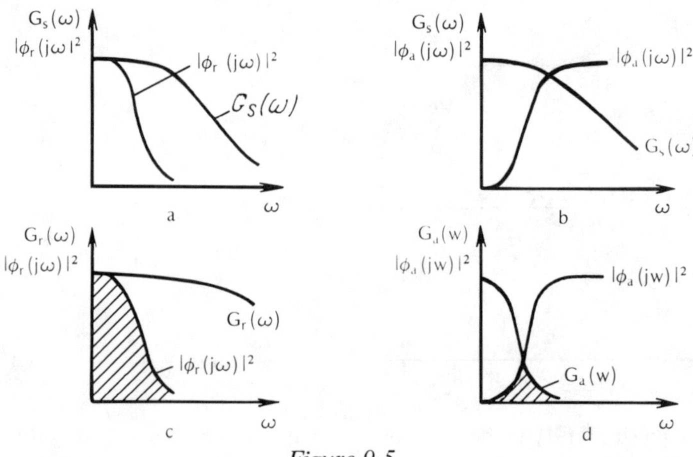

Figure 9.5

Let useful signal s be a stationary function, characterized by spectral density $G_s(\omega)$, the graphs of which, illustrated in Figure 9.5a, b, show that each of the channels can reproduce the useful signal with distortions, the character of which varies: a radio channel does not reproduce the high-frequency components of the useful signal, and the channel of an autonomous sensor does not reproduce low-frequency components.

Hence it follows that in a combined system the low-frequency components of the useful signal go preferentially to the adder output through the radio channel, and the high-frequency components through the channel of the autonomous sensor. In this case the amplitude and phase relations for each spectral component of the useful signal at the adder inputs are such that their geometric sum coincides with the same component at the input of the combined system.

In fact, by representing the i-th component of the spectrum of the sensor input signal as $s_i(t) = \text{Re } U_i \times \exp\left[-j(\omega t + \phi_i)\right]$, we obtain for its value at the output of the combined sensor

$$s_{im}(t) = \text{Re}\left[\phi_r(j\omega) + 1 - \phi_r(j\omega)\right] U_i \exp\left(-j\omega t + \phi_i\right) \tag{9.2.15}$$

$$= \text{Re } U_i \exp\left(-j\omega t + \phi_i\right).$$

Thus the useful signal in a combined two-channel system is split, in a manner of speaking, into two channels, and is then restored in the adder. Exactly the same proof could be arrived at for other types of transfer functions, for example for (9.2.9) and (9.2.10). It need only be considered here that after conversion to the frequency range the expression

$$Y(j\omega) = \omega s\,(\omega)\,\exp\,(j\pi/2)$$

corresponds to the expression $Y = Ds$. Splitting the useful signal into two propagation (measurement) channels provides an opportunity to improve the accuracy of measurement of s with a combined system. The above examination could also be applied to more than two channels. Each channel should make a contribution in the part of the spectrum of the output signal that is least covered by the noise spectrum.

The disturbances that act upon each sensor have the same propagation path between the corresponding input and output of the combined sensor. Therefore, assuming that invariance condition (9.2.2) is satisfied and that the spectral densities of disturbances $\text{Int}'_r(t)$ and $\text{Int}'_a(t)$, expressed in terms of the inputs of the sensors, are $G_r(\omega)$ and $G_a(\omega)$, respectively, we write the expression for the dispersion of the resulting error of the complex:

$$\sigma_z^2 = \sigma_r^2 + \sigma_a^2 = \frac{1}{2\pi}\int_0^\infty G_r(\omega)\,|\,\phi_r(j\omega)\,|^2\,d\omega + \frac{1}{2\pi}\int_0^\infty G_a(\omega) \times |\,\phi_a(j\omega)\,|^2\,d\omega,$$

$$\tag{9.2.16}$$

where σ_r^2, σ_a^2 are the dispersions of error components $z_r(t)$ and $z_a(t)$, attributed to the action of $J'_r(t)$ and $J'_a(t)$, respectively. The graphical interpretation of expression (9.2.16) is presented in Figure 9.5c, d.

Only the components of the spectra $G_r(\omega)$ and $G_a(\omega)$ which fall within the hatched regions will pass to the output of the combined sensor. It also follows from examination of the indicated figures that the greater the difference between the spectral densities of the disturbances in terms of shape (width), the greater the gain in precision can be achieved as a result of combining. For example, if $G_a(\omega)$ approaches the δ-function in the region $\omega = 0$, an arbitrarily large gain can be achieved, since $\phi_a(0) = 0$, and, conversely, the closer $G_r(\omega)$ and $G_a(\omega)$

are in shape, the smaller the gain offered by combining. Actually, assuming $\phi_a(j\omega) = 1 - \phi_r(j\omega)$ and considering that $G_r(\omega) = G_a(\omega)$, we obtain from (9.2.16)

$$\sigma_z^2 = \frac{1}{2\pi} \int_0^\infty G_a(\omega) \, | \, 1 - \phi_r(j\omega) + \phi_r(j\omega) \, |^2 \, d\omega$$

$$= \frac{1}{2\pi} \int_0^\infty G_a(\omega) \, d\omega,$$

i.e., the dispersion of the error of the complex is equal to the dispersion of the error of each sensor, operating independently. This is a somewhat unexpected result. It is well known, indeed, that the storage of a signal by the simple method of adding independent measurement results yields a $\sqrt{i}$-fold gain in the signal-to-noise ratio, where i is the number of signals added. Here, however, there are no contradictions. The fact is that to convert to simple summation it must be assumed that $N = 0$, $\phi_r = \phi_a$, $s_r(\omega) = s_a(\omega)$, i.e., that two identical sensors with independent and identical errors are used. Then, by virtue of the independence of Int_r' and Int_a' and of the equation $G_r(\omega) = G_a(\omega)$, the mean square total error is $\sigma_{av} = \sqrt{2}\sigma_r$ and the useful signal is doubled. Hence it follows that adding provides a $\sqrt{2}$-fold gain in the adder output signal-to-noise ratio.

It should not be forgotten, however, that in addition to the fluctuation error, caused by jamming and disturbances, sensors also have dynamic errors, and in view of the assumption that the equality $\phi_r = \phi_a$ is satisfied, invariance condition (9.2.2) is not. At the same time it is well known that to the minimum summary error corresponds the choice of parameters of the system such that dynamic error σ_d is approximately equal to fluctuation error σ_f. Thus, in order to determine the total error it is necessary also to consider these components of the errors of the sensors. Hence it follows that by simply adding signals distorted by independent noise it is possible to achieve a $\sqrt{i}$-fold gain in accuracy as a result of combining the signals only when dynamic errors are ignored. Thus, the mandatory condition for achieving any significant gain in precision by combining sensors is to make sure that there is a difference in the spectral composition of the disturbances acting upon them. This conclusion is also valid for complexes in which more than two sensors are combined. As a rule, the errors of autonomous nonelectronic sensors are nonstationary functions of time with variances σ_a^2 that increase in time. However, the transfer functions for the errors of autonomous sensors are such that the indicated nonstationarity can be eliminated by combining. In fact, let $J_a(t)$ change by the law

$$J_a(t) = J_0 + \int_0^t J_a'(t)\, dt = J_0 + \frac{1}{D}\, J_a'(t).$$

Then, for error component $z(t)$, caused by the action of $J_a'(t)$, we obtain, in consideration of (9.2.8),

$$z(t) = \frac{TD}{TD+1}\left[J_0 + \frac{1}{D}\, J_a'(t)\right].$$

Hence it follows that combining in accordance with Figure 9.3 excludes the influence of mathematical expectation J_0 on measurement results, eliminates the nonstationarity of the errors of the autonomous sensor and provides additional opportunities to filter them.

Before proceeding to examination of the practical feasibility of combined sensors in accordance with the diagram in Figure 9.3, we should like to point out once again that auxiliary filters ϕ_1 and ϕ_2 need be included only when the structural diagrams of the electronic and autonomous nonelectronic sensors are given and conditions (9.2.2) are not satisfied for $F_r(D) = 1$ and $F_a(D) = 1$. In other words, filters ϕ_1 and ϕ_2 should supplement transfer functions W_r and W_a in such a way that condition (9.2.2) will be satisfied.

Having made the above comment let us proceed to the question of the circuit implementation of the channel filters and their practical feasibility. We shall start the examination with the radio channel, since it specifically determines the noise immunity of the complex. It is clear from the above examination that the maximum suppression of jamming requires that the passband of the radio channel be narrowed. The radio channel includes an electronic tracking system, the transfer function of which is expressed as $W_r(D)$.

The realization of (9.2.7), (9.2.9) or (9.2.11) as an electronic system poses no difficulty. Therefore it may be shown that the inclusion of an auxiliary filter, the transfer function of which is expressed through F_r, is absurd in any case. One may use simply one electronic tracking system and add its output signal with the output signal of an autonomous sensor, passing through filter ϕ_2, the structure of which is selected in accordance with condition (9.2.2). However, a combined sensor with such a structure may not be suitable, since it does not provide the required radio signal tracking reliability in terms of the parameter being measured. We have already indicated the need to take into consideration the nonlinearity of the sensitive element, and consequently of the entire electronic tracking system in relation to parameter $s(t)$ of the radio signal being tracked.

When selecting transfer function $W_r(D)$ it is often necessary to proceed from the need to minimize the automatic tracking interruption probability. Also, the parameters of a combined sensor are determined by the requirement to minimize error $z(t)$.

These two conditions usually cannot be satisfied without including auxiliary filter ϕ_1 (Figure 9.3). Furthermore, the gain of the sensitive element decreases (see Figure 9.1) as a result of jamming, the condition of invariance is violated and the dynamic error of the combined system increases.

Thus the use of this system is advantageous only in cases when the capacity of an electronic sensor to perform its functions under conditions of jamming is limited by the tolerable error, and not by interruption of the automatic tracking mode. In other words, the system is suitable in cases when it is not necessary to achieve any appreciable gain in noise immunity.

Although these deficiencies prevent the complete realization of all the potential capabilities of combining, combined systems designed in accordance with the diagram in Figure 9.3 do find application. Sensors so designed are built and operated independently, and they are combined by means of the combined processing of the output signals of the individual sensors. The lack of memory for the corrected errors of the sensors is an important deficiency of this method of combining. If a radio channel breaks down as a result of jamming the complex ceases to function. Therefore, systems that are conceptually close to the one examined above,　but which have memory for corrected errors, are used in practice.

So-called error compensation systems are most frequently employed. A dynamic structural diagram of such a single-coordinate complex is illustrated in Figure 9.6. Here a filter, with the transfer function $F_k(D)$, is the combining

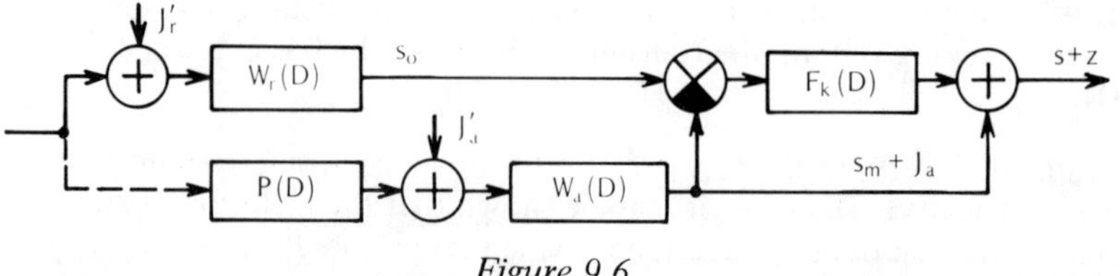

Figure 9.6

filter. It simultaneously performs functions of the filtering of the errors of both sensors and can store errors when required. In view of the fact that

$$s_r = s_m + J_r, \quad s_a = s_{am} + J_a,$$

where $J_r = W_r(D)J_r'$ and $J_a = W_a(D)J_a'$; s_m and s_{am} are the measured values of parameter s at the outputs of the components with the transfer functions $W_r(D)$ and $W_a(D)$, respectively, we may write for the output signal of the sensor

$$s + z = [1 - F_k] (s_{am} + J_a) + F_k (s_m + J_r). \qquad (9.2.17)$$

Hence it is obvious that the error of measurement of s, in addition to the component

$$z_f = [1 - F_k] J_a + F_k J_r, \qquad (9.2.18)$$

caused by the action of jamming J_r and disturbances J_a, contains a dynamic error, attributed to the time constants of the sensors. In fact, assuming $Y = D^N S$, we find the expression for the dynamic error:

$$z_d = [1 - W_a (1 - F_k) D^N - W_r F_r] s,$$

which is equal to zero only for widebound sensors and when coordinates with the same dimensionalities are combined (N = 0). This condition, naturally, is peculiar to an electronic sensor and can be only approximately satisfied.

Comparing (9.2.17) with (9.2.21), we see that when $F_r = F_k$ and $F_a = 1 - F_k$ the combined systems with the diagrams shown in Figures 9.4 and 9.6 will have the identical jamming filtering. The same is true with respect to dynamic error. Thus the described systems are equivalent in terms of filtering properties, but the one shown in Figure 9.6 is simpler, since it has a single filter, common to both channels. The structure of the combining filter is determined by the relationship between the input signals $Y = P(D)s$. It is easiest to combine the signals of two sensors, which measure the same coordinate s = Y, with the filter, the dynamic structural diagram of which is shown in Figure 9.7[1].

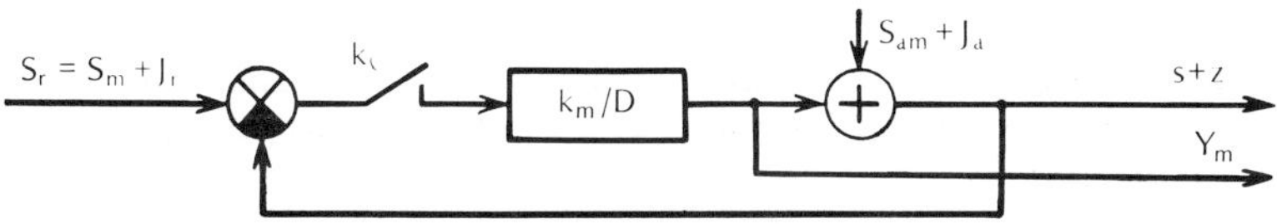

Figure 9.7

The output signal of the complex with switch (K) closed is

$$s_{out} = \left[\frac{k_m}{D+k_m} W_r(D) + \frac{D}{D+k_m} W_a(D) \right] s + \frac{k_m}{D+k_m} J_r + \frac{D}{D+k_m} J_a \, .$$

$$(9.2.19)$$

---

[1] In the diagram in Figure 9.7 signals are processed in a gyromagnetic compass. As the headings measured with magnetic and gyroscopic sensors it uses the values s_r and s_a. This is apparently the first combined sensor incorporating the joint processing of two signals.

The system becomes invariant only when wideband sensors are used, but the dynamic error of the complex

$$\Delta z_d = \left[1 - \frac{k_m}{D+k_m} W_r(D) - \frac{D}{D+k_m} W_a(D) \right] s$$

regardless of the choice of gain k_m of the integrator of the combining filter, cannot be smaller than dynamic error

$$\Delta s_{dr} = [1 - W_r(D)]\, s$$

regardless of which electronic meter is operating.

Assuming for definition $W_r(D) = k_v/(D + k_v)$ and that the autonomous meter operates without dynamic error, we have

$$\Delta z_d = \frac{k_m D}{D^2 + (k_m + k_v)D + k_m k_v}\, s. \qquad (9.2.20)$$

The dynamic error of the electronic meter is given by the expression

$$\Delta s_{dr} = \frac{D}{k_v + D}\, s. \qquad (9.2.21)$$

Comparing (9.2.20) and (9.2.21), we see that even with a broadband autonomous sensor the condition $\Delta z_t \geqslant \Delta s_{dr}$ is satisfied. Here the errors will be equal only when $k_m \to \infty$.

The complex can operate in the periodic or sporadic nonelectronic sensor signal correction mode. The key is used for this purpose. When the key is closed the correction signal

$$Y_m = W_k(W_r - W_a)s + W_k(W_r J_r' - W_a J_a'), \qquad (9.2.22)$$

where $W_k = W_k(D) = k_m/(D + k_m)$, is established at the integrator output of the combining filter.

When the key is open signal (9.2.22) is established at moment of time $t = t_k$, so that

$$s + z = s_{out} = W_a s + W_a J_a' + Y_m(t = t_k). \qquad (9.2.23)$$

If the autonomous sensor operates with a constant or slowly changing error (in comparison with the correction period), then the error will be cancelled, since signal $Y_m(t = t_k)$, established at the integrator output, will have a component that represents the error of that sensor with the opposite sign. As for errors of the electronic sensor, signal (9.2.23) will represent their values at moment of time $t = t_k$.

Because of the fact that the radio channel can operate in the periodic correction mode with a sufficiently long period, commensurable with the correlation time of the errors of the autonomous sensor, and is turned on only for a period of time equal to the transient time of the complex, the security of the radio link is improved and reconnaissance of the parameters of the radio signal is thereby impeded. However, as in the system in which signals are added (Figure 9.3), combining the sensors through a combining filter does not improve the noise immunity of the electronic sensor itself. The passband of the system that tracks the parameter s(t) must be selected, as before, with a view toward minimizing the mean square error. This kind of combining provides no gain in radio signal tracking reliability. Of course, the fact that jamming experiences additional filtering in the combining filter eases to some extent the requirements on the minimum tolerable input signal to noise ratio of the electronic sensor.

The operation of a complex, in which coordinate s(t) and its derivative are combined in the sensor, when the combining filter jointly processes the signals

$$s_r = W_r(s + J'_r) \text{ and } s_a = W_a(Ds+J'_a).$$

is examined below.

The structure of the filter illustrated in Figure 9.8 is most economical and best satisfies conditions (9.2.9) and (9.2.10).

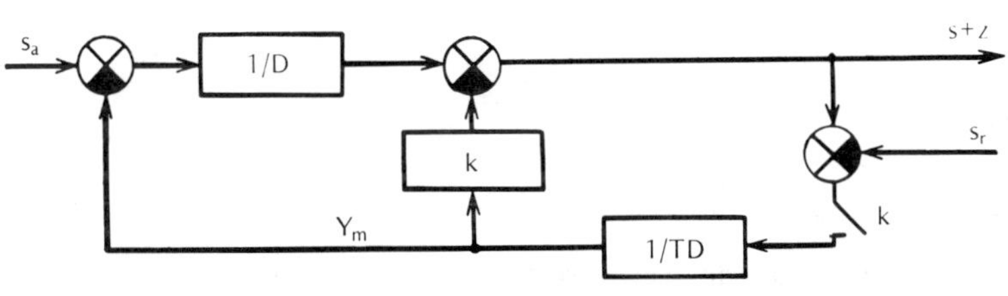

Figure 9.8

The useful component of the signal is

$$s_{out} = W_a \frac{TD^2}{M(D)} s + W_r \frac{kD+1}{M(D)} s, \qquad (9.2.24)$$

where $M(D) = TD^2 + kD + 1$

With broadband sensors ($|W_a(j\omega)| = 1$ and $|W_r(j\omega)| = 1$ for all frequencies of the spectrum of the measured signal), measurements are made without dynamic error. The error attributed to jamming is described by the expression

$$z_f = \frac{TD}{M(D)}J_a + \frac{kD+1}{M(D)}J_r. \tag{9.2.25}$$

Comparing (9.2.25) with (9.2.9) and (9.2.10), we see that the resulting filter structure completely satisfies and above-formulated conditions. It corrects systematic and low-frequency errors of the nonelectronic velocity sensor and performs additional smoothing of the errors of the electronic system. It is not hard to find the structure of the filter that satisfies conditions (9.2.11) and (9.2.12), derived on the assumption that $P(D) = D^2$, i.e., one that jointly processes the signal and its second derivative. Modifications of filters that memorize the constant component J_a of an error, its linearly increasing component, etc., can be found. The same is true with respect to the dynamic error of the electronic system.

The filter illustrated in Figure 9.8 enjoys extensive application for the periodic correction of an autonomous coordinate sensor (range computer) on the basis of air speed data.

This operating mode is examined in greater detail below. Let an autonomous sensor determine target velocity $v = Ds$ with an error that increases in time by $J_a = a_0 + a_1 t$, and let an electronic sensor have a transfer function of the form

$$W_p(D) = k_v/(D + k_v), \tag{9.2.26}$$

i.e., it is a first order system and a broadband nonelectronic sensor. Then

$$s_a = Ds + a_0 + a_1 t$$

$$s_m = \frac{k_v}{D+k_v}s \tag{9.2.27}$$

where s_m is the measured value of s.

The expression for the dynamic error of the complex will be

$$\Delta z_d = \frac{D(kD+1)}{(TD^2+kD+1)(k_v+D)}s. \tag{9.2.28}$$

The dynamic error of the electronic sensor is

$$\Delta s_{dr} = Ds/(k_v + D). \tag{9.2.29}$$

Comparing (9.2.28) with (9.2.29), we see that for a signal of the form $s =$ $= s_0 + vt$ the steady state values of Δz_d and Δs_{dr} coincide and are equal to v/k_v. Thus, radio correction of the autonomous coordinate sensor, which includes a velocity sensor and integrator (range computer) can be accomplished to an accuracy limited by the dynamic error of the electronic sensor.

The system illustrated in Figure 9.8 also has memory for the errors of the autonomous sensor. In fact, the signal Y_m is described by the expression

$$Y_m = \frac{s_a}{M(D)} - \frac{Ds_r}{M(D)} , \qquad (9.2.30)$$

where, as before, $M(D) = TD^2 + kD + 1$.

Under the assumption made above concerning the character of the error of the autonomous sensor the steady state correction signal $Y_{ms} = a_0 + a_1 t$ contains the exact values of the errors of the autonomous sensor, and consequently, after switch K is opened (the correction system is turned off), the influence of the constant error will be completely excluded, the constant error will be memorized by the integrator, and the linearly changing error will be taken into account in the entire preceding period of time.

Thus combined systems of the examined type solve the problem of improving jamming immunity only by virtue of the fact that the radio system can be turned on for comparatively short periods of time of the correction of signals obtained from autonomous systems. Sometimes this can be very important. In nonautonomous radio navigation systems, for example, operating in conjunction with surface responder beacons, mutual interference often is the basic kind of jamming. The periodic switching of the radio correction systems of autonomous sensors sharply increases the transmission capacity of a beacon. Such a combined system will also be useful under conditions of radio signal fade, for example as a result of brief receiver overloads, switching off of the source of the radio signals, etc.

The problem of improving jamming immunity is effectively solved by systems in which electronic tracking systems are corrected by signals from autonomous nonelectronic sensors.

9.3. Systems with Corrected Electronic Tracking Sensors

Combined coordinate sensors, in which the electronic tracking system has two inputs, one ordinary, in the form of an estimator, and the other for the input of a signal from a nonelectronic instrument (this signal is called a correcting signal), are used extensively. A simplified diagram of such a system is illustrated in Figure 9.9.

Here, as before, P(D) represents the relationship between the coordinates measured by the electronic and nonelectronic systems. It is assumed, for the

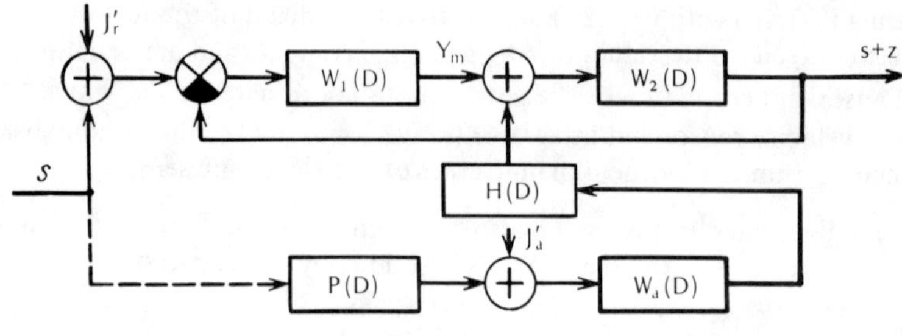

Figure 9.9

purpose of simplifying the calculations, that both sensors operate in the same coordinate system. Statement P(D) may acquire the values 1, D or D^2. If P(D) = 1 the system has position correction, and when P(D) = D and P(D) = D^2 the systems have velocity correction and acceleration correction, respectively.

The component with the transfer function H(D) matches the scales of the signals from both sensors. Its transfer function is determined from the condition z = 0 for $J'_r = J'_a = 0$.

The electronic system consists of two components with transfer functions W_1 and W_2. The measuring (sensitive) element is a part of the component whose transfer function is expressed through $W_1(D)$.

The measured signal is

$$s_{out}(t) = s + z = \frac{W+HPW_2 W_a}{1+W}\, s + \frac{W}{1+W}\, J'_r + \frac{HW_2 W_a}{1+W}\, J'_a$$

$$(9.3.1)$$

where $W = W_1 W_2$.

The problem of selecting the optimum sensor parameters is solved in accordance with the criterion of minimum variance σ_z^2 of error z. The latter consists in the general case of two components: one (dynamic error) is caused by the constant of the sensor, and the other is caused by the action of jamming J'_r and disturbances J'_a.

According to optimum systems theory minimization of variance σ_z^2 involves making a compromise between the magnitudes of the indicated components. However, the optimality of the system is violated as soon as the statistical characteristics s(t), $J'_r(t)$ or $J'_a(t)$ are changed.

The difficulty of finding the optimum dynamic structure of the system and its relative cost gave rise to a somewhat different approach to the design of combined sensors. The structure is selected from the condition of the absence of dynamic errors (called the invariance condition above), and the parameters are selected from the condition of the minimization of errors caused by $J_r'(t)$ and $J_a'(t)$. The invariance condition will be satisfied if the transfer functions of the components are selected in accordance with the equality

$$HPW_2 W_a = 1. \tag{9.3.2}$$

The transfer function of the nonelectronic sensor W_a is assumed to be given. The point of input of the correcting signal is determined by the type of correction. For example, in position correction ($P = 1$) invariance condition (9.3.2) may be written as

$$H = 1/W_a W_2. \tag{9.3.3}$$

An invariance system can be designed only when the nonelectronic sensor is broadband. Assuming $W_a = k_a$ and $W_2 = k_2$, we obtain $H = 1/k_a k_2$. In reality an autonomous sensor always has a time constant. In the simplest case its transfer function may be written as $W_a(D) = k_a/(T_a D + 1)$. The invariance condition requires that the matching component have the transfer function $H(D) = (T_a D + 1)/k_2 k_a$.

The infeasibility of designing a component with such a transfer function confirms that it is not possible to build an absolutely invariant system. However, the transfer function for the useful component of the signal of an inertial correcting signal sensor is more favorable from the standpoint of dynamic error than for a separately operating sensor. As a matter of fact, assuming in (9.3.1) $J_r' = J_a' = 0$, $H = 1/k_2 k_a$, $W_a(D) = 1/(T_a D + 1)$, $W_1 = k_1/(T_r D + 1)D$ and $W_2 = k_2$, which corresponds to a first-order astatic system, we obtain for the output signal of a combined system

$$s(t) + z(t) = s_{out}(t)$$

$$= \frac{T_r D^2 + (k_v T_a + 1)D + k_v}{T_a T_r D^3 + (T_r + T_a)D^2 + (1 + k_v T_a D + k_v)} \, s(t). \tag{9.3.4}$$

Dynamic error is described by the equation

$$\Delta s_d = s - s_m$$

$$= \frac{T_a T_r D^3 + T_a D^2}{T_a T_r D^3 + (T_r + T_a)D^2 + (1 + k_v T_a)D + k_v} \, s(t). \tag{9.3.5}$$

Let us do a comparative analysis of the dynamic properties of an electronic tracking system, corrected by the position signal of an inertial sensor, and of the same sensor without correction.

In the absence of correction the dynamic error may be found from the equation

$$\Delta s_{dr} = \frac{T_r D^2 + D}{T_r D^2 + D + k_v} s(t).$$ \hfill (9.3.6)

Comparing expressions (9.3.5) and (9.3.6) we see that combining an electronic sensor and an intertial autonomous sensor is equivalent to increasing the order of the loop by one, i.e., a corrected first-order system operates without velocity errors. This result has a simple physical explanation. Referring to Figure 9.9 we see that by virtue of the presence of a second channel for the propagation of the signal s(t), the input of the sensor actually receives two signals, one of them passes through the main input, and the other through the inertial component (autonomous sensor).

Let

$$s(t) = s_0 + vt \qquad \text{for } t > 0,$$

$$s(t) = 0 \qquad\qquad \text{for } t \leqslant 0.$$ \hfill (9.3.7)

After transient time $t \gg T_a$ of the inertial component its output signal also will be a linear function of time with lag τ_d, so that $s_{am}(t) = s_0 + v(t - \tau_d)$. Consequently the tracking electronic system actually tracks not signal (9.3.7), but error

$$\Delta s_a(t) = s(t) - s_{am}(t) = v\tau_d.$$ \hfill (9.3.8)

However, the first order system reproduces (in the steady state mode) a constant signal (more accurately a jump) without error. For a signal of the form

$$s(t) = s_0 + vt + \frac{at^2}{2}$$ \hfill (9.3.9)

the combined system will have steady state error

$$\Delta s_d = aT_a/2k_v.$$

At the same time, for a noncorrected system of the same type, in accordance with (9.3.6), the dynamic errors will be determined as follows:

for a signal of the form (9.3.7)

$$\Delta s_{dr} = v/k_v,$$

for a signal of the form (9.3.9)

$$\Delta s_{dr} = \frac{v}{k_v} + \frac{at}{2T_r k_v}.$$ \hfill (9.3.10)

Now let us examine the transient mode of measurement, assuming that $P(D) =$
$= 1$. Under this condition $H = 1/W_2 W_a$, $J_a = W_a J'_a$. Then

$$s_m(t) = s(t) + \frac{W}{1+W} J'_r + \frac{1}{1+W} J_a, \qquad (9.3.11)$$

and the measurement error is

$$z(t) = \frac{W}{1+W} J'_r + \frac{1}{1+W} J_a. \qquad (9.3.12)$$

At the same time we know that the error of a tracking system without correction
is

$$\Delta s_r = \frac{W}{1+W} J'_r + \frac{1}{1+W} s. \qquad (9.3.13)$$

Comparing (9.3.12) and (9.3.13) we see that they coincide, but in (9.3.12)
signal $s(t)$ is replaced by error J'_a. Hence follows the important conclusion that
an electronic tracking system actually tracks not input signal $s(t)$, but error
J'_a, including dynamic error, of the autonomous sensor.

We proceed to the analysis of the measurement errors of signal $s(t)$, caused by
jamming. According to (9.3.12) the transfer function for jamming remains
unchanged. However, it would be premature to conclude on this basis that
the noise immunity of corrected and uncorrected tracking systems will
remain unchanged.

The fact is that the bandwidth of the tracking system, as we have already
mentioned, is selected on the basis of a compromise between the dynamic
error and the error caused by the action of jamming. But since the expressions for the dynamic errors of these two classes of systems do not
coincide the quantitative values of the errors will also be different, regardless
of what hypothesis is stated concerning the statistics of input signal $s(t)$.

We will show that a system with position correction compensates constant
and slowly changing errors of an autonomous sensor. For this purpose we
return to formula (9.3.12). For the measurement error component $z(t)$,
caused by $J'_a(t)$, we write

$$z_a = \frac{1}{1+W} W_a J'_a,$$

where $W = W_1 W_2 = k_v/(T_r D^2 + D)$. After substituting W we obtain

$$z_a = \frac{D(T_r D+1)}{Y D^2 + D + k_v} W_a(D) J'_a(t).$$

Hence it follows that the constant component of error $J'_a(t)$ is compensated by the tracking system and has no influence on measurement results in the steady state mode. Thus signal Y_m (Figure 9.9) contains a component that is proportional to mathematical expectation J'_a, which makes it possible to use the periodic ETS operation mode.

Before examining the influence of the above-established properties of a corrected system on the solution of the problem of improving noise immunity we shall analyze the dynamic properties of the system for correction by a signal proportional to the rate of change of coordinate $s(t)$. The velocity signal from the output of the autonomous sensor should reach the input of the integrator of the tracking system.

Assuming that $P(D) = D$, and $W_2(D) = k_2/D$, we establish on the basis of (9.3.2) that the condition of matching of the scales of the signal, as before, requires that the condition $H = 1/k_a k_2$ be satisfied. Such systems are called combined electronic sensors with velocity correction.

It is not difficult to show, using (9.3.1), that formulas (9.3.11) and (9.3.12), derived above, are also valid for velocity correction systems. Here, however, it is important to make one comment concerning the character of disturbances J'_a. It may be assumed that disturbance J'_a, determined for position correction systems, is $J''_a(t) = DJ'_a(t)$ for velocity correction. We will prove this by way of example of gyroscopic sensors.

Drift $\Delta\psi_g(t)$ of a position gyroscope in relation to disturbing moments, for example to the moment of friction $M_{fr}(t)$ in the suspension, may be written as

$$\Delta\psi_g = k \int_0^t M_{fr}(t)\,dt,$$

where k is a constant, determined by kinetic momentum.

The error of the velocity gyroscope, caused by friction, is

$$D\psi_g(t) = kM_{fr}(t).$$

In consideration of the above comment we find an expression for component z, caused by error J'_a of an autonomous sensor on the assumption that $W_a(D) = 1$. On the basis of (9.3.1) we have

$$\Delta z_a = \frac{T_r D + 1}{T_r D^2 + D + k_v}\,J'_a(t). \tag{9.3.14}$$

It follows from (9.3.14) that slowly changing and constant errors of an autonomous velocity sensor are not corrected in a combined sensor, but are only reduced by a factor of k_v, i.e., the steady state error is

$$\Delta z_{ass} = J'_a/k_v \qquad (9.3.15)$$

This means that the stated error component will introduce a discrepancy in the measured parameter at the input of the sensor of the electronic system.

In a tracking system with acceleration correction, when $P(D) = D^2$, the signal from the acceleration sensor (for example an accelerometer) in the case of the correction of first-order systems, must be integrated before input into the electronic coordinate sensor. If a second-order system is corrected, the signal is sent directly to the input of the first integrator.

In the above examples it was possible to achieve an invariant combined sensor by feeding a simple signal, proportional to the measured coordinate or to one of its derivatives, to the tracking system. This can be done in reality if no constraints are imposed on the point of input of the correcting signal. This means that each component of the tracking system is elementary: amplifier, integrator, etc. However, real components of tracking systems often have the transfer functions of nonelementary components. An electric motor, for example, has the transfer function

$$W_m(D) = \frac{k_m}{D(T_m D+1)} . \qquad (9.3.16)$$

If it is required to make invariant a system which uses an electric motor as the output system, it will be necessary, in accordance with (9.3.2), to supply to its input a signal, which represents the weighted sum of the first and second derivatives of the measured coordinate. Here, obviously, it will be desirable to have two autonomous sensors. It may turn out to be advantageous to preprocess signals of two and more autonomous sensors before input into the tracking system using combining filters, the dynamic structural diagrams of which are shown in Figure 9.7 or 9.8.

By using a priori data about the change of tracked parameter (coordinate) $s(t)$ it is possible to greatly narrow the passband of an electronic system and to reduce its function to the tracking of error $J_a(t)$ of an autonomous non-electronic sensor.

This is the main reason why it is possible, by correcting electronic tracking systems, to greatly increase their noise immunity. The more complete the information about the measured coordinate, obtained from an autonomous sensor, the greater the gain in noise immunity will be.

In application to the problem of measuring target coordinates (Figure 9.2), the following may be combined: angle or angular velocity sensor, automatic range finder and closing speed sensor.

As the first example we shall examine combined sensors for measuring angle γ and angular velocity $\dot{\epsilon}$ of the line of sight. Let a moving object carry a first-order tracking angle measuring system, tracking angle γ. A gyroscope, the measuring axis of which is aligned with the longitudinal axis $0x_1$ of the object, is used to measure the proper angular motions of the object. This combined measuring system obviously is one with incomplete information, since the component of angle γ, attributed to the motion of the target, is not measured by the gyroscope.

To exclude the influence of the change of angle ϑ on the operation of the tracking system it is necessary, as ϑ changes by $\Delta\vartheta$, to rotate the antenna relative to the air frame by angle $-\Delta\vartheta$. To do this the gyroscope signal is supplied to the input of the antenna rotation motor. Since the motor, in simplified terms, may be assumed to be an integrating component, the motor is enclosed in rigid negative feedback in order to match the dimensionalities of the main and correcting signals. Accordingly, and also in consideration of the diagram in Figure 9.9, we obtain for the combined angle measuring system the diagram illustrated in Figure 9.10. An extra component (an

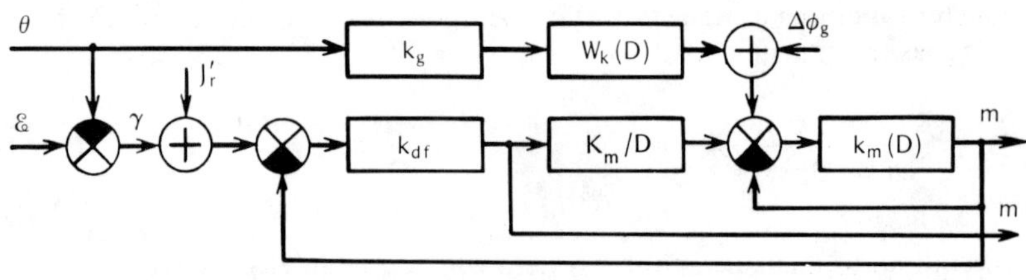

Figure 9.10

integrator with gain k_m) is included to make the closed system astatic. The component with transfer function $W_k(D)$ matches the scales of the signals from the gyroscope and direction finder. Changes of the component of angle γ, caused by the motion of the target, are tracked as a result of the appearance of a signal at the output of the direction finder, the gain of which is k_{df}. If axis $0x_1$ of the target deviates (angle ϑ changes), then the axis of the gyroscope stays in position and a signal appears at its output. This signal passes through the input of the motor with transfer function k_m/D, which rotates the antenna so that the input signal becomes equal to zero.

We denote the transfer function of the drive motor in consideration of feedback through

$$\phi(D) = k_m/(D + k_m).$$

In the absence of jamming and errors $\Delta\psi_g$ of the gyroscope, the output signal of the sensor is

$$\gamma_m = \frac{W_e}{1+W_e}\,\epsilon - \frac{k_g W_k \phi + W_e}{1+W_e}\,\vartheta, \tag{9.3.17}$$

where $W_e = W_1\phi$, $W_1 = k_g k_m/D$; k_g is the gain of the gyro.

The error signal (dynamic error) at the direction finder input is described by the expression

$$\Delta\gamma_d = \epsilon - \vartheta - \gamma_m$$

$$= \frac{1}{1+W_e}\,\epsilon - \frac{1-k_g\phi W_k}{1+W_e}\,\vartheta, \tag{9.3.18}$$

where $\epsilon - \vartheta = \gamma$.

It follows from (9.3.18) that there are at least two possibilities of making a system invariant in relation to changes of angle ϑ. The first way is to build a tracking drive (often called an internal loop) with a wide passband. In fact, the invariance condition will b exactly satisfied when $|\,W_k(j\omega)\phi(j\omega)\,| = 1$ through the entire frequency spectrum of angle ϑ.

The other way is to select the right transfer function $W_k(D)$. Then invariance condition (9.3.2) will be satisfied when

$$W_k(D) = (D + k_m)/k_g k_m.$$

It is not feasible to design a filter with such a transfer function. Invariance is attainable when two gyros, one position and the other velocity, are used, i.e., a system with combined correction. The choice is usually limited to $W_k(D) = 1/k_g$, and high-quality antenna stabilization is achieved by increasing coefficient k_m.

The reproduction error caused by jamming J'_r and by gyroscope errors $\Delta\psi_g$ will be described by the expression

$$\Delta\gamma_f = \frac{k_v}{D^2 + k_m D + k_v}\,J'_r + \frac{k_m D}{D^2 + k_m D + k_v}\,\Delta\psi_g \tag{9.3.19}$$

As can be seen by formulas (9.3.18) and (9.3.19), an effort to reduce the influence of changes of angle ϑ on reproduction error results in an increase of the error caused by gyro drift $\Delta\psi_g$. But if this error can be ignored, which is often done in practice, particularly if the operating time of the complex is short, then it is necessary to take into account the error caused by changes of angle ϵ. The spectrum of ϵ is virtually always many times narrower than the spectrum of ϑ. The bandwidth of the combined tracking system for jamming can be narrowed in comparison with a simple system by approximately the same number of times as the spectral width of ϑ exceeds the spectral width of ϵ. The effectiveness of noise type jamming, employed against angle measuring channels, decreases by the same amount.

Target tracking reliability is also improved by combining. The antenna is aimed at the target for a certain amount of time by the stabilization system in the presence of profound or short-term signal fade.

The diagram in Figure 9.10 is also suitable for measuring the angular velocity of the line of sight. This signal is represented by error $\Delta\gamma_d$ (9.3.18), and consequently it may be taken from the direction finder output (from the input of an auxiliary integrator).

As the second example of the same problem of homing we shall examine the operation of a combined automatic range finder. The electronic tracking system tracks target range r (more accurately delay $\tau_d = 2r/c$), determined by expression (9.1.4)..We write (9.1.4) as the sum

$$r = r_0 + r_s(t) + r_t(t),$$

where $r_t = \int\limits_0^t v_t \cos q_t dt; \; r_s = \int\limits_0^t v_s \cos qdt$, and we assume that the target carries an instrument that measures the projection v_{sr} of velocity vector v_s (Figure 9.2) onto axis O_sO_t (for example an air or ground speed indicator and a computer, which calculates $v_{sr} = f(v_s, \gamma, \alpha)$, where α is the angle of attack).

For the sake of simplicity we shall assume that the condition $v_{sr} \approx v_s$ is satisfied, i.e., the target is guided by the direct guidance method. We also assume that the automatic range finder is first-order. Accordingly the dynamic structural diagram of the combined automatic range finder is represented in the form as illustrated in Figure 9.11.

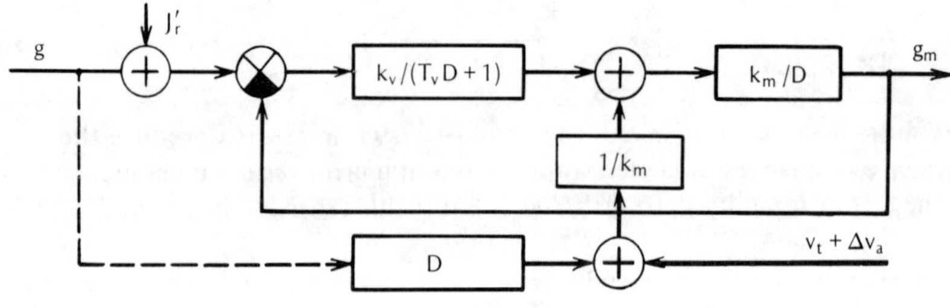

Figure 9.11

Since the PMS does not measure the component $v_t = Dr_t$, the complex is a system with incomplete information and v_t is considered to be a component of error $J'_a = v_t + \Delta v_a$, where Δv_a is the PMS instrument error. On the basis of (9.3.11) we write for the output signal of the meter

$$r_m = r + \frac{k_r}{N(D)} J'_r + \frac{T_v D + 1}{N(D)} (v_t + \Delta v_a), \qquad (9.3.20)$$

where $N(D) = T_B D^2 + D + k_v$, $k_v = k_B k_m$ is the gain of the system in terms of velocity. As can be seen by (9.3.20), the system is static in relation to J'_a. Here the steady state static range measurement error $\Delta r_{ss} = v_t/k_v$ (for zero mathematical expectation of the random value Δv_a) increases as k_v decreases.

The example at hand particularly graphically shows that the more completely PMS information describes the measured coordinate the larger the gain in noise immunity. In fact, assuming $v_t = 0$, we find that the reduction of k_v, and consequently of the band of the tracking system will be limited only by a measurement error Δv_a. By combining an automatic range finder with different kinds of PMS it is possible not only to improve noise immunity in relation to wideband noise jamming, but also to solve to a great extent the problem of protection against decoy jamming.

The drift rate of an automatic range finder is limited by the maximum tolerable dynamic tracking error Δr_d of the decoy pulse. If jamming pulse power is much greater than the pulse power of the useful signal, then drift may take place at the rate

$$v_{ss} \leqslant \Delta r_{dm} k_v.$$

The value of Δr_{dm} for which tracking is not yet interrupted is approximately one-half the linear part of the discriminator response curve.

In a combined sensor, especially a navigation system that receives complete information from the PMS, the value of k_v may be reduced tens of times in comparison with a simple automatic range finder. Therefore the tolerable drift rate may be so low as to possibly render such a type of jamming useless. But even if drift or interruption of automatic tracking does take place due to jamming, with a PMS, the signals of which are corrected by a radio range finder, it is possible to store range information, and in second-order systems information about the rate of change of range, for a sufficiently long period of time. This provides an opportunity to sharply narrow the signal search band.

If the search time is fixed, then, obviously, it is possible with a combined system to increase the time of storage in the detector by a factor of $n = r/\Delta r_e$, where Δr_e is the maximum error of the corrected PMS signal.

For given false alarm and signal miss probabilities this makes it possible to reduce the required ratio $q = u_{ef}/\sigma_n$ by a factor of $k = \sqrt{T_{co}/T_{sim}}$, where u_{ef} is the effective voltage of the signal and T_{co} and T_{sim} are the storage times in the detector of combined and simple systems, respectively.

Systems in which electronic tracking systems are corrected by signals from autonomous nonelectronic PMS enjoy extensive application [16, 18].

It is preferable in navigation systems to use more sophisticated complexes, in which the mutual correction principle is employed. The primary reason for this is that such systems include nonlinear converters (computers), which convert signals from the coordinate system in which the PMS operates to the coordinate system of the electronic sensor, and then often the reverse process. Such complexes are essentially nonlinear useful signal-noise mixture converters. All these nonlinearities are described by "smooth," but as a rule multivalued (for example trigonometric) functions.

Three extremely undesirable phenomena take place in the computer during the joint conversion of a mixture of jamming and signal:

"suppression" of the converted signal by the accompanying noise, and consequently violation of the condition of scale matching of the sensors of the processed signals, i.e., the invariance condition;

possible violation of the requirement of single-valuedness of the converted signal;

conversion of a combined sensor to a system with variables that are changed randomly in time by the parameters in view of the random nature of the converted signals.

These extremely undesirable phenomena necessitate that the structure of a complex be so selected that all signals reaching the computer inputs be satisfactorily filtered and that systematic sensor errors be excluded from them. The stated conditions are often satisfied by systems with mutual correction.

9.4. Combined Navigation Systems with Mutually Correcting Sensors

This class of systems essentially combines PMS radio correction systems (Figure 9.8) and electronic tracking sensor correction systems (Figure 9.9). The essence of the idea of this kind of combining consists in the fact that before being fed into the computer the signals of both sensors are corrected, prefiltered to correct systematic errors, etc.

Combined systems with mutually correcting sensors include nonelectronic coordinate sensors, nonautonomous or autonomous electronic navigation systems, computers, often digital, analog-code and code-analog converters, etc. Their theoretical analysis is an exceedingly difficult task.

However the basic concepts of such combining can be explained and such systems can be subjected to qualitative analysis by using simple examples of mutually correcting single-coordinate sensors.

It is convenient to conduct the analysis for a specific example, which is the dynamic structural diagram of an instrument that measures coordinate s, illustrated in Figure 9.12[1]. This sensor includes the following components:

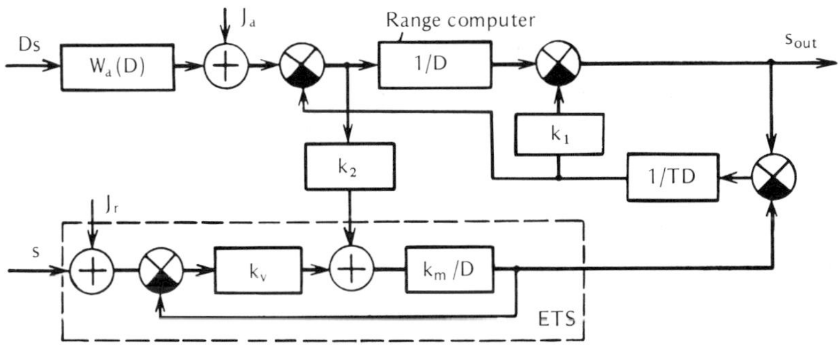

Figure 9.12

autonomous velocity sensor with transfer function $W_a(D)$, electronic coordinate sensor, which is a first-order electronic tracking system (ETS), and a range computer.

To the range computer input is supplied the radio-corrected signal s(t) of the rate of change of the coordinate. This signal, after being converted to the coordinate system in which the electronic tracking system operates (the computer in the diagram is replaced by a proportional component with gain k_2), is fed to the input of an integrator with gain k_m.

The signal of the electronic system, in turn, corrects the range computer after being converted to the coordinate system of the autonomous system. If both computers have negligible time constant the output coordinate $W_a = 1$ will be determined by the expression

---

[1] When analyzing the dimensionalities of the terms of the transfer function it should be borne in mind that the one in the transfer function of the range computer has the dimensionality 1/s.

$$s_{out} = \frac{TD^3 + (Tk_0 + k_1 k_2 k_m)D^2 + k_1 k_0 D + k_0}{N(D)} s(t)$$

$$+ \frac{TD^2 + (Tk_0 + k_1 k_2 k_m)D}{N(D)} J_a(t) + \frac{k_1 k_v D + k_v}{N(D)} J_r'(t), \qquad (9.4.1)$$

where $k_v = k_B k_m$ is the gain of the electronic tracking system in terms of velocity;

$$N(D) = TD^3 + (k_1 + Tk_v)D^2 + (1 + k_1 k_v - k_2 k_m)D + k_v \qquad (9.4.2)$$

is the characteristic equation of the combined system.

It is easy to see that the condition of invariance of the useful component of output coordinate s_{out} in relation to the parameters of the system is satisfied when

$$k_2 = 1/k_m. \qquad (9.4.3)$$

In consideration of (9.4.3) we obtain for the output signal

$$s_{out} = s + \frac{TD^2 + (Tk_v + k_1)D}{N'(D)} J_a + \frac{k_1 k_v D + k_0}{N'(D)} J_r', \qquad (9.4.4)$$

where $N'(D) = TD^3 + (k_1 + Tk_v)D^2 + k_1 k_v D + k_v$. A number of qualitative conclusions may be drawn from analysis of relation (9.4.4).

First, the constant or slowly changing errors of the autonomous sensor in the steady state mode ($D \to 0$) have no influence on measurement results. Second, the constant errors of the electronic sensor pass undistorted to the output of the system, while the fluctuation components experience considerable smoothing, since the order of the polynomial of the numerator in the transfer function for jamming is three times smaller than the order of the polynomial of the denominator.

Thus the system satisfies all the above requirements on the combined signal processing system.

Presented below is a comparative analysis of the dynamic properties of two types of combined sensors: of the just examined complex with mutually correcting sensors, and of the same sensor, but without a corrected electronic tracking system.

To convert from the first to the second type it is obviously necessary and sufficient to assume $k_2 = 0$ in expressions (9.4.1) and (9.4.2).

The output of the system without a corrected electronic sensor is

$$s'_{out} = \frac{TD^3 + Tk_v D^2 + k_1 k_v D + k_v}{N''(D)} s(t)$$

$$+ \frac{TD^2 + Tk_v D}{N''(D)} J_a(t) \quad + \frac{k_1 k_v D + k_v}{N''(D)} J_r(t), \tag{9.4.5}$$

where $N''(D) = TD^3 + (k_1 + Tk_v)D^2 + (1 + k_1 k_v)D + k_v$.

It follows from (9.4.5) that the sensor cannot be invariant to the input signal, regardless of the choice of parameters, with the exception of the case when the gain of the electronic sensor in terms of velocity is $k_v \to \infty$. This approach is obviously unsatisfactory: as k_v increases the parameter tracking error of the radio signal, caused by jamming J'_r, increases, and the system becomes unstable, i.e., leaves the automatic tracking mode, due to the nonlinearity of the discriminator.

Let us find the dynamic error of an uncorrected sensor. By definition

$$\Delta s_d = z_d = [1 - W_0(D)] \, s(t), \tag{9.4.6}$$

where $W_0(D)$ is the transfer function for signal $s(t)$.

Substituting $W_0(D)$ from (9.4.5) into (9.4.6), we obtain

$$\Delta z_d = \frac{D(k_1 D + 1)}{N''(D)} s(t). \tag{9.4.7}$$

Comparing (9.4.7) with (9.3.6), we see that the steady state value of the dynamic error of the combined system is equal to the dynamic error of the electronic sensor itself.

This again confirms that autonomous navigation and guidance systems may be corrected with an error not smaller than the dynamic error of the correcting electronic sensor.

It may also be shown that correcting signal $s_c(t)$ does not contain a constant component of the error of the velocity sensor. This means that signal $s(t)$ can be tracked with an electronic system without dynamic error.

It is also important to note the satisfactory smoothing of J'_r by a narrow-band electronic system that is invariant to the input signal, which reduces the probability of interruption of automatic tracking.

Thus, all the positive properties of autonomous systems with radio correction and corrected electronic sensors are manifested in systems with mutually corrected sensors.

BIBLIOGRAPHY

1. Ageyev, D.V., "Fundamentals of Linear Selection Theory," *Nauchno-Tekhnicheskiy Sbornik LIIS* (Scientific-Technical Symposium of LIIS), 1935, No. 10, pp. 35-41.

2. Ayzenberg, G.S., *Antenny Ul'trakorotkikh Voln* (UHF Antennas), Moscow, Svyaz'izdat, 1957.

3. *Antenny Ellipticheskoy Polyarizatsii* (Elliptical Polarization Antennas), Collected Translations, edited by A.I. Shpuntov, Moscow, IL (Foreign Literature Publishing House), 1961.

4. Atlas, D.M., "Advances in Radar Meteorology," *Advances in Geophysics,* H.E. Landsberg and J. Van Miegham, eds., New York, Academic Press, 1964.

5. Atrazhev, M.P., V.A. Il'in and N.P. Mar'in, *Bor'ba s Radioelektronnymi Sredstvami* (Electronic Countermeasures), Moscow, Voyenizdat, 1972.

6. Amiantov, I.N., *Izbrannyye Voprosy Statisticheskoy Teorii Svyazi* (Selected Problems in Statistical Communications Theory), Moscow, Soviet Radio Publishing House, 1971.

7. Babanov, Yu. N., Ya. G. Rodionov, V.F. Ryabkov and A.M. Shabalin, "Development of Means of Protecting Radio Communications Systems from Pulse Interference," *Radiotekhnika* (Radio Engineering), 1967, Vol. 22, No. 11, pp. 28-36; (translation: *Telecomm. and Radio Engrg., Pt. 2,* Vol. 22, No. 11, November 1967, pp. 82-88).

8. *Voprosy Statisticheskoy Teorii Radiolokatsii* (Problems in Statistical Radar Theory), I.A. Bakut, I.A. Bol'shakov, B.M. Gerasimov, *et al.,* Moscow, Soviet Radio Publishing House, Vol. 1, 1963; Vol. 2, 1964, (translation available from NTIS as AD 608462 and AD 645775)

9. Barton, D., *Radar System Analysis,* Artech House, 1976.

10. Bass, F.G. and I.M. Fuks, *Rasseyaniye Voln na Statisticheski Nerovnoy Poverkhnosti* (Scattering of Waves by Statistically Irregular Surfaces), Moscow, "Nauka," 1972; (translation available from JPRS 66061-1 and 66061-2).

11. Beketov, V.I., "Analysis and Synthesis of Polarization Ellipses," *Radiotekhnika,* 1964, Vol. 19, No. 10, pp. 22-25; (translation: *Telecomm. and Radio Engrg., Pt. 2,* Vol. 19, No. 10, Oct. 1964, pp. 81-84).

12. Baghdady, E.J., *Lectures on Communication Systems Theory,* McGraw-Hill, 1961.

13. Belousova, N.V. and V.L. Lebedev, "Tracking Interruption in Automatic Frequency Control System," *Radiotekhnika,* 1963, Vol. 18, No. 10, pp. 35-42; (translation: *Telecomm. and Radio Engrg., Pt. 2,* Vol. 18, No. 10, October 1963, pp. 33-41).

14. Blattner, D.J., "Electronic Countermeasures the Art of Jamming," *Electronics World,* Vol. 62, No. 6, December 1959, pp. 47ff.

15. Benenson, L.S., *Antennyye Reshetki* (Antenna Arrays), Moscow, Soviet Radio Publishing House, 1966.

16. Boldin, V.A., "Combined Radio Navigation Systems," in the book: *Radiotekhnika* (Radio Engineering), Vol. 3, edited by R.G. Mirimanov, Moscow, VINITI (All-Union Institute of Science and Technical Information (of the State Committee of the USSR Council of Ministers for Science and Technology and of the USSR Academy of Sciences)), 1972, pp. 259-332.

17. Bobnev, M.P., *Generirovaniye Sluchaynykh Signalov* (Random Signal Generation), Moscow, "Energiya," 1970.

18. Bobnev, M.P., B. Kh. Krivitskiy and M.S. Yarlykov, *Kompleksnyye Sistemy Radioavtomatiki* (Combined Radio Automation Systems), Moscow, Soviet Radio Publishing House, 1968.

19. Boloshin, I.A., *Elementy Maloshumyashchikh Priyemnikov Sverkhvysokikh Chastot* (Elements of Low-Noise SHF Receivers), Moscow, VVIA (Air Force Engineering Academy) imeni N. Ye. Zhukovskiy, 1961.

20. Brayson, A. and Kho-yu-shi, *Prikladnaya Teoriya Optimal'nogo Priyema* (Applied Theory of Optimum Reception), Moscow, "Mir," 1972.

21. Burenin, N.I., *Radiolokatsionnyye Stantsii s Sintezirovannoy Antennoy* (Synthetic Aperture Radar), Moscow, Soviet Radio Publishing House, 1972; (available in translation from Joint Publications Research Service, Arlington, VA as JPRS-59391, 29 June 1973).

22. Van Der Warden, B.L., (Mathematical Statistics), Berlin, Springer, 1957; trans. available: New York, Springer-Verlag, 1969.

23. Valeyev, V.G. and Yu. G. Sosulin, "Detection and Simultaneous Measurement of Parameters of Quasideterministic Signal," *Teoriya i Tekhnika*

Radiolokatsii (Theory and Technology of Radar), edited by A.G. Saybel',
No. 207, Moscow, "Mashinostroyeniye," 1970, pp. 172-189; (trans.
JPRS 56143).

24. Vakin, S.A. and L.N. Shustov, *Osnovy Radioprotivodeystviya i Radio-
    tekhnicheskoy Razvedki* (Principles of Jamming and Electronic Recon-
    naissance), Moscow, Soviet Radio Publishing House, 1968; (available in
    translation from NTIS as AD 692642, -43).

25. Vakman, D. Ye., *Slozhnyye Signaly i Printsip Neopredelennosti v Radio-
    lokatsii* (Complex Signals and Indeterminancy Principle in Radar), Moscow,
    Soviet Radio Publishing House, 1965; (available in translation: New York:
    Springer, 1968).

26. Van Trees, G., *Detection, Estimation and Modulation Theory,* Vol. 1,
    Wiley, 1968.

27. Varakin, L.D., *Teoriya Slozhnykh Signalov* (Theory of Complex Signals),
    Moscow, Soviet Radio Publishing House, 1970.

28. Velis, D.D. and D.I. Reynolds, *Golografiya* (Holography), Moscow, Voy-
    enizdat, 1970.

29. Velichkin, A.I., *Teoriya Diskretnoy Peredachi Nepreryvnykh Soobshcheniy*
    (Theory of Digital Continuous Data Transmission), Moscow, Soviet Radio
    Publishing House, 1970.

30. Vinitskiy, A.S., *Ocherk Osnov Radiolokatsii pri Nepreryvnom Izluchenii
    Radiovoln* (Fundamentals of CW Radar), Moscow, Soviet Radio Publishing
    House, 1961; (available in translation: JPRS 23120).

31. Viterbi, A.J., *Communication,* McGraw-Hill, 1966.

32. Vinitskiy, A.S., *Modulirovannyye Fil'try i Sledyashchiy Priyem ChM*
    (Modulated Filters and FM Tracking Reception), Moscow, Soviet Radio
    Publishing House, 1969.

33. Vishin, G.M., *Selektsiya Dvizhushchikhsya Tseley* (Moving Target Indica-
    tion), Moscow, Voyenizdat, 1966.

34. Volkov, V.M., *Logarifmicheskiye Usiliteli* (Logarithmic Amplifiers), Kiev,
    Gostekhizdat UkrSSR, 1972.

35. Volkov, V.M., *Logarifmicheskiye Usiliteli na Tranzistorakh* (Transitor
    Logarithmic Amplifiers), Kiev, Tekhnika, 1955.

36. Volkov, V.M., *Funktsional'nyye Usiliteli s Shirokim Dinamicheskim
    Diapazonom* (Functional Amplifiers with Wide Dynamic Range), Kiev,
    Tekhnika, 1967.

37. Herscovici, S. and A. Detape, "Performance of a Diversity Radar," *Annales de Radioelectricité 12*, No. 50, October 1957 (in French).

38. Goldman, S., *Frequency Analysis, Modulation, and Noise,* New York, McGraw-Hill, 1948.

39. Gonorovskiy, I.S., *Radiotekhnicheskiye Tsepi i Signaly* (Radio Engineering Circuits and Signals), Moscow, Soviet Radio Publishing House, 1967.

40. Gorgonov, G.I., "Precision Analysis of Multivariate Optimum Filtering," *Izv. AN SSSR. Ser. Tekhnicheskaya Kibernetika* (News of the USSR Academy of Sciences. Engineering Cybernetics Series), 1972, No. 4, pp. 200-202.

41. Goryunov, M.V., "WLN Systems in Which Maltuned Pulsed Jamming is Standardized Independently of Its Input Amplitude and Duration," *Trudy GPI* (Proceedings of Gor'kiy Polytechnic Institute imeni A.A. Zhdanov), 1972, Vol. 28, No. 7, pp. 5-8.

42. Gorbachev, A.A., "Improving Noise Immunity of AM Receiver through Application of Nonlinear Feedback in HF Part of Receiver," *Radiotekhnika,* 1963, Vol. 18, No. 2, pp. 37-42; (translation: *Telecomm. Radio Engrg.,* Vol. 17, No. 7, July 1962, pp. 9-12).

43. Grasso, G. and P. Guarguaglini, "General Detection Procedure for a Multi-frequency Radar," *Alta Frequenza 37,* No. 2, February 1968.

44. Grebene, A.B. and H.R. Camenzind, "An Integrated FM Multiplex Receiver/Generator," *IEEE,* 1969 NTC REC.

45. Grigor'yev, L.N., "Action of Multiplicative Stationary Interference on AGC System," *Voprosy Radioelektroniki. Ser. Obshchetekhnicheskaya* (Problems in Electronics. General Engineering Series), 1970, No. 11, pp. 100-104.

46. Grigor'yev, L.N., "Action of Random Process on AGC System with Nonlinear Characteristics," *Voprosy Radioelektroniki. Ser. Obshchetekhnicheskaya,* 1969, No. 9, pp. 83-87.

47. Gudmen, D., *Primeneniye Golografii* (Application of Holography), Moscow, "Mir," 1973.

48. Gudzenko, L.I., "On Periodically Nonstationary Processes," *Radiotekhnika i Elektronika,* 1959, Vol. IV, No. 6, pp. 1062-1064; (trans: *Radio Engrg. and Electron. Phys.,* Vol. 4, No. 6, 1959, pp. 220-24).

49. Gelb, A. and A. Sutherland "Software Advances in Aided Intertial Navigation," *Navigation,* Vol. 17, No. 4, Winter 1970-71, pp. 258-69.

50. Gusev, K.G., *Polyarizatsionnaya Modulyatsiya* (Polarization Modulation), Khar'kov, 1968.

51. Gutkin, L.S. and Yu. P. Borisov, *Radioupravleniye Reaktivnymi Snaryadami i Kosmicheskimi Apparatami* (Radio Control of Missiles and Space Vehicles), Moscow, Soviet Radio Publishing House, 1968.

52. Gutkin, L.S., *Teoriya Optimal'nykh Metodov Radiopriyema pri Fluktuatsionnykh Pomekhakh* (Theory of Optimum Methods of Radio Reception with Fluctuation Interference), Moscow, Soviet Radio Publishing House, 1972.

53. Garrett, P.H., "Adaptive Maximum-Likelihood Receiver for Direct Range Measurement," *IEEE Trans.,* Vol. AES-7, No. 5, September 1971, pp. 863-74.

54. Delano, R., "A Theory of Target Glint or Angular Scintillation in Radar Tracking," *Proc. IRE 41,* No. 12, December 1953, pp. 1778-84.

55. Dech, R., *Nelineynyye Preobrazovaniya Sluchaynykh Protsessov* (Nonlinear Conversions of Random Processes), Moscow, Soviet Radio Publishing House, 1965.

56. Derusso, P., R. Roy and Ch. Klouz, *Prostranstvo Sostoyaniy v Teorii Upravleniya* (The Space of States in Control Theory), Moscow, "Nauka," 1970.

57. Dolukhanov, M.P., *Rasprostraneniye Radiovoln* (Propagation of Radio Waves), Moscow, Soviet Radio Publishing House, 1972.

58. Donchenko, V.K. and L.G. Ryndyk, "Estimation of Losses in 'WLN-Analog Storage' System as Functions of Band Ratio of WLN System," *Trudy GPI,* 1971, Vol. 27, No. 15, pp. 20-23.

59. *Dokumenty X Plenarnoy Assamblei MKKR. Otchet No. 322. Raspredeleniye po Zemnomu Sharu Atmosfernykh Pomekh i Ikh Kharakteristiki* (Documents of 10th Plenary Assembly of IRCC. Report No. 322. Global Distribution of Atmospheric Interference and Its Characteristics," Moscow, "Svyaz'," 1965.

60. Donal'son, D.D. and F. Kh. Kishi, "Survey of Theory and Methods of Adaptive Control Systems," *Sovremennaya Teoriya Sistem Upravleniya* (Modern Theory of Control Systems), Moscow, "Nauka," 1970, pp. 264-319.

61. Davenport, W.B. and W.L. Root, *An Introduction to the Theory of Random Signals and Noise,* New York, McGraw-Hill, 1958.

62. Dillard, G.M. and C.E. Antoniak, "A Practical Distribution-Free Detection Procedure for Multiple Range Bin Radars," *IEEE Trans.*, Vol. AES-6, No. 5, September 1970.

63. Zhuravlev, A.G., "Performance of Phase Automatic Frequency Control System in Sinusoidal Interference," *Radiotekhnika*, 1963, Vol. 18, No. 9, pp. 38-46; (translation: *Telecomm. Radio Engrg. Pt. 2*, Vol. 18, No. 9, September 1963, pp. 31-39).

64. Sorkin, C. S. and B.D. Steinberg, "Conical Scan Tracking in Ground Clutter with Noncoherent and Coherent MTI," *Proc. IRE East Coast Conf. on Aero. and Nav. Electronics*, 1958. Reprinted in Johnston, **Radar ECCM**, Artech House, Dedham, MA, 1979.

65. Zubkovich, S.G., *Statisticheskiye Kharakteristiki Radiosignalov, Otrazhennykh ot Zemnoy Poverkhnosti* (Statistical Characteristics of Radio Signals Reflected from the Ground), Moscow, Soviet Radio Publishing House, 1968; (available from NTIS as AD 783046).

66. Zyuko, A.G., *Pomekhoustoychivost' i Effektivnost' Sistem Svyazi* (Noise Immunity and Efficiency of Communications Systems), Moscow, "Svyaz'," 1972.

67. Ivantsevich, N.V. and Yu. I. Nikitenko, "Evaluation of Noise Immunity of Radio Links in Relation to Atmospheric Interference," *Voprosy Radioelektroniki. Ser. Obshchetekhnicheskaya*, 1969, No. 16, pp. 49-58.

68. Ivakhnenko, A.G., *Samoobuchayushchiyesya Sistemy Raspoznavaniya i Avtomaticheskogo Upravleniya* (Self-Teaching Recognition and Automatic Control Systems), Kiev, "Tekhnika," 1969.

69. Enloe, L.H., "Decreasing the Threshold in FM by Frequency Feedback," *Proceedings IRE*, Vol. 50, No. 1, January 1962, pp. 18-30.

70. Itskhoki, Ya. S. and N.I. Ovchinnikov, *Impul'snyye i Tsifrovyye Ustroystva* (Pulsed and Digital Systems), Moscow, Soviet Radio Publishing House, 1972.

71. Katsenbogen, M.S., *Kharakteristiki Obnaruzheniya* (Characteristics of Detection), Moscow, Soviet Radio Publishing House, 1965.

72. Kalinichev, B.P., "Amplitude Distribution of Atmospheric Interference" *Elektrosvyaz'*, 1968, No. 2, pp. 76-77; (trans. *Telecomm. and Radio Engrg.*, *Pt. 1*, Vol. 22, No. 2, February 1968, pp. 52-54).

73. Kalman, R.E., and R.S. Bucy, "New Results in Linear Filtering and Prediction Theory," ASME Trans., *J. Basic Eng.*, Vol. 83, Series D, No. 1, March 1961, pp. 95-108.

74. Kanareykin, D.B., N.F. Pavlov and V. A. Potekhin, *Polyarizatsiya Radiolokatsionnykh Signalov* (Polarization of Radar Signals), Moscow, Soviet Radio Publishing House, 1966; (portions available in translation: JPRS 46624).

75. Kanevskiy, Z.M., *Peredacha Soobshcheniy s Informatsionnoy Obratnoy Svyazyo* (Information Feedback Data Transmission) Moscow, "Svyaz'," 1969.

76. Knyazev, A.D. and V.F. Pchelkin, *Problemy Obespecheniya Sovmestnoy Raboty Radioelektronnoy Apparatury* (Problems of Compatibility of Radio Electronic Systems), Moscow, Soviet Radio Publishing House, 1971.

77. Kantor, L. Ya., *Metody Povysheniya Pomekhozashchishchennosti Priyema ChM Kolebaniy* (Methods of Improving Noise Immunity of FM Receivers), Moscow, "Svyaz'," 1967.

78. Klass, A., "Radar with Increased Range, Using Frequency Difference Method," *Voprosy Radiolokatsionnoy Tekhniki,* 1958, No. 3, pp. 11-15.

79. Kovit, B., I. Holahan and P. Dax, "Electronic Countermeasure and Counter-Countermeasure Techniques," *Space Aeronautics 33,* No. 4, April 1960. Reprinted in Johnston, S.L., **Radar ECCM**, Artech House, Dedham, MA, 1979.

80. Kock, W.D., *Lasers and Holography: An Introduction to Coherent Optics,* Doubleday, 1969.

81. Kondratenkov, G.S., *Obrabotka Informatsii Kogerentnymi Opticheskimi Sistemami* (Data Processing with Coherent Optical Systems), Moscow, Soviet Radio Publishing House, 1972.

82. Korado, V.A., "Two-Channel Detection of Random Signals on Background of Interference with Unknown Intensity," *Izv. AN SSSR. Ser. Tekhnicheskaya Kibernetika* (News of the USSR Academy of Sciences. Engineering Cybernetics Series), 1970, No. 4, pp. 152-159.

83. "Extraterrestrial Radio Interference (A Survey)," *Zarubezhnaya Radioelektronika,* 1960, No. 7, pp. 3-30.

84. Krivetskiy, A.A., "A Method of Controlling Pulsed Jamming of Telegraphic Communications," *Elektrosvyaz',* 1961, No. 11, pp. 25-29. (Trans: *Telecomm. and Rad. Eng.,* No. 11, 1961, pp. 22-26).

85. *Radiopriyemnyye Ustroystva* (Radio Receiver Systems), edited by V.I. Siforov, Moscow, Soviet Radio Publishing House, 1974.

86. Krivitskiy, B. Kh., *Avtomaticheskiye Sistemy Radiotekhnicheskikh Ustroystv* (Automatic Radio Engineering Systems), Moscow, "Energiya," 1962.

87. Krivitskiy, B. Kh., "Pulse Transmission by an Amplifier with Fast Automatic Gain Control," *Radiotekhnika*, 1958, Vol. 13, No. 4, pp. 68-76; (translation: *Radio Engrg.*, Vol. 13, No. 4, 1958, pp. 93-105).

88. Krivitskiy, B. Kh., "Dynamic Properties of Automatic Gain Control Systems in Consideration of Limitation," *Izv. Vuzov SSSR, Ser. Radio-elektronika* (News of Higher Institutes of Learning USSR, Radio Electronics Series), 1967, Vol. X, No. 3, pp. 254-263.

89. Kriksunov, L.Z., O.I. Naygovzin and V.I. Mekhryakov, *Chastotno-uremennyye i Prostraistvenno-Chastotnyye Kharakteristiki Moduliru-yushchikh Ustroystv* (Frequency-Time and Space-Frequency Characteristics of Modulating Systems), Moscow, "Mashinostroyeniye," 1972.

90. Croney, J. and P. Wallis, "A Sidelobe Suppression System for Primary Radar," *The Radio and Electronic Engineer*, Vol. 28, No. 4, October 1964, pp. 247-60.

91. Croney, J., "The Reduction of Sea and Rain Clutter in Marine Radars," *J. Inst. Nav. 7*, April 1954.

92. Krylov, G.M. and A.V. Vishiyevskaya, *Proyektirovaniye Logarifmiches-kikh Usiliteley s Nepreryvnym Detektirovaniyem Signala* (Planning of Logarithmic Amplifiers with Continuous Signal Detection), Moscow, "Energiya", 1970.

93. Kuz'min, S.Z., *Tsifrovaya Obrabotka Radiolokatsionnoy Informatsii* (Digital Radar Data Processing), Moscow, Soviet Radio Publishing House, 1967.

94. Kulikovskiy, A.A., *Radiopriyemnyye Ustroystva* (Radio Receiving Systems), Moscow, VVIA imeni Zhukovskiy, 1966.

95. Cook, C.E. and M. Bernfeld, *Radar Signals*, Academic Press, 1967.

96. Kock, W.E., "Synthetic End-Fire Hologram Radar," *Proc. IEEE,* November 1970, Vol. 58, No. 11, pp. 1858-59.

97. Lazutkin, B.A., *Radiotekhnichn Pristroi z Kompensatsieyu Zavad. Vidavintstvo,* Kiev, Tekhnika, 1972.

98. Levin, B.R. and Yu. S. Shinakov, "Bias System for Simultaneous Detection of Several Signals and Evaluation of Their Parameters," *Radiotekhnika,* 1971, Vol. 26, No. 4, pp. 16-21; (trans: *Telecomm. and Radio Engrg., Pt. 2,* Vol. 25, No. 4, April 1971, pp. 67-70.

99. Levin, B.R., *Teoreticheskiye Osnovy Statisticheskoy Radiotekhniki* (Theoretical Principles of Statistical Radio Engineering), Book 1, Moscow, Soviet Radio Publishing House, 1966.

100. Levin, B.R., *Teoreticheskiye Osnovy Statisticheskoy Radiotekhniki,* Book 2, Moscow, Soviet Radio Publishing House, 1968.

101. Leonov, A.I. and K.I. Fomichev, *Monoimpul'snaya Radiolokatsiya* (Monopulse Radar), Moscow, Soviet Radio Publishing House, 1970; (available in translation from NTIS as AD 742696).

102. Livshits, A.R., "On Probability of n-Coincidences," *Radiotekhnika i Elektronika,* 1957, Vol. 2, No. 8, pp. 947-950; (trans: *Radio Engrg. and Electron. Phys.,* Vol. 2, No. 8, August 1957, pp. 1-6).

103. Linde, D.P., *Radioperedayushchiye Ustroystva* (Radio Transmitting Systems), Moscow, "Energiya," 1969.

104. Hansen, V.G., "A Sequential Logic for Improving Signal Detectability in Frequency-Agile Radars," *IEEE Trans. AES-4,* No. 5, September 1968. Reprinted in *Radars: Vol. 6,* D.K. Barton, ed., Artech House, Dedham, MA, 1977.

105. Locke, A.S., *Guidance,* Van Nostrand, 1955.

106. Maksimov, M.V., "Mathematical Models of Multichannel Radio Electronic Systems with Radio Interference," *Izv. AN SSSR. Ser. Tekhnicheskaya Kibernetika,* 1967, No. 3, pp. 157-163; (trans: *Engrg. Cybernetics,* No. 3, 1967, pp. 156-62.

107. Maksimov, M.V. and G.I. Gorgonov, *Radioupravleniye Raketami* (Radio Control of Rockets), Moscow, Soviet Radio Publishing House, 1964.

108. Maksimov, M.V. and G.I. Gorgonov, *Aviatsionnyye Sistemy Radioupravleniya* (Aviation Radio Control Systems), Moscow, VVIA imeni Zhukovskiy, 1973.

109. Maksimov, M.V., "Effect of Noise on Single-Balance and Dual-Balance Converters in Continuous Wave Radars Stations," *Radiotekhnika,* 1973, Vol. 28, No. 5, pp. 17-22; (translation: *Telecomm. Radio Engrg. Pt. 2,* Vol. 28, No. 5, May 1973, pp. 69-73).

110. Maksimov, M.V., *Pomekhoustoychivost' Mnogokanal'nykh Komandnykh Radioliniy Upravleniya* (Noise Immunity of Multichannel Command Guidance Radio Links), Moscow, Soviet Radio Publishing House, 1970.

111. "Ways and Means of Counteracting Antimissile Defense," edited by Sorgiyevskiy, *Zarubezhnaya Radioelektronika,* 1966, No. 1, pp. 3-31.

112. *Adaptivnyye Automaticheskiye Sistemy. Sb. Statey* (Adaptive Automatic
     Systems. Collection of Articles), edited by G.A. Medvedev, Moscow,
     Soviet Radio Publishing House, 1972.

113. *Metody Pomekhoustoychivogo Priyema ChM i FM Signalov. Sb. Statey*
     (Noise-Immune FM and PM Reception Methods. Collection of Articles),
     edited by A.S. Vinitskiy and A.G. Zyuko, Moscow, Soviet Radio Publish-
     ing House, 1972, pp. 45-71.

114. Middleton, D., *An Introduction to Statistical Communications Theory,*
     New York, McGraw-Hill, 1960.

115. Momot, Ye, G., *Problemy i Tekhnika Sinkhronnogo Priyema* (Synchronous
     Reception Problems and Technology), Moscow, Svyaz'izdat, 1961.

116. Muhammed Abd Al'-Vakhab Ismail, *Radiolokatsionnyy Vysotomer s
     Dvoynoy Chastotnoy Modulyatsiyey* (Double-Frequency Modulation
     Radar Altimeter), Moscow, IL, 1957.

117. Nemirovskiy, M.S., *Pomekhoustoychivost' Radiosvyazi* (Noise Immunity
     of Radio Communications), Moscow, "Energiya," 1966.

118. Obrezkov, G.V. and V.D. Razevig, *Metody Analiza Sr'va Slezheniya*
     (Methods of Analyzing Tracking Interruption), Moscow, Soviet Radio
     Publishing House, 1972.

119. O'Neill, E., *Introduction to Statistical Optics,* Addison-Wesley, 1963.

120. Ostrovityanov, R.V., "On the Problem of Angle Noise," *Radiotekhnika i
     Elektronika,* 1966, Vol. XI, No. 4, pp. 592-601; (available in translation:
     Radio Eng. and Electron. Phys. 11, No. 4, 1966, pp. 507-15).

121. Paliy, A.P., *Radiovoyna* (Radio Warfare), Moscow, Voyenizdat, 1963.

122. Papaleksi, N.D., *Radiopomekhi i Bor'ba s Nimi* (Radio Interference and
     Its Control), Moscow, "Gostekhizdat," 1942.

123. Papulis, A., *Systems and Transforms with Applications in Optics,* McGraw-
     Hill, 1968.

124. Peters, L. and F. Weimer, "Radars for Complex Targets," *Proc. IEE 110,*
     No. 12, December 1963, pp. 2149-62.

125. Peterson, E.L., Statistical Analysis and Optimization of Systems
     Wiley, 1961.

126. Peterson, E.L., *Error-Correcting Codes,* MIT Press, 1964.

127. "Polarization of Radar Signals Reflected from Aircraft, Atmospheric Precipitation and Terrain," *Ekspress-Infromatsiya. Ser. Radiolokatsiya, Televideniye, Svyaz'* (Express Information. Radar, Television, Communications Series), 1964, No. 10, pp. 1-12.

128. Ray, H., "Improving Radar Range and Angle Detection with Frequency Agility," *Microwave J. 9*, No. 5, May 1966, pp. 63-68. Reprinted in *Radars: Vol. 6*, D.K. Barton, ed., Artech House, Dedham, MA, 1977.

129. *Primery i Zadachi po Statisticheskoy Radiotekhnike* (Examples and Problems in Statistical Radio Engineering), edited by V.I. Tikhonov, Moscow, Soviet Radio Publishing House, 1970. Authors: V.T. Goryainov, A.G. Zhuraviev and V.I. Tikhonov.

130. *Radioelektronika v 1972 g. Obzor po Materialam Inostrannoy Pechati. T. 3. Radiolokatsionnaya Tekhnika* (Radio Electronics in 1972. A Survey of Foreign Literature. Vol. 3. Radar Technology), pp. 10-15.

131. "Radar Station with Step by Step Frequency Change," *Ekspress-Informatsiya. Ser. Radiolokatsiya, Televideniye, Svyaz'*, No. 5, pp. 1-6.

132. "Radar Stations with Changing Carrier Frequency," *Ekspress-Informatsiya. Ser. Radiolokatsiya, Televideniye, Svyaz'*, 1970, No. 6, pp. 1-13.

133. "Radio Electronics in 1971. Survey of Foreign Literature," *Radio-protivodeystviye i Radiotekhnicheskaya Razvedka* (Electronic Countermeasures and Electronic Reconnaissance), Vol. 7, pp. 23-28.

134. *Radiolokatsionnyye Ustroystva* (Radar Systems), edited by V.V. Grigorin-Ryabov, Moscow, Soviet Radio Publishing House, 1970. Authors: V.V. Vasin, O.V. Vlasov, V.V. Grigorin-Ryabov, et al.

135. Reed, H.R. and Russell, C.M., *Ultra High Frequency Propagation*, Wiley, 1953.

136. *Radiolokatsionnyye Stantsii Bokovogo Obzora* (Side-Looking Radar), edited by A.P. Reutov, Moscow, Soviet Radio Publishing House, 1970. Authors: A.P. Reutov, B.A. Mikhaylov, G.S. Kondratenkov, and B.V. Boyko; (translation available from NTIS as AD 787070).

137. Rice, S., "Theory of Fluctuation Noise," *Teoriya Peredachi Signalov pri Nalichii Pomekh* (Theory of Data Transmission in Interference), edited by I.A. Zheleznov, Moscow, IL, 1963

138. Remizov, L.T. and S.A. Zaytsev, "Observation of Weak Changes of Field Intensity on Superlow Frequencies," *Radiotekhnika i Elektronika*, 1970, Vol. XV, No. 8, pp. 1563-1567; (trans: *Radio Engrg. and Electron Phys.*, Vol. 15, No. 8, 1970, pp. 1351-54).

139. L.T. Remizov, I.V. Oleynikova, A.N. Korolev and I.G. Vyskrebtsov, "Spectrum of Fluctuation Component of Atmospheric Radio Interference in Superlong Wave Range," *Radiotekhnika i Elektronika,* 1972, Vol. XVII, No. 2, pp. 291-294; (trans: *Radio Engrg. and Electron. Phys.,* Vol. 17, No. 3, February 1972, pp. 724-26.

140. Richmond, J.H., "Digital Computer Solutions of the Rigorous Equations for Scattering Problems," *Proc. IEEE,* Vol. 53, No. 8, August 1965, pp. 796-804.

141. Rodionov, Ya. G., *ChM Priyem s Obratnym Upravleniyem* (FM Reception with Feedback), Moscow, Soviet Radio Publishing House, 1972.

142. Safronov, G.S. and A.P. Safronova, *Vvedeniye v Radiogolografiyu* (Introduction to Radio Holography), Moscow, Soviet Radio Publishing House, 1973.

143. Svet, V.D., *Opticheskiye Metody Obrabotki Signalov* (Optical Signal Processing Methods), Moscow, "Energiya," 1971.

144. Segal', S.G., "System for Automatic Control of Transmission Level. Patent No. 97578," *BI,* 1954, No. 3.

145. Sedyakin, N.M., *Elementy Teorii Impul'snykh Potokov* (Elements of Pulse Stream Theory), Moscow, Soviet Radio Publishing House, 1965.

146. Seleznev, V.P., *Navigatsionnyye Ustroystva* (Navigation Systems), Moscow, Oborongiz, 1961.

147. Skolnik, M., *Introduction to Radar Systems,* New York, McGraw-Hill, 1962.

148. Solov'yev, N.P., "Reflective Properties of Ground Covers," *Trudy Rizhskogo Instituta Inzhenerov GVF* (Proceedings of Riga Institute of Engineers GVF (Civil Air Fleet)), 1963, No. 27, p. 31.

149. Solodovnikov, V.V., *Statisticheskaya Teoriya Lineynykh Sistem Avtomaticheskogo Regulirovaniya* (Statistical Theory of Linear Automatic Control Systems), Moscow, "Nauka," 1960; (translation: Dover Press, 1960).

150. Smogilev, K.A., I.V. Voznesenskiy and L.A. Filippov, *Radiopriyemniki SVCh* (SHF Radio Receivers), Moscow, Voyenizdat, 1967.

151. Sigmen, A., *Lazery* (Lasers), Moscow, "Mir," 1966.

152. Sergiyevskiy, B.D., "Reaction of Receiver with Quadratic Detector to Waves Modulated by Phase or Frequency Fluctuations," *Radiotekhnika i Elektronika,* 1962, Vol. VII, No. 5, pp. 782-786; (translation: *Radio Engrg. and Electron. Phys.,* Vol. 12, No. 5, 1962, pp. 741-51).

153. "Superminiature Traveling Wave Tubes for Antenna Phased Arrays in ECM Systems," *Elektronika SVCh* (SHF Electronics), 1971, No. 12, pp. 88-89.

154. Surikov, B.T., *Raketnyye Sredstva Bor'by s Nizkoletyashchimi Tselyami* (Missile Systems for Combatting Low-Flying Targets), Moscow, Voyenizdat, 1973.

155. *Spravochnik po Radioelektronike* (Handbook on Radio Electronics), edited by A.A. Kulikovskiy, Moscow, "Energiya," 1969.

156. *Spravochnik po Osnovam Radiolokatsionnoy Tekhniki* (Handbook on Principles of Radar Technology), edited by V.V. Druzhinin, Moscow, Voyenizdat, 1967; (partial translation available as JPRS 46664 and 47864).

157. Soroko, L.M., *Osnovy Golografii i Kogerentnoy Optiki* (Principles of Holography and Coherent Optics), Moscow, "Nauka," 1971.

158. "Statistical Characteristics of Power Ratio of Components with Different Polarization in Reflected Radar Signal with Circular Polarization," *Ekspress-Informatsiya. Ser. Radiolokatsiya, Televideniye, Radiosvyaz'*, 1970, No. 6, pp. 14-27.

159. Stratonovich, R.L., *Printsipy Adaptivnogo Priyema* (Principles of Adaptive Reception), Moscow, Soviet Radio Publishing House, 1973.

160. Strett, D. (Reley), *Volnovaya Teoriya Sveta* (Wave Theory of Light), Moscow, Gostekhizdat, 1940.

161. Starr, A.T., *Radio and Radar Technique*, London: Pitman, 1953.

162. Stein, S. and J. Jones, *Modern Communication Principles with Application to Digital Signaling*, New York, McGraw-Hill, 1967.

163. Stroke, G.W., *An Introduction to Coherent Optics and Holography*, New York: Academic Press, 1966.

164. Tartakovskiy, G.P. and V.G. Repin, "Statistical Synthesis of Adaptive Systems for Checking Hypotheses with Estimation of Distribution Parameters Related to These Hypotheses," *Radiotekhnika*, 1971, Vol. 26, No 4, pp 8-12.

165. Tartakovskiy, G.P., *Dinamika Sistem Avtomaticheskoy Regulirovki Usileniya* (Dynamics of Automatic Gain Control Systems), Moscow, Gosenergoizdat, 1957.

166. *Teoreticheskiye Osnovy Radiolokatsii* (Theoretical Principles of Radar), edited by V. Ye. Dulevish, Moscow, Soviet Radio Publishing House, 1964. Authors. V. Ye. Dulevish, A.A. Korostelev, Yu. A. Mel'nik, et al; (translation available from NTIS as AD 673149).

167. *Teoreticheskiye Osnovy Radiolokatsii* (Theoretical Principles of Radar),
      edited by Ya. D. Shirman, Moscow, Soviet Radio Publishing House, 1970.
      Authors: Ya. D. Shirman, V.N. Golikov, I.N. Busygin, et al.

167a. Shirman, Ya. D., *Razresheniye i Szhatiye Signalov* (Resolution and Pulse
      Compression), Moscow, Soviet Radio Publishing House, 1974.

168. Teterich, N.M., *Generatory Shuma i Izmereniye Shumovykh Kharakter-
      istik* (Noise Generators and Measurement of Noise Characteristics),
      Moscow, "Energiya,", 1968.

169. Teplyakov, I.M., "Noise Immunity of a Receiver with a Noise Limiter
      Circuit in the Presence of Fluctuation Interference," *Radiotekhnika,*
      1961, Vol. 16, No. 4, pp. 72-74; (translation: *Radio Engrg.,* Vol. 16,
      No. 1, January 1961, pp. 76-79).

170. Timakhov, O.N. and V.K. Lyubchenko, *Selektory Impul'sov* (Pulse
      Selectors), Moscow, Soviet Radio Publishing House, 1966.

171. Tikhonov, V.I., "Phase Locked Automatic Frequency Control Operation
      in Presence of Noise," *Avtomatika i Telemekhanika*, 1960, Vol. XXI,
      No. 3, pp. 301-309; (trans: *Automation and Remote Control,* Vol. XXI,
      No. 3, 1960, pp. 209-214).

172. Tikhonov, V.I., *Statisticheskaya Radiotekhnika* (Statistical Radio Engineer-
      ing), Moscow, Soviet Radio Publishing House, 1966.

173. Tikhonov, V.I. and K.B. Chelyshev, "Statistical Dynamics of Phase-Locked
      Automatic Frequency Control," *Radiotekhnika i Elektronika,* 1963,
      Vol. VIII, No. 2, pp. 331-334; (trans: *Radio Eng. and Electr. Phys.,* Vol.
      VIII, No. 2, 1963, pp. 287-294).

174. Tikhonov, V.I. and V.T. Goryainov, "Action of Normal Noise on Limiter,"
      Elektrosvyaz', 1961, No. 11, pp. 13-24; (trans: *Telecommunications,*
      No. 11, 1961, pp. 10-22.

175. Trofimov, K.N., *Pomekhi Radiolokatsionnym Stantsiyam* (Jamming of
      Radar Stations), Moscow, Voyenizdat, 1962.

176. White, W.D., "Circular Radar Cuts Rain Clutter," *Electronics 27,* No. 3,
      March 1954, pp. 158-60.

177. Udalov, A.P. and B.A. Supkun, *Izbytochnoye Kodirovaniye pri Peredache
      Informatsii Dvoichnymi Kodami* (Redundant Coding in Binary Code Data
      Transmission), Moscow, "Svyaz'," 1964.

178. Widrow, B., R. Mantey, A. Griffiths and B. Gud, "Adaptive Antenna
      Systems," *Proc. IEEE,* Vol. 55, No. 12, December 1967, pp. 2143-59.

179. Utkina, G.I. and P.S. Davydov, "Evaluation of Effectiveness of Certain False Alarm Stabilization Systems in Nonstationary Gaussian Noise," *Teoriya i Tekhnika Radiolokatsii* (Theory and Technology of Radar), edited by A.G. Saybel', No. II, Moscow, "Mashinostroyeniye," 1967, pp. 116-126.

180. Fal'kovich, S. Ye., *Otsenka Parametrov Signala* (Signal Parameter Analysis) Moscow, Soviet Radio Publishing House, 1970.

181. Fal'kovich, S. Ye. and Z.N. Muzyka, *Chuvstivitel'nost' Radiopriyemnykh Ustroystv s Tranzistornymi Usilitelyami* (Sensitivity of Receivers with Transistor Amplifiers), Moscow, "Energiya," 1970.

182. Francon, M., *Holography*, Berlin: Springer-Verlag, 1972 (in German); (trans: Academic Press, 1974).

183. Fink, L.M., *Teoriya Peredachi Diskretnykh Soobshcheniy* (Digital Data Transmission Theory, Moscow, Soviet Radio Publishing House, 1963.

184. Helstrom, K., *Statistical Theory of Signal Detection*, Pergamon Press, 1961.

185. Kharkevich, A.A., *Bor'ba s Pomekhami* (Combatting Interference), Moscow, Fizmatgiz, 1963.

186. Gustafson, B.G. and B.O. As, "System Properties of Jumping-Frequency Radars," *Philips Telecom. Rev.,* Vol. 25, No. 1, July 1964, pp. 70-76. Reprinted in *Radars,* Vol. 6, D.K. Barton, ed., Artech House, Dedham, MA, 1977.

187. Hansen, R.C., *Microwave Scanning Antennas,* Academic Press, Vol. I (1964), Vol. II (1966), Vol. III (1966).

188. Hansen, V.G. and B.A. Olsen, "Nonparametric Radar Extraction Using a Generalized Sign Test," *IEEE Trans.,* Vol. AES-7, No. 5, September 1971, pp. 942-50.

189. Tsibul'ko, L. Ye., "Analysis of Phase Automatic Frequency Control System by Mathematical Modeling," *Radiotekhnika,* 1970, Vol. 25, No. 3, pp. 55-61; (trans: *Telecomm. and Radio Engrg., Pt. 2,* Vol. 25, No. 3, March 1970, pp. 100-104.

190. Tsivlin, I.P., *Elektronnyy Dal'nomer s Dvumya Integratorami* (Electronic Range Finder with Two Integrators), Moscow, Soviet Radio Publishing House, 1964.

191. Tsypkin, Ya. Z., *Osnovy Teorii Obuchayushchikhsya Sistem* (Fundamentals of Theory of Self-Teaching Systems), Moscow, "Nauka," 1970.

192. Tsvetkov, A.G., *Printsipy Kolichestvennoy Otsenki Effektivnosti Radio-elektronnykh Sredstv* (Principles of Qualitative Analysis of Effectiveness of Radio Electronic Systems), Moscow, Soviet Radio Publishing House, 1971.

193. Chelpanov, I.B., *Optimal'naya Obrabotka Signalov v Navigatsionnykh Sistemakh* (Optimum Signal Processing in Navigation Systems), Moscow, "Nauka," 1967.

194. Shastova, G.A., *Kodirovaniye i Pomekhoustoychivost' Peredachi Tele-mekhanicheskoy Informatsii* (Coding and Noise Immunity of Tele-mechanical Data Transmission), Moscow, "Energiya," 1966.

195. Shapovalov, V.I., "On Construction of Signal Detection Characteristics of Frequency Diversity Radar Stations," *Voprosy Radioelektroniki. Ser. Obshchetekhnicheskaya* (Problems in Radio Electronics. General Engineering Series), 1965, No. 3, pp. 51-56.

196. Shabalin, A.M. and B.D. Zabegalov, "Wide-Band-Limiter-Trap System for Pulsed Interference Detection," *Trudy GPI,* 1971, Vol. 27, No. 15, pp. 82-85.

197. Shabalin, A.M., "Method of Suppressing High-Frequency Pulsed Inter-ference in Receiver by Interrupting High-Frequency Wide-band Channel and Using Aperiodic Trap Filters," *Trudy GPI,* 1962, Vol. 18, No. 2, pp. 74-90.

198. Shabalin, A.M., "On Noise Immunity of WLN System," *Trudy GPI,* 1964, Vol. 20, No. 5, pp. 36-40.

199. Shakhgil'dyan, V.V. and A.A. Lyakhovkin, *Sistemy Fazovoy Avtopod-stroyki Chastoty* (Phase Automatic Frequency Control System), Moscow, "Svyaz'," 1971.

200. Shirokov, V.V., "Effect of Signal Fluctuations on a Receiver with Auto-matic Gain Control," *Radiotekhnika i Elektronika,* 1961, Vol. VI, No. 9, pp. 1452-1459; (translation: *Radio Engrg. and Electron. Phys.,* Vol. 6, No. 9, 1961, pp. 1288-95).

201. Shirokov, V.V. and V.G. Repin, "The Influence of Interference on an Automatic Gain Control System," *Radiotekhnika,* 1959, Vol. 14, No. 4, pp. 65-72; (translation: *Radio Engrg.,* Vol. 14, No. 4, 1959, pp. 88-98).

202. Schwartz, M., "Statistical Decision Theory and Detection of Signals in Noise," *Modern Radar,* edited by R.S. Berkowitz, Wiley, 1965.

203. Schlesinger, R.J., *Principles of Electronic Warfare,* Prentice-Hall, 1961.

204. Shaw, E. and D.E.N. Davies, "Theoretical and Experimental Studies of Resolution Performance of Multiplicative and Additive Aerial Arrays," *Radio and Electronic Engr.,* Vol. 28, No. 4, October 1964, pp. 279-91.

205. Shchelkunov, S.A. and H.T. Friis, *Antennas: Theory and Practice,* Wiley, 1952.

206. *Elektronnyye Metody Sozdaniya Pomekh i Protivodeystviye Pomekham. Tekhnicheskaya Informatsiya (TsAGI)* (Electronic Methods of Jamming and Electronic Countermeasures. Technical Information (TsAGI)), No. 23, 1966.

207. Hareg, L.W., "Radio Frequency Interference from Electric Power Systems," *Proc. IEEE,* 1970, Vol. 31, No. 8.

208. Larson, R., J. Zelenka and E. Johansen, "A Microwave Hologram Radar System," *IEEE Trans.,* Vol. AES-8, No. 2, March 1972, pp. 208-17.

209. Hausz, W., "Angular Location, Monopulse and Resolution," *Microwave J.,* February 1964, Vol. 7, pp. 60-65 (reprinted: *Radars,* Vol. 1, Artech House, 1974).

210. Baur, K., "Der Weelenanalisater eine Einrichtung zur glichreitungen Peilung mehreren einfallender Wellenzune," *Frequenz,* 1960, Vol. 14, II, No. 2, pp. 18-25.

211. Brooks, F.E., Jr., "A Receiver for Measuring Angle-of-Arrival in a Complex Wave," *Proc. IRE,* April 1951, Vol. 39, No. 4, pp. 407-11.

212. Eckler, A.R., "The Construction of Missile Guidance Codes Resistant to Random Interference," *Bell System Techn. J.,* Vol. 39, July 1960, pp. 973-94.

213. Urkowitz, H., C. Hauer and J. Koval, "Generalized Resolution in Radar Systems," *Proc. IRE,* October 1962, Vol. 50, No. 10, pp. 2093-2105; (reprinted: *Radars,* Vol. 4, Artech House, 1975).

214. Kuck, J.H., "Means for Tracking Multiple Target Formations by Radar," U.S. Patent No. 3130402 of 12 February 1957.